PLANNING SUSTAINABLE CITIES

An infrastructure-based approach

DIRECTED AND EDITED BY
SPIRO N. POLLALIS

Routledge
Taylor & Francis Group
NEW YORK AND LONDON

First published 2016
by Routledge
52 Vanderbilt Avenue, New York, NY 10017

and by Routledge
2 Park Square, Milton Park, Abingdon, Oxon OX14 4RN

First issued in paperback 2020

Routledge is an imprint of the Taylor & Francis Group, an informa business

Library of Congress Cataloging in Publication Data

ISBN 13: 978-0-367-66995-9 (pbk)
ISBN 13: 978-1-138-18842-6 (hbk)

Typeset in FF DIN OpenType TTF Std
by Monotype Imaging Inc.

Publisher's Note
This book has been prepared from camera-ready copy provided by the editor.

CONTENTS

PREFACE

This publication fills a gap between the professions of planners and engineers, both contributing to the fundamental process of planning and building infrastructure for cities. It responds to the urgency of integrating these practices, with the objective of creating sustainable urban environments. As cities collectively start to respond to the magnitude of global urban changes and especially to the climatic risks (seen, for example, in the impact of Superstorm Sandy on New York City), there is an overall awareness that key professionals need to work in an integrated manner in order to achieve sustainable urban solutions. With this awareness, there is a need as well as a demand for a framework that can create a collaborative platform for different stakeholders involved in cities to work in unison. This need is widely acknowledged by more and more professionals working in the planning and engineering sectors.

Due to the size and complexity of infrastructure systems, multiple stakeholders with different issues and agendas are involved during planning. This book is intended to form the foundation of a common collaborative platform, primarily among public authorities, planners, and engineers,enabling those who have traditionally functioned in silos to work closely together in infrastructure planning. Establishing shared knowledge that responds to the language of these professionals is integral to our approach. The content of this book should be resourceful and broad enough to create an informed knowledge of the infrastructure systems of the city, and goes in depth to a level that is comprehensible to professionals engaged in practices of public policy, planning, and engineering, as well as to community organizations.

Professor Spiro N. Pollalis
Zofnass Program Director
Professor of Design, Technology and Management at the Harvard Design School

ACKNOWLEDGMENTS

This research was made possible by the generous support of Paul Zofnass and Joan Zofnass, whose determination and commitment to sustainability led to the establishment, in 2007, of the Zofnass Program at Harvard University (research.gsd.harvard.edu/Zofnass, or www.zofnass.org). The founding objective of the Program was the creation of a sustainability rating system for infrastructure projects in close collaboration with the Harvard professional schools and with industry.

In 2007, Professor Spiro Pollalis was appointed the first Director of the Zofnass Program and assumed the task of formulating and implementing the vision. A team of researchers and students was assembled to carry out the research and the Planning Guidelines is the direct or indirect outcome of a multi-year effort on sustainability of the researchers throughout the years who have been part of the Zofnass Program. The Zofnass Program has also been supported by Harvard University faculty advisors from the School of Design, the Center for the Environment, the School of Public Health, the Business School, and the Law School, who bring a wide range of expertise to the Program.

In parallel, the Zofnass Program established the Sustainable Industry Advisory Board (SIAB). SIAB members support the Program financially and, equally important, with the commitment of the firms' leadership share their professional expertise and provide valuable feedback to the Program's research. ARCADIS, Autodesk Inc., exp., Golder Associates, Granite Construction, CH2M, HNTB Corporation, MWH Global, NV5, Power Engineers, and Stantec are the SIAB members.

The Program organizes quarterly workshops, by invitation only. The researchers of the Program, Harvard faculty advisors, SIAB members, and guests attend these workshops, led by public authorities and private parties involved in large-scale projects. The workshops allow intense interaction on complex issues that require utmost expertise and understanding of state-of-the-art practice. We attribute to these workshops the results of our research, the successful outcome of both our rating system and our Planning Guidelines.

The Zofnass Program maintains a close ongoing relation with the Institute for Sustainable Infrastructure (ISI, www.sustainableinfrastructure.org). ISI was established in 2012 as a nonprofit organization with three founding members: the American Society of Civil Engineers (ASCE), the American Council of Engineering Companies (ACEC), and the American Public Works Association (APWA). After its founding, ISI adopted our Zofnass rating system, giving it a new life after the research phase, and has transformed it into Envision®, a commercial-grade product for self-evaluating the sustainability of infrastructure projects, for rating them, and for training Envision® certified professionals (ENV SP). Our interactions with ISI have had an impact on our work.

The Planning Guidelines were presented, discussed, and scrutinized in all our workshops over the past two years. The Harvard faculty advisors and SIAB members provided feedback on the research and reviewed the manuscript of this book. SIAB members contributed introductory essays for each chapter, sharing their valuable professional perspectives.

It should be noted that the Water Infrastructure chapter was partially supported by a two-year grant of the Surdna Foundation to the Zofnass Program (www.surdna.org).

BOOK CREDITS

The work presented in this book is the result of the research effort of the Zofnass Program at the Harvard Graduate School of Design in the period 2013-15. The entire team seamlessly contributed to the methodology, graphics and texts, with specific tasks:

Principal Investigator of the Zofnass Program
Director of the Planning Guidelines Research
Professor Spiro N. Pollalis

Core Research Team
Evgenia Hagistavrou
Dimosthenis Lappas
Eleonora Marinou
Yannis Orfanos
Vicky Sagia
Richa S. Vuppuluri
Olga Tzioti

Editor
Professor Spiro N. Pollalis

Publication Coordination
Richa S. Vuppuluri
Vicky Sagia

Design and Layout Coordination
Eleonora Marinou

Planning Guidelines' Tables Coordination
Vicky Sagia

Chapter Coordination
Chapter 3. General Framework:
Vicky Sagia
Chapter 4. Landscape as Infrastructure:
Evgenia Hagistavrou
Chapter 5. Transportation Infrastructure:
Evgenia Hagistavrou, Dimosthenis Lappas
Chapter 6. Water Infrastructure:
Dimosthenis Lappas, Richa S. Vuppuluri
Chapter 7. Energy Infrastructure:
Evgenia Hagistavrou
Chapter 8. Solid Waste Infrastructure:
Olga Tzioti
Chapter 9. Information as Infrastructure:
Eleonora Marinou, Yannis Orfanos
Chapter 10. Food as Infrastructure:
Dimosthenis Lappas, Yannis Orfanos
Chapter 11. New City in Asia:
Spiro N. Pollalis, Olga Tzioti
Chapter 12. City Expansion in Europe:
Spiro N. Pollalis, Angela Kouveli

Cover Design:
EMBLANK (Alexandra Papagianni and Chrysostomos Tsimourdagkas)

ADVISORS TO THE ZOFNASS PROGRAM

Harvard University Faculty

Amy C. Edmondson, Novartis Professor of Leadership and Management, Harvard Business School (HBS)
Richard T.T. Forman, Research Professor of Advanced Environmental Studies in the Field of Landscape Ecology; Associate of Harvard Forest, Harvard Graduate School of Design (GSD)
José A. Gómez-Ibáñez, Derek Bok Professor of Urban Planning and Public Policy, Harvard Graduate School of Design (GSD)
Wendy B. Jacobs, Clinical Professor of Law; Hilt Grant Faculty Director; Special Advisor on Learning and Practice, Harvard Law School (HLS)
John D. Spengler, Akira Yamaguchi Professor of Environmental Health and Human Habitation; Contin Ed/Spec Prog Instructor; Faculty & Stud Serv Sr Specl, Harvard School of Public Health (HSPH)
Katherine E. Konschnik, Policy Director; Lecturer on Law, Harvard Law School, Environmental Policy Initiative (HLS)
Hanif Kara, Professor in Practice of Architectural Technology, Harvard Graduate School of Design (GSD); Design Director at AKT II
John Macomber, Senior Lecturer, Gloria A. Dauten Real Estate Fellow, Harvard Business School (HBS)
Ali Malkawi, Professor of Architectural Technology; Director of Harvard Center for Green Buildings and Cities, Harvard Graduate School of Design (GSD)
Richard Peiser, Michael D Spear Professor of Real Estate Development, Harvard Graduate School of Design (GSD)
Daniel P. Schrag, Sturgis Hooper Professor of Geology, Professor of Environmental Science and Engineering; Director of the Harvard University Center for the Environment, Harvard Center for the Environment (FAS)

Sustainable Industry Advisory Board (SIAB)

Arcadis: Douglas M. Owen, PE, BCEE, ENV SP, Executive Vice President, Chief Technical Officer for the Water Division
Autodesk: Terry D. Bennett, LS, LPF, MRICS, ENV SP, LEED AP, Senior Industry Strategist, Civil Infrastructure
CH2M: Erin Mosley, PE, ENV SP, Vice President, Director of Management Consulting
exp: Alida Saleh, Head, Environment & Sustainable Development
Golder Associates: Hannah Hamling, Global Sustainable Development Leader
Roberto Mezzalama, MSc, Global ESIA Technical Community Leader
Granite Construction: Geoff Boraston, Vice President Environmental Affairs
HNTB: Jim Grant, Energy and Fueling Director
Inter-American Development Bank (IDB): Ana Maria Vidaurre, Principal Investment Officer, Infrastructure Division of Structured & Corporate Finance Department
MWH Global: Loren A. Labovitch, Vice President and Director, Emerging Markets
NV5: Dickerson Wright, CEO
Laura Bonich, PE, LEED AP, Director Business Development
Stuart Lesser, President, Principal-In-Charge of Joslin Lesser & Associates Inc. (an NV5 company)
Power Engineers, Inc: Richard Corolewski, PE, Electrical Engineer / Federal Division Director
Stantec Corporation: Rich Allen, Senior Vice President, Chief Operating Officer
Marty Janowitz, MES, ENV SP, Vice President & Practice Leader Sustainable Development

PLANNING SUSTAINABLE CITIES

An infrastructure-based approach

PART 1

SUSTAINABILITY, INFRASTRUCTURE, AND CITIES

Part 1 introduces the Zofnass Program, its activities, research scope, and approach, as well as key factors affecting the sustainability of urban infrastructure systems and aspects of infrastructure planning.

Zofnass Program Events

Sustainability, Infrastructure, and Cities

the Zofnass Program for Sustainable Infrastructure

Chapter 1

1.1 | THE PROGRAM

Paul Zofnass
Zofnass Program Founder
President, Environmental Financial Consulting Group, Inc. (EFCG)

Zofnass Program Workshop, October 2014.

Having grown up in a beautiful suburb of Boston, surrounded by forests and farms, I developed at a very young age a passion for preserving our natural world. I have also gained an appreciation for how effectively measuring performance or quality can influence and incentivize human behavior. Give performance a rating, and we compete to do better. Consider, for example, how school grades push students to study harder; or how athletic teams or individuals strive to improve their performance to the extent that it can be measured. Corporate goals and metrics incentivize revenue or earnings growth. In hybrid cars, by making the miles per gallon achieved by the driver (in five-minute increments!) clearly visible, manufacturers have leveraged our innate human desire to improve to encourage better fuel economy. Consider how the US Green Building Council's LEED® rating system for sustainability in building design has sensitized people to the meaning, importance, and benefits of "green building" and encouraged them to build more sustainably.

For the past 25 years I have been working with the engineering/consulting industry as a financial and strategy advisor. Observing the effectiveness of LEED, I began to wonder why a similar type of rating system had not been developed for the design and planning of public infrastructure. One reason may be that the scope and range of issues that need to be taken into consideration for public infrastructure are generally far more complex than for a single building. (Consider for example the impact of a new airport, railroad, highway, urban development project, sewer system, bridge or tunnel complex, waterway, power grid, or even new community or city, any of which can affect hundreds of thousands, in some cases millions, of people. These all have major effects not only on the environment but on many other aspects of society.) If we are going to make the world more "sustainable," don't we need to develop and apply a set of standards by which to analyze, define, and measure "sustainability" as it applies to these large infrastructure programs? And don't we need to continue improving our understanding of the underlying sciences and meaning of sustainability? But where do we begin?

I noted that many of my engineering clients were developing their own tools and methods to measure "sustainability" as it might apply to the kinds of projects they were designing. But there was no commonality in their approaches to these metrics. And in fact, as each firm leveraged its proprietary tools and systems to demonstrate expertise and drive for sustainability in the marketplace, project owners found it difficult to compare the benefits between one tool and another. It was also hard, if not impossible, for a wider consensus to be developed or agreed upon. Yet it was clear to me that such a consensus approach could offer a very beneficial, productive path forward.

It also occurred to me that an approach driven primarily from an engineering perspective would likely fail to take into consideration many other factors in which engineers are not well versed, but which are still critical in understanding the broader scope of "sustainability." Consider, for example, impacts on human health, social communities, and natural ecology, as well as the role of politics, economic development and human well-being, social justice, and obligations and goals of government. These issues go far beyond the resources or knowledge borders of any single engineering firm, or even group of engineering firms. They also cut across all aspects of the social and physical sciences, as well as our overall body of scientific knowledge.

These thoughts suggested to me that to create an effective measurement system for infrastructure sustainability, it needed to be developed by a wide range of sources. These included:

a) A broad and diversified base of participating design and engineering firms,

b) A globally recognized and respected, and multidisciplined, educational institution that could bring to bear all the academic disciplines required to define the concept of "sustainability,"

c) An array of senior government infrastructure directors who could help us understand their needs and issues, so that they could support funding for and use of sustainable design,

d) The support of the major professional associations affiliated with the engineering and design industries, to encourage the use of sustainable design in their work.

In 2007, I approached Harvard University (my alma mater) with an offer to fund 18 months of research to see if they could develop a conceptual framework by which to define sustainability as it might apply to major infrastructure projects, and create an initial set of metrics by which to measure this concept. The Zofnass Program for Sustainable Infrastructure was housed in the Harvard Graduate School of Design, under the direction of Professor Spiro Pollalis and leadership of Professor Andreas Georgoulias, and brought together roughly 30 professors from six different schools within Harvard (Design, Business, Public Health, Government, Sciences, and Law) and over 100 graduate researchers. Over the next two years they met quarterly with sustainability leaders of 30 major engineering/consulting (e/c) firms, and roughly 50 senior government infrastructure staff from different government sectors (federal, regional, state, and city) throughout the country, to attempt to develop a consensus-based set of sustainability parameters.

By 2009, the Program had developed a set of guidelines and metrics to define and measure sustainability as it relates to infrastructure that was sufficiently defensible to warrant further research and development. Twelve organizations (including 10 leading e/c firms) committed to provide significant funding for the ongoing development of this system, as members of the Sustainable Infrastructure Advisory Board at Harvard or SIAB. The SIAB members actively participated in the refinement and application of the system; they included: CH2M HILL, MWH Global, HNTB, Stantec, Golder, Power Engineers, Arcadis, exp., NV5, Granite, Autodesk, and the Inter-American Development Bank (IDB).

In 2011, the Program was approached by the Institute for Sustainable Infrastructure (ISI), a joint venture between three leading professional engineering associations, ASCE, ACEC, and APWA, which was on a similar path to develop a set of infrastructure sustainability standards. Given the need for a single consensus standard, the two programs made the decision to combine their individual systems into one, selecting the best characteristics from each. The combined system is called Envision®, and the two groups work in tandem to continue to improve this combined system. The Zofnass Program at Harvard focuses on further research and development for sustainable infrastructure and for the Envision® system, working to keep it current with the advances in the underlying sciences and knowledge. The Program also holds quarterly workshops that bring together the practical expertise and experience of the SIAB firm members with members of Harvard's faculty and graduate schools and with representatives from governments and NGOs committed to making the world's built infrastructure more sustainable over the long term. The Zofnass Program continues to work with the ISI, helping them to integrate this research and knowledge into a constantly improving Envision® rating system. While the Zofnass Program focuses on R&D and working with governments and NGOs, the ISI focuses on the training, credentialing, and marketing of Envision®.

With Envision®'s integration into the infrastructure marketplace accelerating and roughly 3,000 Envision® Sustainability Professionals, or ENV SPs, in the US and abroad, the Zofnass Program's research is now moving toward the next logical step in creating a more sustainable world, integrating each component of sustainable infrastructure into a system applicable to whole-city design. This work, already seeing initial pilot implementation efforts, is very complex and highly interconnected, but offers a promising vision for how infrastructure of the future might be designed and built.

This textbook, the second since the initiation of this Program, presents our city-scale research results to date. It incorporates sustainable project analyses and case studies, covering many different fields and sectors of infrastructure, with contributions by professionals from the SIAB members and members of the Harvard University faculty.

I strongly believe that if we don't start creating a more sustainable world today, we will not have one in the future. Our sincere appreciation goes to all those who have helped develop this system, who will help us continue to improve it, and who will be using the system to create a better world through sustainable infrastructure.

1.2 | PRODUCED RESEARCH

William J. Bertera
President and CEO, Institute for Sustainable Infrastructure (ISI)

Planning for sustainable infrastructure has been practiced for decades, though often in a context in which the word "sustainability" has rarely appeared ... until recently. Planners, engineers, architects, contractors, and public administrators have, nevertheless, been planning, designing, constructing, operating, and maintaining civil infrastructure using best practices that promote efficiency, safety, longevity, cost-effectiveness, and community values and priorities.

Many of these best practices also promote sustainability as we have come to define it, but none assure that sustainability as a strategy, objective, or goal is considered routinely in the development of civil infrastructure. When it has happened historically, it has occurred as a fortunate by-product of good planning or engineering. But the world is changing, and sustainable infrastructure is no longer, if it ever was, only a choice or a happenstance.

Population growth, its uneven distribution, increasingly taxed natural resources, the sensitivity of the natural world to external alterations, unpredictable weather patterns, and the challenges posed by the threat of global warming and attendant climate change all make doing business as we have in the past untenable. Planning civil infrastructure has become more complicated still. Our infrastructure's claim on resources, its effects on the environment, its contribution to the creation of societies where humans can live in comfort and dignity have magnified and made more obvious its importance.

Infrastructure generally, civil infrastructure in particular, provides for personal security, the public health, economic stability, and quality-of-life benefits. It makes possible concentrations of people in dense environments and the transport of materials and people in pursuit of commercial and social goals: in short, it makes modern societies possible. That this infrastructure be sustainable, that the decisions associated with its planning, design, construction, and operation also be sustainable, i.e., justifiable in the context of an increasingly taxed natural world, has become an essential and critical element of the planning function.

The Brundtland Commission articulated a definition for sustainable development in 1987 that embodies the spirit of the challenge; phrased loosely, "do as little harm as possible and remember that our children walk in our footsteps." In truth, sustainability, especially as it is applied to our civil infrastructure, is a way of thinking about the future more than a precise formula of how to create or reach a precise goal. In this context, what we build is at least as important as how we build. In a world without limits, without scarcity, without diminishing natural resources, without environmental challenges, there are neither reasons nor incentives to preserve and protect. We no longer live in such a world.

If thinking about sustainability has now become a conscious necessity in the development of civil infrastructure, we need tools to help assure that sustainability is not just a random outcome but a planned outcome. The Zofnass Program for Sustainable Infrastructure at the Graduate School of Design at Harvard University in partnership with the Institute for Sustainable Infrastructure (ISI), a Washington-based not-for-profit organization, have developed a tool called Envision® to help planners, engineers, political scientists, sociologists, contractors, and public officials produce civil infrastructure that is demonstrably and intentionally sustainable.

While Envision® can be used to help make single infrastructure projects more sustainable, its more important application may be in its ability to link whole families of infrastructure projects of different kinds into a community-wide web of sustainability. In this context, Envision® is a very powerful planning tool, especially when applied at the early stages of project or community planning.

Although the tool is most productively applied at the early stages of development, it is applicable at any point in a project's life cycle. This is critical because of the extended life cycles of much of our civil infrastructure. Bridges and sewer systems, designed for much shorter lifespans, not infrequently are in use seventy-five and even one hundred years after construction. During the course of its lifespan, all infrastructure undergoes repairs, alterations, rebuilding, and operational changes. Best practices are improved, new materials and processes are developed, and resource and environmental changes occur. Civil infrastructure is not static, nor are the circumstances or communities in which it is located.

The concept of what constitutes sustainability is changing. What was considered sustainable fifty years ago might not be today; the same will be true fifty years hence. Envision® allows us to take advantage of new information and changing priorities every time we consider altering a piece or system of infrastructure. The tool is not prescriptive, however. It encourages the application of local solutions that reflect community values, priorities, and resources. In this context, it recognizes that infrastructure is about more than engineering and design and construction. It is also about a host of other, sometimes difficult-to-quantify variables.

The triple bottom line is an articulation of some of those most important variables. It includes the trifecta of social, economic, and environmental needs and aspirations that are embedded in every decision associated with an infrastructure project ... but which often are not manifest at the point of procurement for lack of a way to measure their importance and to monetize that evaluation in a set of costs and benefits reflected in a return on investment.

Returns on investment are a clear measure of success in the private sector, and cost/benefit valuations matter as much in public-sector infrastructure projects as in those for the private sector. Sustainability is a positive public good, but doing good is not enough when there are more needs than resources. There have to be commensurate benefits, those benefits must be clear, and they must be financially obvious. Sustainability can cost, but it can also deliver financial rewards. The Envision® tool helps decision makers evaluate both costs and benefits at the critical point of procurement, where it matters most.

Importantly, Envision® is a web-based tool available in the public domain. There is no charge for its use. It is easily accessible to both public- and private-sector users. The program itself has several parts, but the most important aspect of the program is its intention. Both Harvard University and the Institute for Sustainable Infrastructure have a goal of making a usable tool available to decision makers to help them create, construct, and operate more sustainable infrastructure projects.

The tool itself comes in two forms. The first is an easy-to-use checklist that requires little time or experience to use and is best applied to test the sustainability of large projects or for use on small projects where quick, easy, and inexpensive analysis is desired. The second and more comprehensive form is an extensive web-based self-assessment worksheet that can require extensive time and experience for its most productive use. ISI provides a web-based course, for a fee, that helps users better apply Envision® and which includes a web-based exam, the successful completion of which results in a professional credential (ENV SP). In either instance, there is a comprehensive user's manual that is also free of charge and can be downloaded.

Although the tool is intended primarily as a self-assessment evaluation for project planners, ISI also offers a Project Recognition and Awards Program to acknowledge successful levels of achievement for those self-assessments submitted for verification by third-party reviewers employed by ISI. Self-assessments and evaluations alike are conducted using sixty different best practices, though not all of the credits are applicable to every project. An important aspect of the awards program is that it is an evaluation of the application of the Envision® tool to the project, not an evaluation of the project itself.

The self-assessments are conducted within five broad categories: Leadership, Quality of Life, Resource Allocation, Natural World, and Climate and Risk. The intention is not to force any particular prescriptive outcome, but to help planners organize their thinking to consider sustainable concepts at every stage of a project's development, and to cause a reassessment of existing best practices in the context of new circumstances, materials, concepts, and priorities.

Envision® was originally designed as a rating system, a way to systematically evaluate both existing and planned projects for their sustainability or their sustainability potential, and it can be used in that way. But it is also a very powerful planning tool that can be used to embed sustainable thinking into each and every decision point in the development of a project. Beyond even this application, it has significant potential for public administration and even good government.

Infrastructure development, sustainable or not, is a complicated subject. It is difficult to capture the public imagination with projects whose costs and lifespan are out of proportion to the daily human experience of most voters and taxpayers. Envision® creates a language with which to talk about these complicated subjects in a way that those most affected by them can understand and appreciate. In this sense, Envision® has the capability of elevating the public dialogue and ultimately changing the priorities used to allocate scarce public resources. In a global community where infrastructure of all kinds is a critical differentiator in commercial competitiveness, this is no small accomplishment.

Although initially available in the United States and Canada, Envision® was developed for the global marketplace. Its flexibility and reliance on accepted best practices make it universally applicable.

1.3 | EXPANDING ON CITY-SCALE RESEARCH

Spiro N. Pollalis
Zofnass Program Director
Professor of Design, Technology and Management at the Harvard Design School

The Zofnass rating system for sustainable infrastructure and subsequently Envision®, the evolution of our system as developed by ISI,[1] are toolkits for the various stages of a project, including the very first steps of the decision-making process, before even the engineers start designing. The intention has always been the selection of the best and most suitable type of project for solving the problem at hand. Making sustainable the particular design of a project is the next, equally important consideration.

Our rating system gains built-in flexibility by separating the aspects that define the sustainability of a project. These aspects are (a) the impact on the natural world, (b) the use of resources, (c) the relation to the risks of climate change, and (d) the impact on quality of life. A fifth aspect, the commitment of the leadership to sustainability, is evaluated in terms of the development and implementation of sustainable features. All five aspects are individually rated in order to be able to identify the project's sustainability in each section. The sum of the individual ratings provides the total sustainability rating of the project.

Because of its structure, our system serves as a guide in the early planning stage toward making the right decisions to ensure the sustainability of a project. Using it in a reverse process, it provides "planning guidelines" toward a sustainable project and thus achieving the highest rating in our system. Furthermore, the training of specialists to become Envision® Sustainability Professionals (ENV SP) gives ample understanding of fundamental sustainability concepts and guides them toward making the right decisions in a project.

Having completed the rating system for infrastructure projects, the next research effort in the Zofnass Program was the sustainability of entire cities, quite expected for a research program in the Harvard Design School with its legacy in urban planning and urban design. In addition, it was a time when members of the Zofnass Program were actively involved in planning new cities, with a strong desire to plan sustainable cities but without having tools that offered objective and measurable processes.

There is a significant need for expansion of cities and the development of new cities. Both the increase of population and the concentration of people in cities drive the need. Pakistan, for example, sees an annual population increase of more than 3.5 million people. A large part of them will live in cities, and the existing cities lack both amenities and capacity. Even while many people will remain in rural areas, we can conclude that there will be a need for another Karachi in less than ten years!

However, creating new cities and major expansions to old cities is only half the story. The planet cannot support more people living the way we live today, and has even less capacity to withstand the current trajectory of growth. A sustainable way of life is a must, and cities provide amenities at a cost of fewer resources and less damage to the natural world and the environment at large. An increased percentage of the population living in cities is a step in the right direction. The next decisive step is the development of sustainable cities.

Cities are made of infrastructure and buildings, so a first approach to the sustainability of cities might focus on individual sustainable buildings and sustainable infrastructure projects. Such a simplistic approach is not sufficient, however, as it bypasses the systemic relationships not only within infrastructure but also among the infrastructure systems and buildings. Such a systemic approach has been advanced over the last 20 years in discussions with leaders of architectural and engineering firms at Harvard University, focusing on a systems approach for better planning and higher efficiency.

At the Zofnass Program, the completed rating system for sustainable infrastructure projects was set as the point of departure for assessing the sustainability of cities. Its overarching approach addressing the essence of environmental, social, and economic sustainability, combined with its flexibility for application to diverse types of projects, made it suitable for the big picture.

The decision was made to consider the sustainability of each infrastructure type at a systems level. The seven types of hard infrastructure had been defined in the Zofnass Program as Transportation, Energy, Water, Solid Waste, Landscape, Information, and Food. The decision to consider them separately eliminated undue complexity, while not precluding integration later. This should not be seen as a different methodology from Envision® which does not distinguish the types of infrastructure. Envision® applies to a single project, while at a city level we consider

ZOFNASS PLANNING GUIDELINES IN RELATION TO ENVISION® RATING TOOL

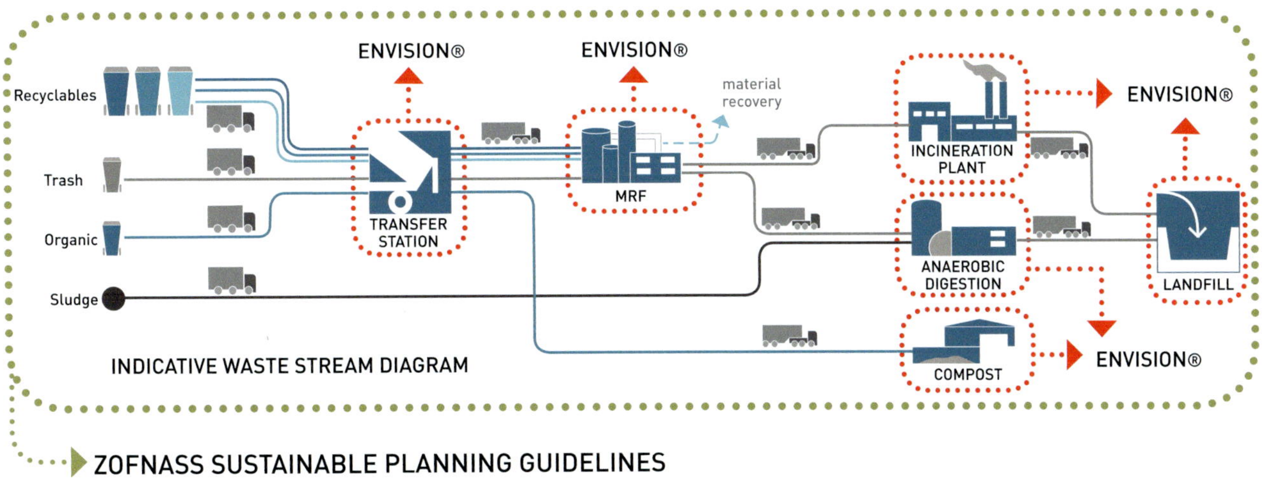

entire infrastructure systems consisting of many projects.

An example of the approach is shown in the graph above for Solid Waste infrastructure. Envision® applies to the individual projects: transfer stations, material recovery facilities, incineration plants, anaerobic digestion, compost, and landfills. However, the sustainability of the city is based not only on the sustainability of each project (and building) but also on the overall planning of the entire infrastructure system and its own sustainability.

In the planning of infrastructure systems, we identified four "system levels," consistently applicable to all seven types of infrastructure. The first level addresses the system's end users and their demand. The second level refers to core strategic decisions. The third and fourth levels address the facilities and operations as prescribed by the strategic plans of the second level. The larger facilities, functioning as nodes and feeding into the infrastructure networks, constitute the third level, while the networks that cover the spatial distribution needs of the infrastructure are the fourth level.

For each infrastructure type and for each of its four system levels, we analyzed the sustainability dimensions and we proposed objectives, planning guidelines, and actions.

After the systematic analysis of the infrastructure systems, we focused on the synergies and interdependencies among them. Energy and Water are directly related. Transportation and Energy are also related. Information is related to all other types of infrastructure. Thus, we concluded this part of the research by providing synergistic planning guidelines leading to an integrated model of the entire infrastructure of a city, as it should be.

The detailed methodology is presented in section 3.2, while the seven infrastructure types are presented in chapters 4 to 10. This book adopts an integrated model, so the synergies are part of each infrastructure system and not an add-on. The objectives, planning guidelines, and actions have been applied to the two cities planned by members of the Zofnass team, with their planning not immediately related to the research itself. These two cities serve as examples of the applicability of the methodology (chapters 11 and 12).

The Planning Guidelines link the buildings to the infrastructure of the city, not only as consumers, as in the current planning guidelines, but also as suppliers of services (i.e., nodes in level three). In the Program's future research, buildings will also be subject to rating using an adapted version of the Envision® credits on how well they fulfill their sustainable role, very much as the infrastructure facilities in the diagram above are rated using the current version of Envision®.

NOTES

1. For the remainder of this essay, we will not distinguish between the Zofnass rating system and Envision®

Photograph by Eleonora Marinou

Planning Infrastructure

Chapter 2

2.1 | CHALLENGES IN INFRASTRUCTURE PLANNING AND IMPLEMENTATION

Spiro N. Pollalis
Zofnass Program Director
Professor of Design, Technology and Management at the Harvard Design School

Our approach considers sustainable infrastructure as a starting point for sustainable cities. The overall city planning must support sustainability, and the individual buildings must be sustainable to help lower the demand on the infrastructure systems. The challenges in starting with infrastructure lie in its interrelations to city planning and to the city's buildings, in the need for sustainability both in new and existing cities, in addressing resilience, and in securing financing.

Sustainable Infrastructure and City Planning

The planning of infrastructure is an integral part of the overall planning. Infrastructure requires space. The size of infrastructure systems depends on the demand from the city's occupants, which translates to the demand from buildings and from the other infrastructure systems. For the cities of the past, it can be argued that this was an ordinary linear process. However, it becomes an iterative nonlinear process in planning sustainable cities, as shown in the next page graph. Preliminary planning data and sustainability considerations drive the initial decisions related to the required infrastructure. Then the infrastructure data and sustainability decisions become part of the planning process and more detailed data are extracted and feed the planning of infrastructure, always under the prism of sustainability and finance. The synergies among the infrastructure systems introduce another level of iterations for even more optimal planning. After the needed iterations and the necessary changes in assumptions and decisions, the system reaches equilibrium, meeting planning, sustainability, and financial objectives. The process is complex and is driven by the expertise and values of the stakeholders. The Zofnass Planning Guidelines do not directly address spatial planning. However, they can serve as a framework and a tool to help the stakeholders work together.

In the outcome, infrastructure systems take a large area of the city, so process integration is a must. The road network takes an average of almost 30% of the city's area. Public open space is rarely lower than 10% and often much more. Energy and water treatment facilities take significant space. Infrastructure networks also take underground space and elevated space. Crossing networks requires special attention. Equally important, the location of infrastructure facilities has an impact on city planning, based on their required proximities to buildings and other infrastructure facilities for efficiency and effectiveness.

The planning of infrastructure systems in a city is like what engineering is in architecture. In buildings, structural and mechanical systems define and require space and cannot be seen as separate from architectural design. In ordinary buildings they were beginning to seem an afterthought. Recently, on the other hand, with an increasing demand for sustainable buildings, engineering systems and especially mechanical systems have been integrated with architectural design and often drive design decisions, both for mechanical solutions as well as in using building physics.

Similarly to the way a truly sustainable design is an integral part of the architectural design, truly sustainable city planning is an integral part of city planning. City planning sets the upper boundary of what is possible for all five sections of the Envision® assessment: (a) Quality of Life, (b) protection of the Natural World, (c) responding to the risks of Climate Change, (d) Resource Allocation especially during operations and (e) Leadership.

Planning Guidelines in Existing Cities

In the western world, new cities are infrequently planned. Instead, the focus is on city expansions and on upgrading existing aging infrastructure systems. The Zofnass Planning Guidelines are equally valid for making decisions to improve the sustainability of existing cities. However, the existing conditions define the latitude of options for the planners. The four levels of infrastructure systems provide an insight for each infrastructure type into what is possible and under what circumstances.

The first system level defines the demand for each type of infrastructure system. The demand for energy and water is higher for an existing stock of buildings of older technology and, in the large scale, upgrading the buildings can slowly reduce this demand. Demand for transportation is also difficult to reduce once the choices of prior urban planning have been made. The demand for solid waste handling is more flexible as it depends less on fixed assets. Analogous observations can be made for all seven types of infrastructure systems.

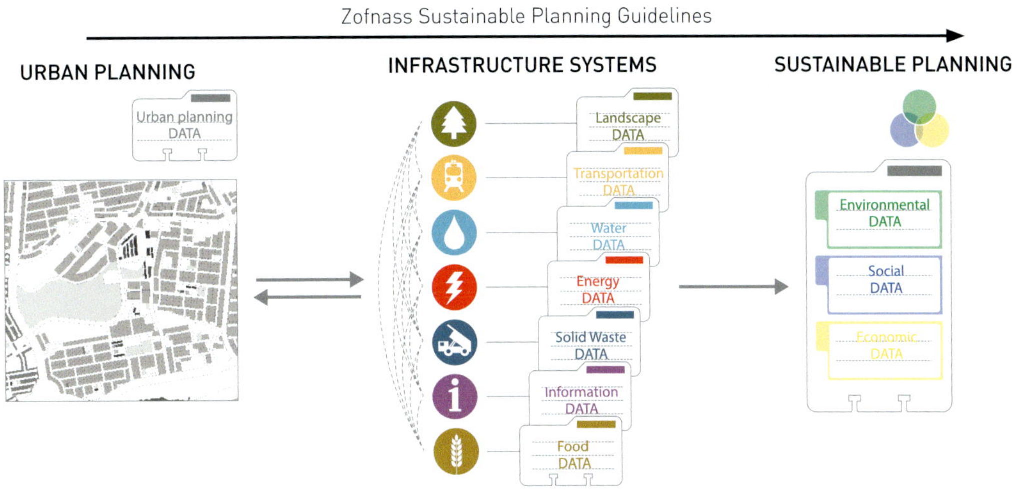

The second level defines the strategic decisions for the infrastructure systems. Plenty of interventions can be made at this level. Decisions for the development of new energy sources, for more efficient water treatment plans, for switching to public transportation can be made with manageable consequences for the existing urban fabric. Strategic decisions on sustainable infrastructure can also be implemented within the urban fabric or in the hinterland of the city. Within the urban fabric, they are bound by the available space, and implementation usually has a higher cost than in new cities. On the other hand, for nodes in the hinterland, the difficulty lies with remoteness and the added requirements of transport.

The third level defines the nodes of the infrastructure systems. Usually nodes can be upgraded as they take up limited space. IT centers, transportation stations, power production units, and material recovery facilities are in this category.

The fourth level defines the network of the infrastructure type. These are usually more complicated to upgrade as access may be difficult or there may be no available space for expansion. Water pipes, electric power distribution networks, railroad tracks, and roads are in this category.

Despite the external constraints on interventions to make existing cities more sustainable, the Planning Guidelines still provide a basis for considerations and actions. On the positive side, there are beneficiaries in existing cities who demand that their cities be more sustainable, and they can undertake the financial burden, either through taxation or with pay-as-you-go models of public-private partnerships.

Planning Resilient Cities

Defining resilience as the ability of a system to retain processes and performance by adapting to change and dealing with undesirable events, the Planning Guidelines ensure the resilience of the infrastructure systems in the sustainable city.

The Climate and Risk section of the Envision® credits addresses the risks associated with climate change, namely extreme weather conditions leading to both scarcity of water and flooding and sea level rise. The Planning Guidelines, based on Envision®, caution and specify actions to face such conditions and address the security of supply to retain processes and performance by adapting to change and dealing with undesirable events. However, resilience is also needed for non-climate-induced disasters, such as earthquakes, oil spills, and fires. The Envision® credits address these at a higher level of abstraction. The "innovation credit," the last credit in each section, pushes the stakeholders to think proactively and address potential disasters before they happen.

The Planning Guidelines address social resilience indirectly through the sections on Quality of Life, the Natural World, and Leadership. Prudent planning reduces social stresses. Envision® and the Zofnass Planning Guidelines have the ingredients to effectively address resilience. However, further research is needed to provide a complete framework of guidelines on ensuring resilience to cities.

Financing Sustainable Infrastructure Systems

The Planning Guidelines, as presented in this book, do not address financing, which is central to any project. The overarching approach is to define the technical approach first and then introduce financial considerations. In any case, the question is not whether we should plan sustainable infrastructure systems but their optimal form based on a series of considerations including financing, especially since preliminary studies show that sustainable planning increases a project's longevity and costs less over the life cycle of a project. The link of the Planning Guidelines to financing is a subject for further research.

2.2 | FINANCING SUSTAINABLE INFRASTRUCTURE

Ana Maria Vidaurre
Principal Investment Officer, Infrastructure Division of Structured & Corporate Finance Department, Inter-American Development Bank (IDB)

With governments cutting back on infrastructure spending due to budgetary constraints and banks tightening on credit, infrastructure investment has declined sharply. According to data from McKinsey & Company and the World Economic Forum, an estimated $57 trillion will be needed to finance infrastructure development worldwide between now and 2030.[1] This amount does not take into account the cost of repairing existing infrastructure, renewal backlogs, or increasing infrastructure's resilience to the impacts of climate change. In Latin America and the Caribbean (LAC), the shortfall is enormous, with an annual need of over $250 billion of investment for the next few years.[2] According to research from the OECD, the existing amount of paved road and railroads in the region is just 25 percent of what it should be expected given the region's socioeconomic characteristics.[3]

Latin America has become the most urbanized of the emerging market regions, with 80 percent of the population already living in cities and the proportion expected to rise to 90 percent by 2050.[4] According to the UN Population Division, Bogotá (Colombia) and Lima (Peru) will have more than 10 million inhabitants by 2030, joining the four current megacities of the region: Buenos Aires, Mexico City, Rio de Janeiro, and São Paulo.[5] In this context, existing infrastructure is not only unable to meet current and future needs, but it is under considerable strain causing costly bottlenecks, shortages of services, and environmental problems. Increasing the resilience of the region's urban infrastructure will be critical to improve quality of life and social inclusion in cities.[6] Infrastructure investment in the LAC region exceeded 3 percent of GDP in the 1980s, but since then it has declined sharply.[7] To plug the existing infrastructure gap, it is estimated that the region should invest around 5 percent of GDP per year, compared to the 1.7–2 percent that is currently being invested.[8] This funding shortfall has notable implications for economic growth, competitiveness, and social well-being.

At a time of fiscal restraint and weak debt and equity markets, tapping private capital is essential to bridge the gap between available public funds and necessary infrastructure investments. Public-private partnerships (PPPs) can be a way to generate and harness synergies, leverage financial resources, and use the expertise of the private sector in designing, building, and operating infrastructure across the life cycle. However, regulatory changes under Basel III have established higher capital reserve requirements and tougher liquidity standards, discouraging traditional long-term bank lending to infrastructure projects. This poses a major challenge for infrastructure financing, but it also creates an opportunity for new players to fill the funding gap. Long-term, nontraditional lenders such as pension funds may be a good match, given their long-term liabilities and the long-term cash flows of infrastructure projects. According to *The Economist*, these institutional investors are well resourced and have over $50 trillion to invest, but only 0.8% of their assets are currently set aside for these projects.[9]

To attract capital from these lenders, it is essential to understand their risk-averse nature and the need for stable and reliable cash generation at a project level. Some incentives that can entice institutional investors into funding projects include a clear and predictable regulatory framework that is able to outlive a particular government, greater transparency in the allocation of resources, bankable key project contracts with clear allocation of responsibilities, as well as visible and reliable information on performance expectations. Financing structures can also be adjusted based on timing of investor's participation in order to best address its risk appetite. In this sense, guarantees could be considered during construction phases, or during operations, covering certain risks. Also, projects could be structured with governments taking on the riskier elements and leaving the more stable part for the private sector to finance.

Sustainability may also be part of the equation. According to the international financial services holding corporation State Street, increased awareness of environmental, social, and governance concerns is influencing more and more the investment decisions of institutional investors.[10] Although it is perceived to make infrastructure projects more costly, sustainability can de-risk infrastructure projects, enhance performance, and provide better risk-adjusted returns in the long term.[11] To underpin tomorrow's economic growth, we have to adapt today's infrastructure to the impacts of climate change. For this reason LAC countries should not only invest to increase their stock of infrastructure, but also in innovative solutions to promote cross-sector synergies, increase resilience, improve the quality of infrastructure services, and boost their positive impacts on social inclusion and poverty reduction.[12] Unpredicted extreme weather and geophysical events pose a major challenge to the region's highly vulnerable infrastructure. Unless current trends are reversed, carbon dioxide emissions in the LAC will increase by 60 percent through 2050, entailing a cost of up to 2.5 percent of GDP.[13] To mitigate climate change, the region has to develop a sustainable economy with a low-carbon infrastructure network that reduces greenhouse gas emissions and ensures energy efficiency. Projects should be designed and implemented based on the concept that infrastructure is an asset that must be properly built, managed, and maintained.[14] Mainstreaming sustainability into infrastructure planning, design, construction, and operation may also reduce project risks through increased transparency, stakeholders' involvement, inclusive practices, innovation, and leadership. Hence it may contribute to further attract much-needed long-term sources of funding.

NOTES

1. McKinsey Global Institute, "Infrastructure Productivity" (2013), http://www.mckinsey.com/~/media/McKinsey/dotcom/Insights%20and%20pubs/MGI/Research/Urbanization/Infrastructure%20productivity/MGI_Infrastructure_Full_report_Jan2013.ashx.
2. Inter-American Development Bank, "Investment, Integration, Infrastructure," *OFID Quarterly* (July 2014), http://www.iadb.org/se4all/documents/OFID_quarterly_Pres._Moreno_interview.pdf.
3. OECD Latin America and Caribbean Investment Initiative, "Bridging Infrastructure Gaps through Smart Investment" (2014), http://www.oecd.org/daf/inv/investment-policy/Agenda-LAC-Peru2014.pdf.
4. Mike Murkowski, Gabriel Goldschmidt, and Mary Rose Brusewitz, "Will Latin America Be Able to Solve Its Infrastructure Problem?", *Latin American Advisor* (October 15, 2014), http://www.thedialogue.org/page.cfm?pageID=32&pubID=3686.
5. UN Population Division, "World Urbanization Prospects" (2014), http://esa.un.org/unpd/wup/Highlights/WUP2014-Highlights.pdf.
6. Murkowski, Goldschmidt, and Brusewitz, "Will Latin America Be Able to Solve Its Infrastructure Problem?"
7. T. Serebrisky, "IDB Infrastructure Strategy: Sustainable Infrastructure for Competitiveness and Inclusive Growth" (2014), Inter-American Development Bank, Washington, DC, http://www10.iadb.org/intal/intalcdi/PE/2014/14088en.pdf.
8. CG/LA Infrastructure Inc., "Strategic Top 100 Latin American Infrastructure Report" (2014).
9. "Infrastructure Financing, a Long and Winding Road: The World Needs More Infrastructure. How Will It Pay for It?" Economist, print edition, March 22, 2014, http://www.economist.com/news/finance-and-economics/21599394-world-needs-more-infrastructure-how-will-it-pay-it-long-and-winding.
10. State Street, "Sustainable Investing: Positioning for Long-Term Success" (2010), http://www.statestreet.com/wps/wcm/connect/1f9fae8044624a16aa02eb27a2dfc506/Vision+Focus+Sustainable+Investing_FINAL.pdf?MOD=AJPERES&CONVERT_TO=url&CACHEID=1f9fae8044624a16aa02eb27a2dfc506.
11. Global Infrastructure Basel Foundation, "Sustainable Infrastructure as an Asset Class" (2014), http://www.gib-foundation.org/content/uploads/2014/03/Sustainable-Infrastructure-as-an-Asset-Class_V7.1.pdf.
12. Serebrisky, "IDB Infrastructure Strategy."
13. Ibid.
14. Ibid.

2.3 | SYNERGIES AND SUSTAINABLE INFRASTRUCTURE PLANNING

Marty Janowitz
MES, ENV SP, Vice President & Practice Leader Sustainable Development, Stantec Corporation

In the context of urban infrastructure, one plus one often leads to three, four, five, or more. This implies that the conventional way of seeing infrastructure units in relative isolation has underappreciated their connections, interactions, and effects on the larger array of infrastructure components. This long-standing isolationist approach is characteristically conveyed by the typical characteristics of siting decisions and planning or design requests for proposals (RFPs). Until recently, did we ever see wastewater treatment facilities proximate to the energy utilities that could use their cooling discharge water or that considered distances for transport and disposal of sludge? Did conventional highway plans and designs pay much attention to stormwater quantity or contaminant impacts on water treatment facilities, much less the placement, capacity, and future demands for schools, shopping, or energy utilities in newly accessible but undeveloped areas? And what about quality-of-life implications from urban sprawl begetting increased commute times, noisy trucks carrying sludge through residential neighborhoods, or parklands cut off by highways from pedestrian or bicycle access?

In this era of growing interest in and sensitivity to urban sustainability, we could quickly take a narrow approach and focus merely on localized resource or energy efficiencies and reduced environmental impacts. These changes are good things—increased use of recycled or reusable materials, increased energy efficiency, better contaminant containment and treatment, etc. But these are a step removed from comprehensive, integrated infrastructure planning and design that accepts the premise that there are inevitable, inextricable, and multiple interactions and potential connections between the infrastructure systems of Energy, Waste, Water, Transportation, Landscape, Information, and Food. Even further, these infrastructure systems have significant interactions and impacts on other dimensions of urban communities—education, health, recreation, and commerce, to name only a few.

It may be easier to grasp the significance of synergism in infrastructure if for a moment we shift our attention from civil works to the human body. We instinctively understand that the body is an integrated system. All parts are interrelated and interdependent. Each finger is connected and interrelated both materially and functionally to virtually every other part of the human system, with obvious neurological, sensory, cardiovascular, and even psychological connections. Simply put, we can scratch behind our ear, feel and respond to a splinter, understand how cold it is, and shake hands warmly, all through the activity or sensory dimensions of an index finger. Today's spicy meatball and double dose of wine may become tomorrow's headache, indigestion, and heartburn. This morning's missed stretches may be the forerunner of this afternoon's backache and stress. The rapid adoption (at least in western developed cities) of healthy diet and exercise materials, training, and resource "infrastructure" is a direct result of the correlation that has been accepted between healthy lifestyles and longer, more vital lives. I suggest we think about infrastructure in a similar organismic fashion, appreciating that our transportation, waste and wastewater, and other civil infrastructure are basically the circulatory, digestive, and skeletal systems of society. At the level of body we easily experience that health, vitality, and longevity are optimized when each system is healthy and when the relationships between them are in balance. Can we take a similar synergistic approach to infrastructure systems?

Such an approach will ensure that discrete elements within the urban infrastructure function to their highest and most efficient potential individually and in relation to all other infrastructure components. Synergies are seen as two-way associations between different systems. Within infrastructure categories, each infrastructure component is intended to be managed as an integrated element within its extended system. Perhaps even more significantly, persistent attention is paid to the functional relationship between infrastructure types, which are therefore considered subsystems of the city. This necessarily requires extensive cross-disciplinary understanding and dedicated attention to design, planning, and operations that are interwoven, aligned, and complementary.

Infrastructure systems are consumers of resources themselves, but mainly managers and providers of resources and services for their end users. How they fulfill these functions is as important as the fact that they do. Integrating flows and components between the different infrastructure systems is necessary to achieve efficiency, resilience, quality of space, and growth. This is even more pressing given rapid global urbanization. We must reflect on the fact that almost half of today's global population is living in cities, a trend that continues to accelerate. Therefore the role of urban infrastructure systems becomes even more critical to supporting healthy, functional human societies, avoiding harmful pollution and environmental degradation, and strengthening resilience and adaptive capacity against seemingly more frequent and powerful natural calamities. The performance and stability of infrastructure is the backbone of the operation and development of cities, affecting the availability and allocation of resources and health, social, and surrounding natural conditions. This requires integration on multiple levels, transforming practices of urban planning and engineering, the decision-making processes of stakeholders, and harmonization between natural and built systems and of course between infrastructure elements themselves.

By now we have largely accepted the principle of avoiding, minimizing, or mitigating the adverse environmental effects of human works. Environmental impact assessments, protection plans, and strict regulatory or permitting requirements are a fact of engineering and planning life. The further step will be adapting the management of infrastructure systems to the needs of the natural systems themselves by seeking opportunities for restorative, genuinely symbiotic, and even mutually enriching interrelationships. Beyond the aspiration for an ecologically grounded humanity, enhanced adaptation of infrastructure to its related natural systems typically contributes increased resilience and efficiency and reduced need for new infrastructure.

One further dimension is worth mention. While infrastructure may generally reside beneath the consciousness of most citizens, in recent decades stakeholders and citizens have increasingly been active protagonists in infrastructure-related controversies. The spheres of "not in my backyard"—prospective adverse effects on property values, human health, aesthetics, and the natural environment—have all illuminated the important relationship between citizens, governments, engineers and planners, and the laws or regulations we construct to define them. Whereas in the past, the responsibility of proponents and designers may have given nodding acknowledgment to these citizen interests, they have often been framed as "interveners" (read that as "getting in the way") or opponents, with public consultation and engagement generally viewed as merely a necessary hurdle. In the past decade or more, this view has generally and radically evolved, to the point where stakeholders are occasionally if not regularly considered constructive collaborators. This synergy can actually contribute to better projects, to doing projects right and only the right projects.

If a synergetic outlook is now a fundamental of good practice, we have long needed a transparent, commonly communicated and readily understood language and framework for addressing issues, options, and choices. The integrated framework outlined herein goes a long way to meeting this need, in part by recognizing that city planning and infrastructure planning should be performed in parallel. Land use planning determines the end users and their needs for services and resources that infrastructure should supply. Similarly, the requirements and characteristics of infrastructure affect the configuration, form, and structure of the urban fabric. These may both have immediate effects and substantially influence future prospects and directions for development, settlement, and community quality of life. The quality and type of services provided by infrastructure can therefore influence and even determine the financial development and social environment of a city.

In sum, a synergistic approach to infrastructure planning intends to improve or facilitate the function of other systems by reducing the initial demand of other systems, connecting by-products and feedstock needs, optimizing placement of entities, combining entities, and mitigating negative impacts of processes. Planning analysis should prioritize integrated, systemic, and high-level planning while advancing the crucial issues of each system. This typically leads to engagement with a broad range of interrelated issues, since city systems depend upon their interconnectivity to function.

Addressing these complexities will require extensive and detailed cross-disciplinary understanding of multiple issues, across multiple domains. All infrastructure systems are interconnected through synergies, but some have a stronger and inherent synergetic nature which will be explored in subsequent chapters. The important work of the Zofnass Program highlights and is addressing these implications, which are at the core of its objectives and methodologies as it works to understand, map, and create tools to optimize systemic synergies at the city level.

PART 2

SUSTAINABLE INFRASTRUCTURE PLANNING GUIDELINES

Part 2 is the core of the book. Chapter 3 describes the scope and methodology of the research that led to this publication, and serves as a manual guiding the reader through the book. Individual chapters are then dedicated to the infrastructure systems of Landscape, Transportation, Water, Energy, Solid Waste, Information, and Food. Each chapter discusses the basic sustainability challenges and principles for that system, and concludes with Planning Guidelines organized in tables.

General Framework

Chapter 3

3.1 | SCOPE

This book provides an analytical framework for achieving urban sustainability, focusing on the services and the performance of infrastructure systems. The concentration of people and economic processes that defines the urban condition requires the sustaining of continuous and reliable flows of resources; infrastructure is the backbone of these flows. Urban planning and infrastructure development are intertwined. Planning determines the end users and thus the demand for services and resources that infrastructure should respond to. On the other hand, the space requirements and distribution of infrastructure facilities and networks affect the configuration of urban areas and the morphology of the urban fabric. Planning and infrastructure development should proceed in parallel to ensure efficiency, high living standards, and resiliency. Urban concepts like mix and distribution of land uses, density, compactness, and centralization versus decentralization need to be negotiated in terms of the infrastructure concepts of services demand, system hierarchy and distribution, efficiency and effectiveness, supply chains, and routing of networks.

Infrastructure systems are mainly considered to offer *provisioning* services covering specific demands for energy, water, food, mobility, information, etc. However, the complex and heterogeneous processes affected or set in motion by urban infrastructure call for the consideration of *regulating* services as well. Examples include the control of contaminants, the preservation of habitats, and protection against extreme phenomena. At the same time, infrastructure is very much present in the public space, constantly interacting with communities and creating the framework for their development. This signifies the *sociocultural* as a third type of service that should be covered by infrastructure planning. Infrastructure configuration has the ability to affect quality of life, promote different lifestyles, and form consumer habits. For example, the available transportation and telecommunication options create interfaces for the interaction of individuals and groups, while the options for waste management can change how consumers use and dispose of materials and products. It is crucial for infrastructure planning to identify and respond to the needs for all three types of services: provisioning, regulating, and sociocultural.

Urban infrastructure in this book is organized into seven basic systems: Landscape, Transportation, Water, Energy, Solid Waste, Information, and Food. Classification does not mean separation; the approach is intended to be a first step toward high-level, integrated planning. The next chapters approach urban infrastructure as a series of systems that should function in synergy and be directly linked with urban planning. "Infrastructure system" in this context includes not only physical structures but also operations, procedures, administration, bylaws, natural processes, and not least the end user. The aim is a collaborative and integrated planning process, an enabling framework that will help identify choices and opportunities, help guide decisions, and justify the final choices while offering simplicity and flexibility in the context of urban complexity.

LANDSCAPE INFRASTRUCTURE

Landscape infrastructure is the system of open spaces inside and around the city, offering a wide range of services tailored to the needs of the specific context. Landscape differs from the other infrastructure systems as it is not defined by its response to a specific demand. It is an alternative, soft, flexible infrastructure with the inherent potential of multifunctionality. Landscape adds to local identity, offers opportunities for recreation, supports habitats and natural processes, provides water, food, and material resources, mitigates and compensates for the negative impacts of other infrastructure systems, and increases city resilience against natural and man-made threats. In the context of this book, Landscape infrastructure is the key tool for protection against coastal, river, and storm flood risk. It is imperative for sustainable urban planning to manage the system of open spaces as components of an integrated system, configured to offer targeted services. The boundaries of this system are flexible and should be defined after careful analysis and study of natural connections, flows, and the city's needs. For example, the need to improve water quality can expand the Landscape system to include the city's watershed.

TRANSPORTATION INFRASTRUCTURE

Transportation infrastructure includes the facilities and processes in place to serve the mobility of passengers and freight. It is a basic catalyst for development, as it is closely interrelated with urban and rural economies. Its strong physical presence and the resulting large footprint create a range of challenges: it creates connections, but often also creates unintended barriers in the natural world and local communities. Sustainable Transportation infrastructure should therefore aim to avoid, minimize, and compensate for negative impacts on individuals, communities, and ecosystems in the process of its inevitable expansion to meet increased user demand. It is important to stress that sustainable transportation cannot be seen independently of sustainable urban development. The configuration of the road network, in particular, has a direct impact on the morphology of the urban fabric. Collaborative planning efforts should emphasize defining land use distribution, density, and compactness to support optimal transit options.

WATER INFRASTRUCTURE

Water infrastructure provides for the treatment, collection, and distribution of potable water, rainwater, and wastewater. The availability of potable water and the management of wastewater are, historically, key factors in the establishment and development of human settlements. Water infrastructure both extracts water from the natural world and returns it with minimum impact. So, in addition to securing and optimizing the performance of networks and facilities, sustainable Water infrastructure planning should include watershed and natural processes management, aiming to actually reduce the built infrastructure required. Control of contaminant sources and recharging of groundwater are part of the Water infrastructure, along with striving to reduce overall water consumption instead of just providing for it. Offering natural drainage and harvesting strategies for rain and stormwater management is another key function, particularly in urban environments where extensive water-impermeable surfaces are found. The planning for expansion of urban areas within the watershed, the distribution of land uses, and the configuration of urban fabric and surfaces should take into consideration their impact on Water infrastructure performance.

ENERGY INFRASTRUCTURE

Energy infrastructure manages and converts available natural resources into energy carriers like electricity, heat, and fuel, delivering them to end users. Energy security is fundamental for quality of life, economic development, and by extension social stability. Energy generation, however, is currently the leading source of greenhouse gas emissions worldwide. This has brought energy to the forefront of efforts for sustainability and resiliency. Sustainable Energy infrastructure should, therefore, secure constant and equal access to energy carriers for all individuals and businesses, while promoting rational use and efficient practices. It should use natural resources within the environment's regenerative capacity, providing renewable and low-emission alternatives.

SOLID WASTE INFRASTRUCTURE

Solid Waste infrastructure includes all the processes and facilities required for the collection, treatment, and final disposal of waste. Practices that largely depend on landfilling as a means of final disposal, however, have significant social and environmental impacts. Planning should therefore focus on source reduction but also on developing strategies that treat solid waste as a valuable resource. Waste management should be structured in close synergy with the supply chains of other systems, catering where possible to their needs for energy and materials. Waste has strong social and political dimensions as it is closely related to the production and consumption patterns and potential of a society. Engaging consumers in developing habits that will support the objectives of source reduction and diversion from landfills is therefore essential.

INFORMATION INFRASTRUCTURE

Information infrastructure supports the flow and processing of information. It consists of different subsystems that people and organizations use to produce, collect, filter, process, and distribute data. It can be characterized as a "next-generation public infrastructure" supporting human operations, decision making, and action. In recent decades, the availability of solid and high-capacity Information infrastructure has emerged as a cornerstone for growth and high-quality services. This book uses a twofold definition of information as infrastructure: first, it is approached as infrastructure per se, including telecommunication and internet services, and second, as a support system of all other infrastructure systems, to interconnect and assist them in achieving high performance. As such, Information infrastructure is flexible and inherently synergistic.

FOOD INFRASTRUCTURE

Finally, Food infrastructure involves the structures and processes used for the production, processing, distribution, and trade of food. It is quite different from other systems such as Transportation, Energy, and Water in that it has a more decentralized and private character. However, rapid population growth in combination with urbanization trends and changing consumption and diet patterns is expected to lead to an increased demand for food in the near future, also resulting in increased freight transportation. Intensified food production already puts pressures on the planet's natural resources and generates significant greenhouse gas emissions. In turn, climate change strains the capacity of existing food production. These conditions urge cities to make food management part of their overall planning, coupling ornamental landscapes with food production and thus becoming more self-sufficient and resilient. Planning should aim to protect the food-productive landscape near cities, improve the efficiency of the food supply chain, and ensure consumers' awareness.

In the scope of this book, collaboration and integration are pursued along three basic axes. First and foremost is the axis of stakeholders' efforts and objectives. Due to the size and complexity of infrastructure systems, multiple stakeholders with different priorities and agendas are inevitably involved during planning. This book aims to form the foundation of a common collaborative platform, primarily among public authorities, utility companies, financial institutions, planners, and engineers, to closely involve planning practices that have traditionally functioned in silos. Establishing shared knowledge that responds to the language of these professionals is integral to the approach. The second axis of integration runs between infrastructure and natural systems. Infrastructure occupies vast expanses of land, extracts large volumes of natural resources to use as feedstock, and produces by-products that are returned to the natural environment. These interactions can cause the alteration of climate patterns, the degradation of habitats, and the disruption of water and nutrient cycles and other processes. The impact can extend far beyond the boundary of a city, with significant and nonreversible effects. Infrastructure planning should therefore include a thorough understanding of the affected natural systems, in order to preserve or restore them or compensate for their disruption. The third axis of integration involves the infrastructure processes themselves. Planning decisions must identify and support linkages and interactions among infrastructure systems, aiming for efficiency, resiliency, and reduced impact. These linkages are referred to in this book as synergies. They can be used as opportunities for each infrastructure system to assist or be assisted by others, in order to resolve operational challenges, optimize placement of facilities, secure feedstock, resolve the management of by-products, and more. The synergies map how each system relates to the others, providing decision makers with a better understanding of the repercussions planning strategies can have across the city's operations. In this way they highlight potential traps and vulnerabilities and facilitate risk management.

These notions of integrated, systemic, and high-level planning, along with the crucial issues of the seven infrastructure systems, are presented through text and graphics and are encapsulated in the Planning Guidelines at the end of each chapter. Forming the core content of this book, tables of guidelines for Energy, Landscape, Transportation, Solid Waste, Water, Information, and Food are given, supporting a unified and cross-disciplinary process for improving existing urban developments or developing new ones. The guidelines can be used both as a practical tool for decision makers to optimize the contribution and interaction of various experts in the planning process, and as an educational tool for all related stakeholders independent of their degree of involvement with sustainable infrastructure. The result should be sustainable urban planning.

APPLICATION OF PLANNING GUIDELINES

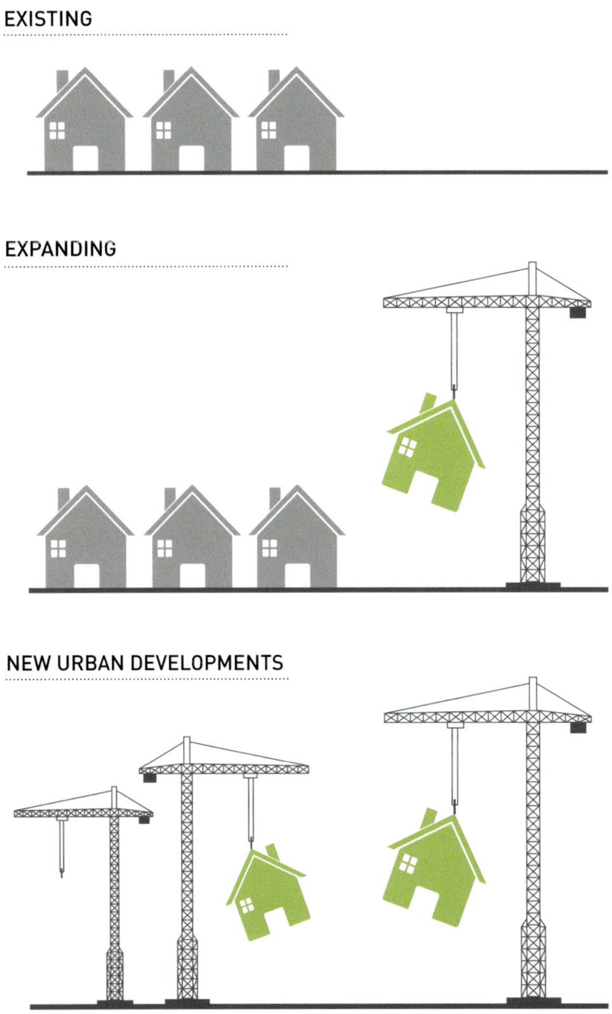

3.2 | **METHODOLOGY**

Cities can vary significantly in terms of their size, age, natural context, social and economic circumstances, and administrative structures. Urban environments are always works in progress, encompassing endless procedures and flows that often change rapidly. For these reasons urban planning projects present certain particular challenges. They have timeframes that often expand across decades, requiring initial plans to be constantly revised and adapted to ever-changing conditions. Numerous stakeholders from different backgrounds need to collaborate and reach consensus. The Planning Guidelines are therefore structured to provide a framework that can be applied to vastly different cities by stakeholders of varying expertise. They offer clear guidance for high-level planning, toward specific objectives, without repeating the complete set of requirements that should be met at project level. They provide flexibility in the face of the ever-evolving status of cities and the changes in available technologies and know-how. Users are free to define the boundaries of the system they are studying and use the guidelines for smaller or larger projects and for long-term or short-term planning.

As a method to streamline and guide the planning process, offering consistency while still highlighting particularities, each of the seven infrastructure systems is decoded in terms of four "system levels." The system levels are used to group the processes and various infrastructure facilities found within the systems and, most importantly, to structure a high-level planning and decision-making process. The definition of levels is consistent among infrastructure systems. The first level represents the main demand or consumption that must be covered by the system. It is crucial to base the design of any system on an understanding of realistic and up-to-date demand estimates and projections. The second level addresses the strategic approach and the resources employed in order to meet the demand as defined in the first level. The third and fourth levels represent the specific facilities and operations that will realize the strategic plans of the second level. Specifically, the third level refers to facilities performing as nodes of the system and the fourth to the networks set in place to cover the spatial needs. The only exception to this structure is the Landscape system. Time has a very strong effect on Landscape entities. While other systems start to decline after implementation, due to wear and tear, a well-planned Landscape system only becomes more established with time, adding value. The fourth system

ZOFNASS PROGRAM CATEGORIES

PLANNING GUIDELINES STRUCTURE

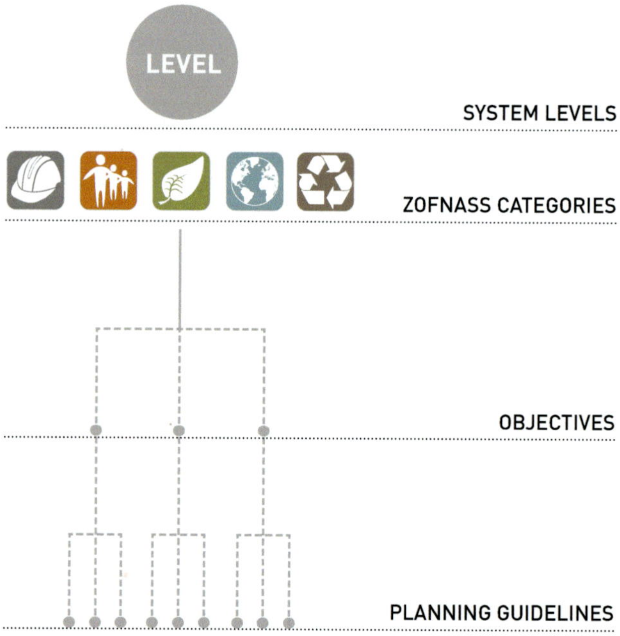

The Planning Guidelines do not describe specific solutions. Their role is to prompt project teams to develop a set of actions adapted to each individual project.

ACTIONS

level of Landscape infrastructure is therefore defined as Maintenance to address the need for adaption to natural cycles and patterns.

For each system level, a set of objectives is defined, indicating the general goals for planning decisions. The objectives for the first system level typically call for a clear analysis of the primary demand on the infrastructure system, and its optimization through the engagement of end users. For the second system level, the main objectives refer to the selection of context-adapted management strategies, diverse and secure supply resources, efficient technologies, and adequate allocation of by-products. In the third and fourth levels, core objectives address the placement, distribution, and management of facilities and networks for optimum, secure, and flexible operation. The objectives are drawn from the five focus areas for sustainability used by the Zofnass Program: Leadership, Quality of Life, Resource Allocation, Natural World, Climate and Risk.

Leadership encourages the development of holistic strategies for high-level planning through innovation, integration, and stakeholder engagement. Quality of Life addresses the physical, economic, and social impact of planning decisions on the affected communities. It includes the concepts of health, well-being, community involvement, and local identity. Resource Allocation refers

INFRASTRUCTURE SYSTEM LEVELS

to the energy, water, and physical materials required to build an infrastructure system and operate it, as well as to the by-products and residues that leave the system. Each of these should be treated respectfully and efficiently. Natural World encompasses the habitats, species, and nonliving natural systems that interact with an infrastructure system. This area focuses on understanding and minimizing the negative impacts of infrastructure while incorporating natural systems in a synergistic way. The scope of Climate and Risk is to minimize greenhouse gas emissions and other pollutants and to ensure resiliency for infrastructure and city against changing conditions and natural and man-made threats, both short-term and long-term. Each planning objective is primarily attributed to one of these focus areas, though it might be relevant to other areas as well.

The Planning Guidelines are given as guidance for reaching each objective. This structure serves several purposes. Objectives encapsulate the core issues for the sustainability of each infrastructure system, allowing for a quick grasp of the challenges ahead. They provide a general framework for assessing the overall direction and priorities of planning and provide flexibility as they apply to all contexts. At the same time they work complementarily to the guidelines, better illustrating the purpose of each. The guidelines do not prescribe specific actions in order to reach the objectives; rather they determine the direction of planning through objectives and they leave specific actions open-ended. Thus, the guidelines approach facilitates the creation of plans customized to each city's unique context, conditions, and priorities. It promotes innovation, allowing for the incorporation of new or improved technologies as they evolve. Project teams can select from a wide spectrum of actions: technical, administrative, educational, regulatory, etc. The decision-making process using the guidelines is not linear but an iterative process. While the system levels structure indicates a sequence in planning, the entire set of objectives and guidelines should be taken into consideration before finalizing decisions.

In addition to guiding planning, the guidelines can be used to assess existing cities and systems, identifying areas for improvement or monitoring the effect of plans already in place. They can also serve as a framework against which to compare the effect of smaller projects on overall city goals. Their modular structure provides further versatility. They can be applied to all seven infrastructure systems, covering the majority of processes within a project, or to one specific infrastructure system, or can even be used to analyze a specific system level. However, it should be stressed that integrated planning requires consideration of all four system levels. The flexibility of the guidelines is further enhanced by tracing and highlighting the synergies among systems. When a guideline is relevant to another system, the connection is highlighted, as shown in the graph below. At the end of each chapter these connections are identified in tables that reference specific relevant guidelines from the other systems. Relevant guidelines facilitate developing integrated and multifaceted action plans, tailored to the needs of each project, by drawing elements from different sectors.

Two types of synergies are identified (TO and FROM), based on whether a guideline can assist another system or be assisted by the other system. Thus, project teams can identify how their decisions impact other infrastructure or whether actions in other systems should be included for integrated sustainable planning. Not all interactions among infrastructure systems are noted; the emphasis is on highlighting strategic connections with significant impact that are relevant to high-level planning. A third type of synergy (WITH) is identified in relation to Information and Landscape infrastructure. (WITH) indicates that a certain guideline might require planning for an Information or Landscape subsystem following multiple guidelines from these systems. Both Information and Landscape constitute highly synergic systems, able to offer a multitude of supporting and regulating services, determined to a significant degree by the needs of other systems. Information in particular can constitute a subsystem of all other infrastructure systems, assisting in operations and management, while Landscape elements can be used to provide feedstock or mitigate the impact of processes on communities and the natural word.

ABBREVIATIONS OF INFRASTRUCTURE SYSTEMS

L → **Landscape**
T → **Transportation**
W → **Water**
E → **Energy**
SW → **Solid Waste**
I → **Information**
F → **Food**

OBJECTIVES AND GUIDELINES IDENTIFICATION

T → **Infrastructure System**
T1 → **Infrastructure System Level**
T1.1. → **System Level Objective**
T1.1.1. → **Planning Guideline**

SYNERGY INDICATIONS

SAMPLE OF PLANNING OBJECTIVES TABLE

T1
TRIP GENERATION

T1. TRIP GENERATION	
T1.1.	Reduce number of unnecessary trips
T1.2.	Reduce length of trips

1. Planning Objectives are given for each System Level. They are color-coded according to the Zofnass Category they reference

SAMPLE OF PLANNING GUIDELINES TABLE

T1
TRIP GENERATION

Objective		
T1.1. Reduce number of unnecessary trips	T1.1.1.	Provide and promote remote access and e-services
T1.2. Reduce length of trips	T1.2.1.	Promote mixed-use development
	T1.2.2.	Include basic amenities within each residential area
	T1.2.3.	Promote urban fabric compactness
	T1.2.4.	Optimize placement of other infrastructure facilities generating significant traffic

2. Planning Guidelines are given for each Objective

3. Synergies with other infrastructure systems are indicated

SAMPLE OF RELEVANT GUIDELINES TABLE

T1
TRIP GENERATION

Guideline			
T1.1.1. Provide and promote remote access and e-services	I1.1.3.	Identify opportunities to improve city administration and social amenities through information infrastructure and applications
T1.2.2. Include basic amenities within each residential area	F3.1.1.	Promote the equal and dense distribution of food retail facilities across the city
T1.2.4. Optimize placement of other infrastructure facilities generating significant traffic	F4.5.2.	Optimize placement of food processing and distribution facilities to reduce number and length of food distribution trips across the city
	SW4.2.1.	Optimize placement of solid waste transfer stations between solid waste generation locations and treatment / final disposal facilities

TO

WITH

FROM

4. Relevant Guidelines from other Systems are cross-referenced based on Synergies

IMAGE CREDITS

P. 27: Infrastructure Systems' Levels icons designed by Alexandra Papagianni.

Landscape

as Infrastructure

Chapter 4

Landscape

is a regulator of endless processes, producer and consumer of raw materials, highly multifunctional and synergistic.

4.1 | INDUSTRY PERSPECTIVES

Landscape as Infrastructure: An Industry Perspective

Landscape as a high-performance and multiobjective infrastructure, and its potential

The concept of "landscape as infrastructure" occurs relatively infrequently in the global megatrends dialogue. In fact, the topic of landscape is almost invisible there, although it is integral to the consequences of megatrend interaction, including population concentration, well-being and quality of life, resource consumption, and private versus shared space. The guidelines presented here explore innovative ways to integrate the needs for energy, water, transportation, food, solid waste management, and information in the context of climate change and accelerating growth of the urban population. Whereas landscape has always had the ability to accommodate change, these accelerating demands are forcing a paradigm shift in how we deliver these needs today without compromising the resources for tomorrow.

Livability and people-first-focused infrastructure mean a rediscovery of the essentials of improving quality of life. The physical attributes of a more sustainable city interplay to create landscapes and cityscapes of a vernacular form and function. As different metrics are increasingly used to measure success (Human Development Index and Gross National Happiness), it is the livable city that will be the most competitive.

Various approaches to defining livable cities cite sustainability and infrastructure issues as essential components and stress the need for coordinated, multidisciplinary approaches.

How the spatial configuration of landscape entities can be connected to specific services

As the urban population increases worldwide, the protection, conservation, and enhancement of high-quality multifunctional landscapes both in our cities and countryside is becoming more important. Indeed the opportunities afforded by integrating the other utilities of sustainable infrastructure (both retrospectively in our existing cities and in planning new ones) have the capacity to address all our future needs.

These values are applicable at a regional scale as well as a city scale. The development of large-scale energy-related infrastructure, such as reservoirs formed to supply water and generate energy, can be more easily integrated into a landscape at corresponding scale. The reservoir in Central Park is of a scale and design appropriate for a city park, while Lake Havasu and the Hoover Dam provide the same function (in addition to power generation) but are of appropriately regional scale. There is additionality with both, of course, through ecological enhancement and recreational activities.

In the more modern past (the last 150 years) landscape has traditionally been provided by city authorities by way of public expenditure via tax or levies. This has not always been the case: many of the megacities of today were planned by industrialists during the Industrial Revolution who realized that healthy workers were more productive workers. The same key 19th-century issues of providing heat and light to homes and factories, providing clean water and removing foul water, and moving goods and services around quickly and in increasing volume all remain key issues today. The issues of urban surface water drainage and adapting to and mitigating the effects of climate change are more recent additions.

The opportunities to take a new approach to integrating landscape as infrastructure are, more often than not, caught in the weeds of politics, land ownership, and bureaucracy. This is both a hindrance but also an opportunity for cities to help themselves. Several cities in Europe (Dresden, Hamburg) rebuilt themselves after World War II to include green space, boulevards, and pedestrian-friendly transport. Maybe cities like Detroit have an opportunity to rebound from economic adversity by attracting inward investment through a new approach to city planning where landscape is the blueprint for development? Maybe handing over large parts of the city to private enterprise and employers, as we experienced 150 years ago, is an option?

How translating landscape in terms of ecosystem services and landscape functions promotes a sustainable integrated regional and urban planning

We can learn from the past to meet the challenges and opportunities presented by the need for more sustainable infrastructure for our society. The ongoing rapid increase in population in our towns and cities is not new. Particularly during the Industrial Revolution, cities experienced very large spikes in population and whole new cities were created. Landscape has been key to addressing the consequences of this phenomenon in the past and can be again in the future. The creation of Central Park in New

York City (from arable land on the outskirts of what was then a smaller city) gives us confidence that the pressing issues of water supply, city cooling, recreational needs, food security, etc. can be planned for. In addition to facing the added challenge of climate change, we have perhaps lost the organizational design to realize such ambitious projects. Perhaps as much as climate change, the issues of leadership, organizational design, and "who pays today" are the big challenge in making infrastructure more sustainable in the next 100 years.

Priorities, challenges, and possible conflicts of current and emerging sustainable landscape practices

As well as serving a range of physical and social functions, landscapes stimulate highly emotional responses. Perhaps our most treasured possessions in any city or town are our public spaces. Nothing provokes such public anger as the threat to bulldoze a local public park or greenbelt; more positively, nothing motivates a community more than the opportunity to create an urban pocket park, community garden, or restore a historic open space that has become part of the fabric of the city. John Ruskin, the eminent 19th-century writer, wrote of the history and importance of open spaces:"The measure of any great civilization is in cities, and a measure of a city's greatness is to be found in the quality of its public places, its parks and squares."

The key matter for a more fundamental rethink is: who pays today to establish the framework for a healthier, wealthier society to come?

Research for the London Sustainable Development Commission: Sustainability Challenge London[1] in 2008 looked at examples of integrated sustainable action in other cities around the world where benefits of more sustainable approaches have been demonstrated and are potentially transferable to London. In addition, questionnaires were sent to key representatives of the business, public administration, and voluntary sectors to gain a preliminary understanding of some of the challenges these sectors are facing in working toward a more sustainable London. Visionary leadership and clear objectives were identified as vital for success. Plans, policies, and procedures (typically associated with 21st-century western governance and process in realizing infrastructure) can only get us so far. It is people (leadership and community) who provide real change.

Landscape infrastructure as the principal mitigation strategy to compensate for adverse impact of other infrastructure systems on the natural world

The highly synergistic nature of landscape infrastructure (including cityscape, townscape, rural, and naturalized as well as natural landscape) provides a wide range of functions that may compensate for the adverse impact of other infrastructure systems on the natural world, thereby making our cities and society more sustainable. For example, high-quality open green spaces within our towns and cities provide environmental, economic, and social benefits—as carbon sinks (one tree over a 30-year period consumes approximately 1.7t of CO_2; a mature tree can save an average of 22kg of CO_2 per year);[2] heat sinks (reversing heat island effects through nonreflective surfaces and shading); stormwater storage; water recycling systems (filtration systems); transport corridors (roads and rail as well as walk and bike); ecological habitat and biodiversity; energy providers (ground energy systems, solar farms, wind turbines, mini-hydropower); and places for food production (reducing food miles) and green waste recycling. Landscapes, new or retrofitted, are at the forefront of efforts to mitigate and adapt to the effects of climate change.[3] Landscape also provides for less tangible benefits such as health and education. In the past 10 years we have begun to quantify the benefits to society of access (physical and intellectual) to open spaces. Good access (typically within a 12-minute walk) has direct benefit on the healthcare system in terms of exercise, rehabilitation, and also mental health.[4]

We require new ways to analyze our need for more sustainable infrastructure and the potential for landscape to help deliver our infrastructure needs. Completing research on the economics of the sustainability triple bottom line is essential, followed by a focus on the long-term return on investment involving the private sector. For instance, an examination of retirement expenditure is likely to reveal that significant amounts of money will be spent on healthcare, recreation, etc. that can be provided through better-quality and better access to landscapes. Without a consistent decision-making framework, will the outcome be as clear when evaluating safeguarding a rural landscape for agricultural use and food security versus using it to house an increasing population in a new or existing city—or determining which lands our floodwaters will cover if our floodplains have already been redeveloped—or deciding whether to redevelop our brownfields when, if naturalized, they could become part of the "green lung" to reduce urban heat island effect?

Why where matters

Around the world, pressing issues continue to drive innovation and implementation of landscape as infrastructure. The city of Chicago is harnessing the potential of trees to reduce the impact of heat waves by developing and implementing a strategic management plan for its urban forest. The need to reduce stormwater runoff in Chicago's alleys led to the Green Alley program, which introduced wide-ranging environmental improvements to neglected spaces. The Green Streets project in Portland, Oregon, "features attractive curbside planters that absorb flash flood water runoff—essential in a city with nearly a meter of rainfall each year."[5] At Bristol Business Park in Bristol, UK, a sustainable urban drainage system (SUDS) prevents the possibility of water runoff flooding areas such as the nearby village of Hambrook. Seonyudo Park in Seoul, South Korea, is an award-winning ecological park, created by the conversion of a former water treatment plant. At the East London Green Grid in London, a strong planning framework and area-based structures have helped develop new and existing green infrastructure to tackle quality-of-life issues in the Thames Gateway growth area. The Manor and Castle Green Estate in Sheffield, UK, is a public-sector/social enterprise partnership using multifunctional green infrastructure to tackle multiple deprivation at the neighborhood scale. "Environmental improvements have transformed Augustenborg in Malmö, Sweden, from a neighborhood in decline to an exemplar of an environmentally adapted urban area, making it once again an attractive place to live and work."[6]

Jonathan Buckley
BA(Hons) DipLA, CMLI, FFB, Chartered Landscape Architect and former Global Urban Development & Infrastructure Leader at Golder Associates
Jeanette M. Southwood
MASc, PE, FEC, FCAE, QPESA, QPRA, Engineer and Global Sustainable Cities Initiative Leader / Canadian Urban Development & Infrastructure Leader / Principal, Golder Associates

NOTES

1. Golder Associates, "Sustainability Challenge London: Sustainable Cities Initiative," commissioned by the London Sustainable Development Commission, 2008.
2. Golder Associates, 2010.
3. Commission for Architecture and the Built Environment (CABE), UK, "Public Space Lessons: Adapting Public Space to Climate Change," 2009. Further reading by following the case studies included at the end of the text.
4. UK Faculty of Public Health in association with Natural England, "Great Outdoors: How Our Natural Health Service Uses Green Space to Improve Wellbeing," 2010, http://www.fph.org.uk/uploads/bs_great_outdoors.pdf.
5. Commission for Architecture and the Built Environment (CABE), UK, "Case studies: Green streets, Portland, USA," accessed January 13, 2014, http://webarchive.nationalarchives.gov.uk/20110118095356/http:/www.cabe.org.uk/case-studies/green-streets.
6. Commission for Architecture and the Built Environment (CABE), UK, "Case studies: Ekostaden Augustenborg, Malmö, Sweden," accessed January 13, 2014, http://webarchive.nationalarchives.gov.uk/20110118095356/http:/www.cabe.org.uk/case-studies/ekostaden-augustenborg.

Green and Blue Spaces: Landscape as Infrastructure

The original infrastructure

The natural landscape that once dominated our environment was the original infrastructure prior to the development of our towns and urban centers. The vegetation, air, and soil worked harmoniously to infiltrate stormwater and recharge groundwater sources, naturally filter stormwater impurities, moderate surface stormwater flow rates to lakes and rivers, stabilize air temperatures, support wildlife diversity and habitat, improve air quality, and maintain soil structure and balance. All of these balancing processes performed by nature were done organically and without human intervention.

So why don't trees, vegetation, and natural systems play a larger role in our current infrastructure?

Isolated systems: green spaces and blue spaces

The incorporation of vegetation in city planning has often been driven by aesthetics and human comfort, through the inclusion of parks and residential yards for recreation. Green spaces are typically sited in relation to land use and human circulation patterns, not considering the value they could also bring in terms of stormwater management, air quality, temperature mitigation, and wildlife habitat. The full benefits of even familiar landscape elements such as trees and vegetation have not been well understood, nor have the longer-term, negative consequences that can arise from changing the existing landscape and natural watershed patterns.

Our built environments have become warmer, drier, more polluted, and void of ecological diversity because water and vegetation are most often viewed, planned, and operated as separate and isolated systems. The natural landscape that once dominated our predevelopment environments held no distinction between what our built environments have now categorized as "blue space" and "green space. "The placement and preservation of these spaces—open space, greenways, conservation areas, water management systems, water bodies and features—need to go beyond simply considering their recreational uses.

Reconnecting green and blue

A relatively basic and increasingly common approach is to replace impervious surfaces with trees and other vegetation, enabling more rainfall to infiltrate into the ground and less to enter a stormwater collection system. Reconnecting the blue and green spaces in this way can recharge groundwater sources; reduce operation, maintenance, and upgrade needs for the existing infrastructure; and reduce the release of polluted water into waterways. Add to that these other benefits attributed to trees and surrounding vegetation:

- capturing and absorbing significant rainfall,
- promoting natural cooling by evaporation, retaining moisture, and reducing reflected heat,
- reducing energy costs for nearby buildings,
- removing air particulates and pollutants,
- absorbing atmospheric carbon dioxide, a driver of global climate change,
- producing oxygen,
- increasing property values,
- reducing crime,
- increasing social interaction among neighbors,
- enhancing the well-being of anyone with a view of trees, including kids in classrooms and hospital patients.

If the benefits of planting trees mitigate so many urban conditions, why not just plant more trees?

The multilayered benefits do have many cities undertaking large tree-planting efforts. Denver, London, New York City, Shanghai, and Los Angeles are increasing their urban forests with initiatives to plant a million trees each. The greater Sacramento area has taken its effort a step further with a goal of planting 5 million trees by 2025. Chicago has mapped areas of elevated surface temperatures associated with urban heat island effect and is using the data to help prioritize tree planting and green roof implementation efforts.

Nevertheless, while large tree-planting efforts increase urban forest canopy and provide so many benefits to communities, trees are living, breathing organisms that endure more stress in cities than in their native grasslands or forests. Because trees are resource-starved in urban environments, they are constantly seeking out nutrients, water, and oxygen. When proper planning is not executed, trees often break up pavement in search of oxygen and damage pipes in search of water. A tree growing in an urban environment typically will live less than half as long as it would under ideal conditions. When one considers that trees typically do not offer maximum benefits until they near maturity, it raises the question of whether mass tree-planting efforts are really helping.

Integrating green and blue

To optimize our investments in urban vegetation, more thought needs to go into the environments in which trees and vegetation are planted. We must take the planning, siting, and preparation to the next level by considering a more vital reconnection of blue and green spaces—in other words, creating specific urban conditions to ensure greater resiliency in urban canopy cover as well as myriad other benefits. These factors come into play:

Space. Trees and vegetation need room to grow, with adequate soil volume for healthy root systems and moisture retention. After roadways are planned, buildings sited, and the necessary utilities placed, trees and vegetation are often expected to "thrive" in the remaining space. It is not unusual to find a street tree planted in a 4-foot-by-4-foot space within a sidewalk, planted in approximately 50 cubic feet of soil. To grow a large, mature tree, optimum soil volumes range from 1,000 to 1,500 cubic feet. Achieving this soil volume in traditional urban landscapes is difficult due to space constraints. Nevertheless, greater soil volumes for trees can be achieved by thinking differently about planning and constructing streetscape elements. Planning and designing to meet a requirement for tree root growth (or the needs of similar landscape elements) should be included in much the same way that our stormwater and drinking water systems are sized for adequate service.

Soils. Urban soils typically are stripped of the natural organisms, organic matter, and other nutrients trees need. Urban soils also tend to be compacted to a point where root systems become stunted and rainfall cannot penetrate.

Species diversity. For trees in particular, no one species should make up more than 15 percent of any forest population. Diseases have repeatedly devastated urban forests and have completely eliminated tree canopy in some town centers and cities. Maintaining species diversity will not stop diseases from spreading, but it will better equip urban forests to maintain canopy cover.

Urban conditions. Abilities to tolerate heat, salt, pollution, and drought are all key to vegetation success in an urban environment.

Through consideration of even these basic vegetation requirements, we set the stage for landscape as infrastructure: thinking not of vegetation and water systems as independent elements of urban areas, but of the vital connectivity of these blue and green spaces in the earliest planning stages along with all other infrastructure systems.

While we can no longer go back to predevelopment conditions and depend on trees and other vegetation as our sole infrastructure in urban environments, optimizing natural systems as "green infrastructure" that works in concert with our built "gray infrastructure" is key in balancing our resource needs with environmental benefits.

Relying on green and blue

Each city faces unique conditions, challenges, and opportunities. For example, in North America eight highly damaging hurricanes have occurred over the last 10 years, leaving coastal cities examining different methods for building greater resiliency against future storms. New York City and New Orleans are among cities applying large-scale, flexible networks of vegetated systems that rely increasingly on natural processes. Common green infrastructure approaches like bioswales and rain gardens, as well as larger-scale efforts such as revegetated waterways and expanded open spaces, not only aim to provide for recreation but are now designed in the spirit of accepting and living with water in the urban environment, while lessening the impacts of coastal, urban, and overbank flooding. Trees and other vegetation act like sponges to take on initial surges of storm and coastal waters, while gray infrastructure systems are enhanced to work congruently with green infrastructure. The green infrastructure in these instances works to lessen the enormous scale and financial resources required to maintain our current gray infrastructure systems by relying more on natural systems.

A 2009 study estimated that the costs of adapting US drinking water and wastewater systems to the effects of climate change will range between $448 billion and $944 billion ("Confronting Climate Change: An Early Analysis of Water and Wastewater Adaptation Costs," NACWA, 2009), and this does not include the adaptation costs for other infrastructure systems. It may seem that with all of these challenges, cities cannot expand or modernize their current gray infrastructure systems fast enough. This is why many cities are thinking differently about their infrastructure systems, incorporating natural processes into regional planning and design. These approaches bring longer-term benefits and greater resilience against the property damage and loss we see so frequently today, with lower overall costs and better management of risks, than does a reliance on gray infrastructure alone. The growing movement toward integrated infrastructure systems is catalyzing changes to existing regulations and policy and inspiring a new way of thinking about landscape as infrastructure.

Onondaga County, NY, has implemented a balance of green and gray infrastructure with a regional stormwater management program called "Save the Rain." Through a combination of well-designed and complementary green and gray infrastructure, and a long-term view of operations and maintenance requirements, this award-winning program plans to divert or capture 95 percent of contaminated stormwater overflow volume previously released into Lake Onondaga and its tributaries by 2018. See www.SavetheRain.us for more information.

Jade Paul, PLA, Senior Landscape Architect, CH2M
Erin Mosley, PE, ENV SP, Vice President, Director of Management Consulting at CH2M
Robert M. Beinstein, PE, ENV SP, EFCG

4.2 | SUSTAINABLE LANDSCAPE PLANNING

LANDSCAPE

"Landscape is infrastructure's most flexible strategy."[1]

Landscape as infrastructure is essential to the framework of an integrated planning process, providing an alternative soft, flexible infrastructure with the inherent potential of multifunctionality and thus high performance. Like other infrastructure systems, Landscape is analyzed here in terms of the provision of certain services. We highlight its potential to perform multiple simultaneous services, as well as its high dependence on the local context that defines these services.

"Green Infrastructure can deliver multiple benefits from the valuable urban space it occupies, compared with traditional single purpose engineering infrastructure,"[2] "which tend[s] to be designed to perform one function."[3]

Our approach decodes Landscape infrastructure on four system levels that supplement each other. It starts by identifying the services landscape provides (ecosystem services), then translates these services into specific program activities. It determines the spatial configuration of landscape entities necessary to support the identified and assigned activities and services, and finally injects the temporal dimension, providing for landscape evolution through maintenance. This integrated approach addresses landscape entities as part of a system and examines the ways they can contribute to overall city performance objectives.

Landscape as infrastructure proposes a methodology for holistic sustainable landscape planning. Within this framework, an open space is not seen as an independent unit that only aims to be sustainably designed; instead, its role within an extended system of open spaces is examined. Rather than focusing on how to make, for example, an urban park sustainable, this approach observes the aggregate of entities that form the city's system of open spaces and deals with sustainability as a synergic effort within which all spaces make different contributions. "When these elements are recognized as parts of larger systems, they can be designed and managed to achieve higher ecological and infrastructural performance."[4]

The approach of Landscape as infrastructure presented here uses the natural processes that are inherent within landscapes as a starting point and primarily proposes a functional decoding to enable an enhancement of performance.

Apart from its cultural and aesthetic value, landscape supports specific infrastructural services, offering biological solutions to technical problems, especially if planned in synergy with other infrastructure systems and as high-performance landscape. The landscape services can be summarized in six main functions as habitat, source, sink, conduit, filter, and barrier.[5] Landscape configuration is highly dependent on the local context. A registration of all landscape entities, in accordance with other infrastructure systems, would be exhausting if not impossible. The proposed approach provides a high-level framework, detached from scale-fixed and site-specific interpretations.

ALL CITY'S LANDSCAPE ENTITIES AS A SYSTEM

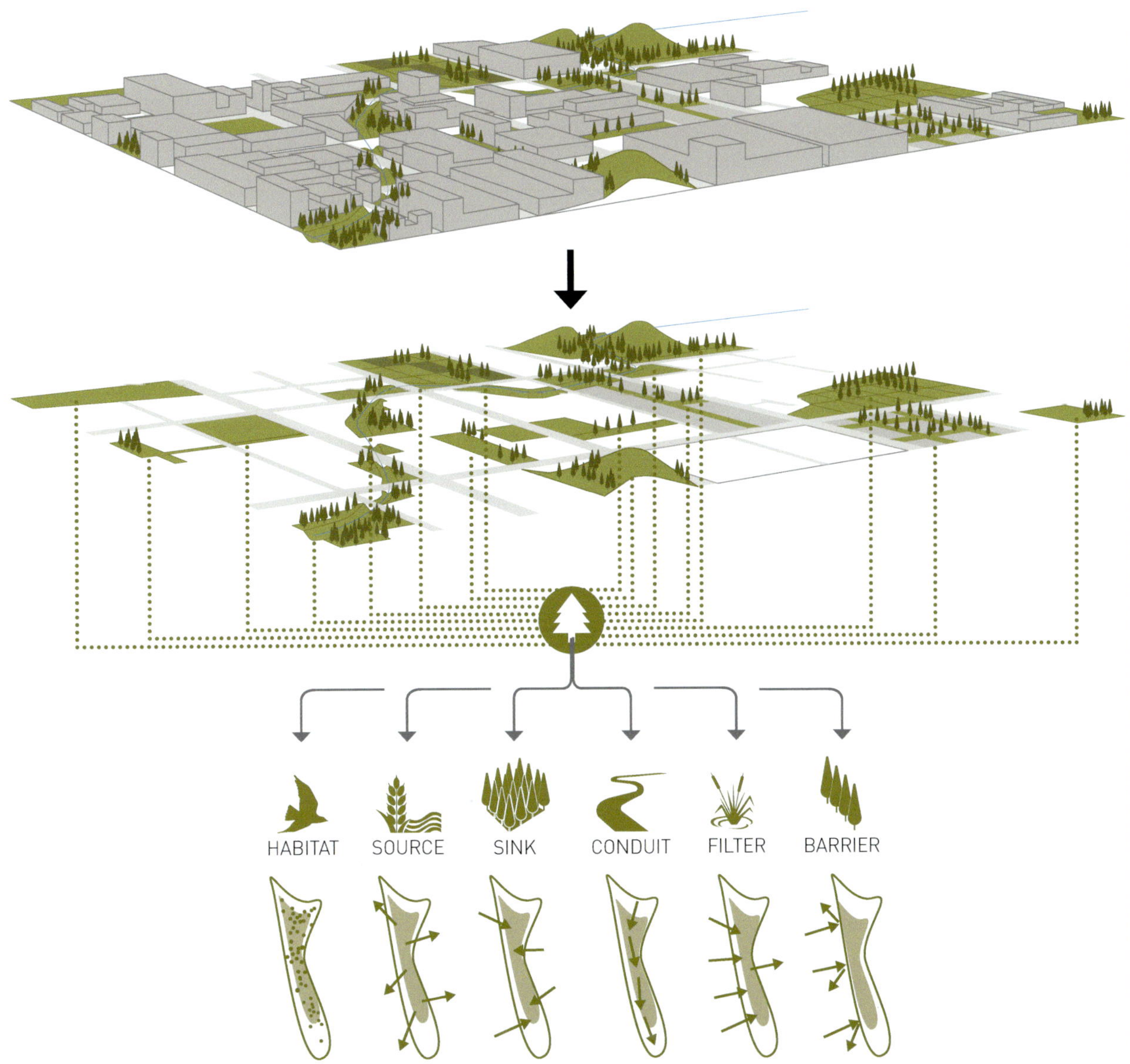

Six main functions: R.T.T. Forman (1995), J.F.Thorne (1993)

CHALLENGES TOWARD A SUSTAINABLE LANDSCAPE SYSTEM

The issues that need to be addressed to achieve a sustainable landscape system can be classified as environmental, social, and economic, but there are complex linkages between them.

The challenge is to align goals and strategies with social and economic benefits, providing services and opportunities for a community in harmony with environmental objectives that affect the community in the long term. It is essential to identify, regulate, and prioritize in order to counterbalance conflicting adopted strategies so that one does not invalidate the other.

"Multifunctional green networks should be used within the planning process to ensure that greenspace creation and management is spatially targeted to achieve optimum gains for social, environmental and economic development."[6]

The adverse environmental impact of other infrastructure systems is less of a factor in Landscape. The major challenge resides in drawing maximum advantage from its inherent potential while making sustainable use of its resources. This highlights the need to approach Landscape infrastructure sustainability in a systematic, overall way and in relation to other infrastructure systems. Landscape's highly synergistic nature is a product of its capacity to provide a range of services and goods that overlap with other systems' services. Its aptness to regulate natural phenomena and processes, as well as its role in mitigating and counterbalancing the impact of other systems on the natural world and the community, make landscape a useful tool in the service of infrastructure. Stormwater management, for example, can be handled through landscape infrastructure's biological solutions to technical problems, through restoring and integrating the natural water cycle in the urban environment, instead of modifying it such as by the use of impervious surfaces and engineered drainage systems.

However, the quantity and quality of services shouldn't be taken for granted. These services are not performed to the same degree by each landscape.

"As human populations grow, so do the resource demands imposed on ecosystems and the impacts of our human activities. For the most part, we have lived our lives, produced and consumed our food, manufactured and used our products, and disposed of our waste as if these ecosystem services are free, invulnerable and infinitely available. However, the impacts of human use and abuse are becoming evermore apparent—air and water quality are increasingly compromised, oceans are being overfished, pests and diseases are extending beyond their historical boundaries, deforestation is eliminating flood control around human settlements. Consequently, society is coming to realize that ecosystem services are not only threatened and limited, but that the pressure to evaluate tradeoffs between immediate and long-term human needs is urgent."[7]

In other words, Landscape is being challenged to maintain and enhance the processes that enable the performance of its services. Resilience, the adaptive capacity of ecosystems to undergo a disturbance and persist maintaining their functions and controls, is constantly being challenged by human activity.[8]

Apart from challenges in relation to other infrastructure and urbanization, Landscape has to meet challenges that arise within its own system. Landscape is usually targeted to the provision of one service, to such an extent and intensity that it comes at the expense of its potential for diverse services.

Productive landscape, being man-made, is an example of a tangible landscape service with direct financial equivalent where human pressure and management for an intensified single function have evident adverse environmental impact. Productive landscapes are asked to deal with agricultural expansion and intensification of production to meet the increasing need for nutrition of the world's constantly growing population. These decision-making strategies aim to solve a socioeconomic problem but have simultaneous environmental impacts, such as extensive land occupation, deforestation, loss of biodiversity, soil acidification and salinization, watershed pollution due to extensive fertilizer use, etc.

"The major challenge for agriculture is to ensure food security, adequate nutrition and stable livelihoods for all, now and in the future, by increasing food production while adopting sustainable and efficient agriculture, sustainable consumption of resources, and landscape-level planning to ensure the preservation of biodiversity."[9]

"The more that powerful groups of humans value a particular service, the more likely they are to drive a landscape toward monofunctionality. Relatively monofunctional landscapes will require high levels of human input to continue delivering their values and functions, and it is likely that a completely monofunctional landscape will cease to be sustainable and will eventually require remediation. Hence, from the point of view of public policy, it will normally be desirable to seek a degree of multifunctionality in all cultural landscapes, and to achieve high levels throughout much of our green infrastructure."[10] A significant challenge lies in understanding the opportunities and vulnerabilities of landscape. Understanding the conditions, cycles, and processes that form and maintain landscape is essential for sustainability, as well as examining and taking into consideration how an intervention at one scale can become a catalyst for events at another.

The causal links between the environment and human health and well-being are complex and often indirect. Landscape responds to all five main aspects of human well-being: materials, health, social relations, security, freedom and choice.[11] "Ecosystem attributes influence the aesthetic, recreational, educational, cultural, and spiritual aspects of human life."[12]

"Ecosystems provide goods and services that support human well-being, and shortages of these can have impact on health and well-being. The regulating functions of ecosystems affect human well-being in a number of ways, including the availability of clean air, freshwater, reduced risk of flooding or drought, stabilization of local and regional climates, and checks and balances on the range and transmission of certain diseases. Without these regulatory functions, the varied populations of human and animal life are inconceivable."[13]

This chapter's way of defining and decoding Landscape infrastructure aims to point out this challenge. Landscape's services are often underestimated and neglected when they do not translate directly into economic values, though they provide numerous contributions to human welfare. This is why there is an ongoing effort, through natural capital accounting and ecosystem services economics and assessment, to understand the benefits provided by ecosystems.

The services of ecological systems and the natural capital stocks that produce them are critical to the functioning of Earth's life support system. They contribute to human welfare both directly and indirectly, and therefore represent part of the total economic value of the planet.[14]

It is important to understand the value of the benefits that nature provides to human society, and to mobilize investments to sustain and enhance them. Other forms of infrastructure can be expensive to build, while nature can often provide cheaper, more durable solutions. Landscape can act as an urban economic and social regenerator that attracts investments supporting a competitive economy as well as balanced development.

Landscape infrastructure offers opportunities to enrich biodiversity, work with natural processes, stimulate natural systems to become self-maintaining, and regenerate resources for continued use by humans and other species.

ENVIRONMENTAL

Challenges for environmental sustainability include:

- Maintaining and enhancing biodiversity
- Regulating natural processes and phenomena
- Mitigating and compensating for the adverse impact of other infrastructure systems on the environment
- Restoring areas degraded by human activity
- Using natural resources required for maintenance in a sustainable manner
- Performing complementarily to, and not against, existing ecosystems

SOCIAL

Challenges for social sustainability include:

- Producing essential materials and goods with sustainable use of available natural resources
- Provisioning regulating services necessary to ensure public health
- Providing equal access to high-quality and safe open public spaces
- Offering a variety of recreational, cultural choices to the community

ECONOMIC

Challenges for economic sustainability include:

- Enhancing the productivity of local land and water bodies in a sustainable manner
- Functioning as an urban economic regenerator attracting investments
- Developing low-cost and durable solutions for maintenance
- Securing funds to cover long-term maintenance needs

STRATEGIES TOWARD A SUSTAINABLE LANDSCAPE SYSTEM

SYSTEM PURPOSE: ENSURE AND PROTECT THE ECOLOGICAL INTEGRITY OF NATURAL AREAS	
	PRESERVE ECOLOGICAL INTEGRITY OF NATURAL AREAS: Preserve biodiversity
	RESTORE: Rectify or compensate for habitat loss, degradation, and fragmentation
	ACHIEVE SELF-SUFFICIENCY: Minimize need for maintenance
	CREATE NEW RESOURCES: Increase "carrying capacity" of landscapes

The primary purpose and service of Landscape as a system is to ensure and protect the ecological integrity of natural areas. This is a prerequisite if it is to be capable of performing its ecological services. Ecological degradation directly affects the quality of landscape's performance, as discussed in the previous section. The strategies listed above suggest basic steps that could be included in the planning of a landscape system. They do not aim to outline all the requirements for a sustainable system, but instead highlight approaches that if implemented could resolve multiple issues simultaneously. The pyramid signifies the order of planning steps based on their potential contribution in resolving main sustainability challenges.

A priority for sustainable landscape management is to preserve biodiversity and protect natural areas of elevated vulnerability. The first step for this strategy is to identify and designate areas to protect. Since natural systems are living systems, ensuring their sustainability involves understanding their structure and dynamics. Conservation designations, statutory designations, natural resources definition, and environmental protection acts are necessary tools in this direction. Setting regulations to assess the compatibility of certain land uses adjacent to protected areas and prohibiting invasive activities such as logging, mining, and sport hunting also introduce ways to control habitat fragmentation, loss, and degradation. An identified habitat for endangered plants or animals is an example of land requiring preservation techniques. In areas where the natural features are an integral part of the community's character, the integration of new development should provide adequate protection. Integration strategies allow natural features to coexist with development while remaining largely undisturbed.

The second strategy is to restore disturbed natural areas to their previous or to a new state of equilibrium. Human activities' adverse impact on ecosystems resilience such as exploitation of natural resources, emissions of pollutants, land-use change, reduction of biodiversity, and anthropogenic climate change, increases the likelihood of regime shifts in ecosystems to less desirable and degraded states affecting their capacity to generate ecosystem services.[15]

Landscape as biodiversity

Landscape infrastructure is assigned the hard task of compensating for the unavoidable impact on biodiversity caused by infrastructure projects. Human intervention alters ecosystems to such a degree that they lose their capacity to tolerate disturbance and restore themselves. Depending on the intensity of their previous use or the magnitude of the disturbance, some sites may be very difficult to return to self-sustaining conditions, due to severely disrupted processes necessary for regeneration or growth. Reforestation and bioremediation of contaminated

Landscape as resource

soils are some strategies to restore lost natural cycles.

The strategies highlighted are all connected with recognizing the inherent dynamics of landscapes: the temporal parameter, the evolution of systems in time. It is essential to take this dimension into consideration when planning a landscape system or entity. Planning should aim for self-sufficiency through generation of resources such as soil, plants, and water on-site, while reducing negative flows off-site such as fertilizer- and pesticide-laden water runoff. Water conservation should be promoted through

Landscape as cycle

selection of plant material appropriate to the local rainfall, soils, and exposure.

Mimicking nature as a planning method is a way of ensuring sustainability of spaces. Mimicking should not be seen as replicating natural forms, but rather as evoking and activating natural processes and regeneration dynamics through design.

Sustainable landscape management finally adopts as a strategy enhancing landscape productivity, the creation of new resources both for humans and for other species, and thus increasing a landscape's carrying capacity. Carrying capacity is defined as "the amount of activity or use that can be handled by a system before it begins to deteriorate": the use that can be absorbed by a given setting.[16]

Landscape has to promote local production of basic goods, introducing alternative methods to establish the adequate conditions when previous planning efforts have not provided for productive activity. Shaping and sizing land to introduce appropriate hydrological regimes can help catalyze and support the success of new plant and animal communities. Examples of practices in this direction include community gardening and urban agriculture as local sources of food. (It is nonetheless important, as already mentioned, to understand their side effects when introducing new functions into a landscape.)

"The sustainability of the production of a particular ecosystem service can refer simply to whether the biological potential of the ecosystem to sustain the yield of that service (such as food production) is being maintained."[17]

4.3 | LANDSCAPE SYSTEM DECODING

LANDSCAPE
DEMAND

LANDSCAPE
PROGRAM

LANDSCAPE
PATTERN

LANDSCAPE
MAINTENANCE

Within the framework of the Zofnass Planning Guidelines, Landscape infrastructure is decoded on the following four system levels: Demand (for services), Program (of activities), Pattern, and Maintenance (of cycle). This decoding provides the ground for setting objectives and guidelines for sustainable landscape planning and management.

In Landscape infrastructure the organizational levels are not ways of classifying entities, but rather different and supplementary ways of analyzing entities. Each level supplements the others, promoting an additional degree of thoroughness in planning.

The first level, Demand for services, encapsulates all the services landscape is able to provide, and thus justifies its denomination as infrastructure. The local context is one of the factors that will define the services that can be provided or need to be introduced within the existing conditions by creating the appropriate conditions to support them.

Demand is followed by Program of activities. Landscape program is the common and familiar process of defining landscapes through the assignment of specific activities. The present approach links activities and services, although not directly. It lists and suggests criteria that create connections between programmatic activities and landscape services.

The third level, Landscape Pattern, represents the system's structure: its entities, their relationships and functions. Landscape pattern is related both to the spatial distribution of entities and to the configuration of physical materials. Function represents the flows of materials, water, and energy that are indispensable to landscape's structure. Landscape pattern, either spatial or functional, entails both the networks and nodes identified in other systems. Landscape corridors, the linear elements of connection, cannot be strictly separated from patches, the equivalent to nodes, as the definition of both is tightly connected to the scale of analysis. Corridors can also be identified within patches according to their spatial configuration, and vice versa, in relation to the level of focus applied.

The fourth system level, Landscape Maintenance, focuses on the practices that provide the living and therefore inherently cyclical system of landscape with continuity, evolution, and regeneration in time.

Though there are dominant connections between the different levels, the present approach encourages the investigation of a multitude of associations and connections. It proposes the reinterpretation of concepts and practices related to landscape planning to achieve optimum social, environmental, and economic gains.

LANDSCAPE INFRASTRUCTURE SYSTEM LEVELS' INTERCONNECTION

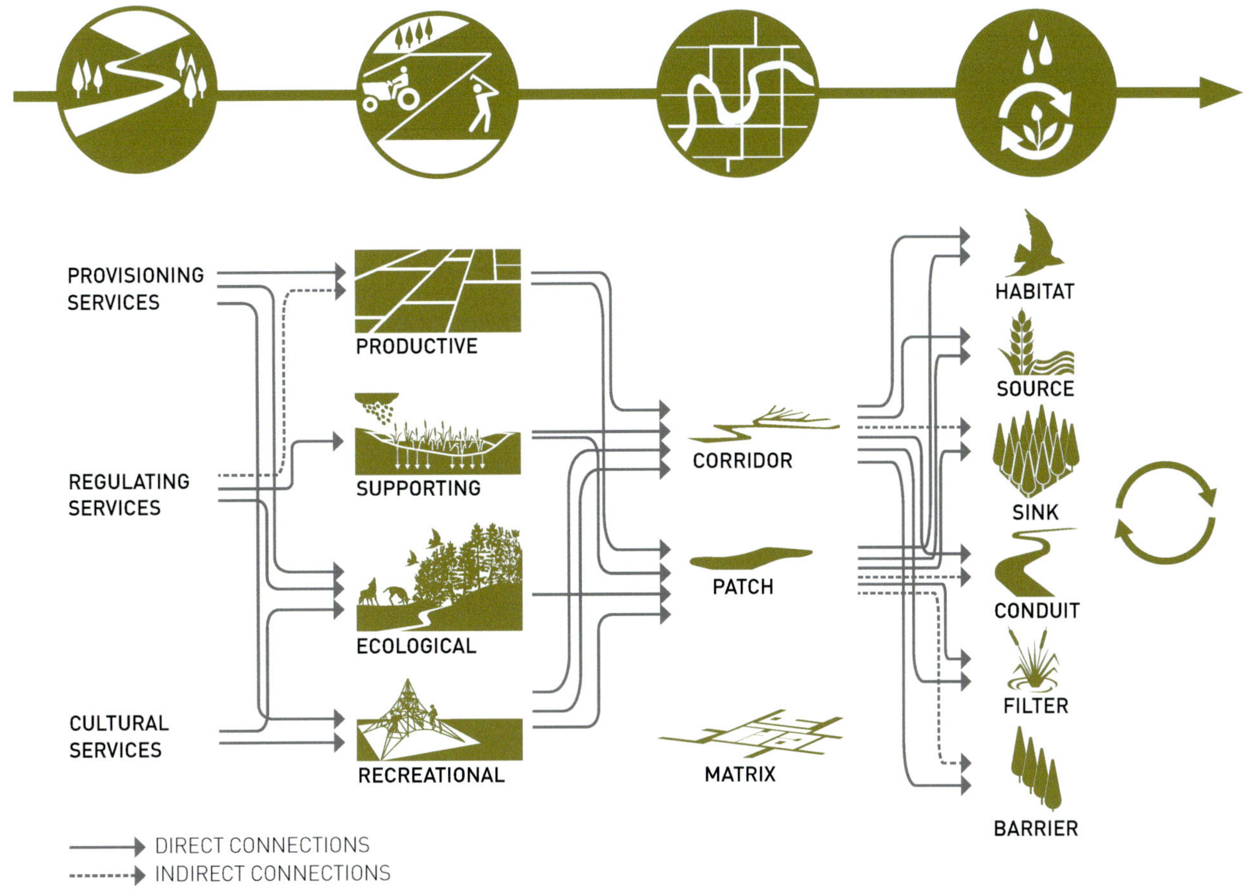

LANDSCAPE DEMAND

Landscape infrastructure Demand aims to define the required and potential services that the system has to provide, such as required protection, restoration, and mitigation services (regulating), potential landscape productive services (provisioning), and potential and required landscape recreational/social services (cultural). The definition of these services for a certain context should be based on extensive analysis leading to the prioritization of services and an initial spatial placement.

The analysis has to explore three basic fields: the natural context, the socioeconomic context, and the macrocontext. Natural context includes the biotic and abiotic conditions of a site and its surroundings, such as: climate, geophysical conditions, surface cover (based on type of vegetative cover in terrestrial ecosystems, or on freshwater, brackish water, or salt water in aquatic ecosystems), species composition, etc.

Socioeconomic context refers to an area's demographics, its population distribution, the users' consumption patterns, the dominant human activities within a certain ecosystem, the resource management systems and institutions, as well as the goals of a certain development plan.

Macrocontext conditions that need to be examined are pressures, both direct and indirect, on the natural world, such as pollution, degradation, fragmentation or loss of habitats, and climate change that exceeds the local context but still needs to be addressed.

Landscape Demand for services is defined through the list extensively documented in the bibliography as ecosystem services, and classified as provisioning, regulating, and cultural.[18,19]

CONTEXT FACTORS THAT DETERMINE LANDSCAPE INFRASTRUCTURE SERVICES

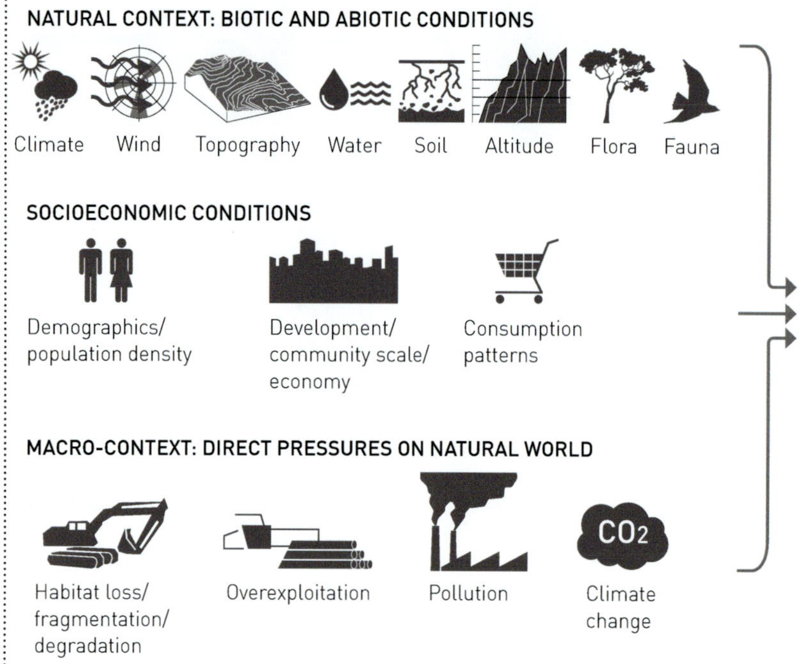

LANDSCAPE INFRASTRUCTURE SERVICES

Provisioning services

- Food (nutrition/animal feeding)
- Freshwater
- Fiber, timber (raw materials)
- Genetic resources
- Biochemicals (medicinal resources)
- Fuel (energy resources)
- Ornamental resources

Regulating services

- Water filtration/purification
- Water regulation (rainwater management/ natural hazard protection)
- Erosion regulation
- Local climate regulation
- Regulation of atmospheric composition
- Crop pollination
- Pest regulation (biological control)
- Disease regulation (biological control)
- Waste decomposition
- Contaminants control
- Primary production/nutrient cycling (habitat protection etc.)
- Soil formation (maintenance of soil fertility)

Cultural services

- Spiritual and religious values
- Aesthetic values
- Recreation and ecotourism
- Knowledge systems and educational values

It is important to clarify that a single service does not necessarily correspond to a single landscape entity, but rather an entity should ideally be able to provide multiple services.

Defining the services that Landscape infrastructure has to deliver is essential for integrated landscape infrastructure planning and for landscape functioning as a system. That is why, parallel to the classification of provisioning, regulating, and cultural, the services have to be specified as "potential" or "required."

Though difficult to quantify, the less tangible landscape services need to be especially evaluated in terms of other infrastructure systems' services in order to approximate an assessment of landscape infrastructure's performance.

The overall goals for the system are set on the Demand level through the services that Landscape as a system has to perform in relation with its context (contextualized overall goals).

FROM SERVICES TO PROGRAM ACTIVITIES

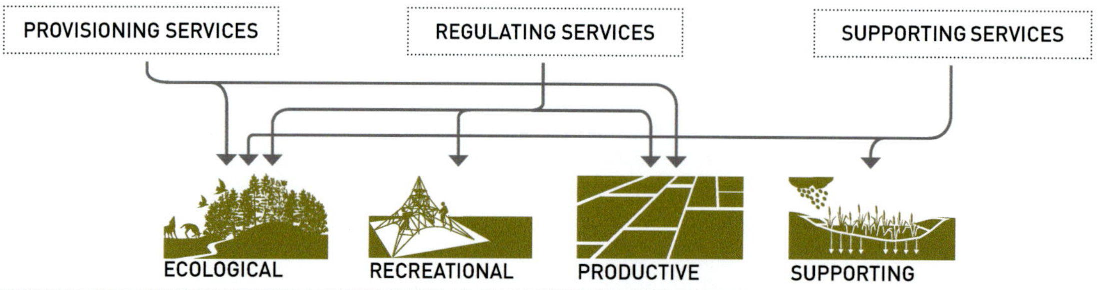

LANDSCAPE PROGRAM

Landscape Program assigns specific activities to entities, linking them with services through suggested criteria, though not directly. In typical practice, Landscape Program and activities generally carry connotations of recreational, social, and productive character rather than of protecting ecological integrity—a tendency that is partly responsible for poor program activities, neglecting the necessary ecological contribution of each landscape activity and leading eventually to poorly performing open spaces.

This approach highlights the need to connect activities with the complete list of landscape (ecosystem) services, cultural, regulating, and provisioning, aiming to broaden the potential of sustainable practices. It encourages the reinterpretation of a programmatic activity to avoid distorted or limited perceptions and elevate its perceived value and its contribution to the community's well-being.

Activities are associated with four general categories: ecological, productive, recreational, and supporting, respectively related to natural processes, natural phenomena, urban conditions, and the performance of other infrastructure systems. This categorization is sufficiently generic to cover different scenarios of site configuration. Different activities are subsumed in these four main categories in regard to a prevailing common role they share.

Program therefore defines the range of space uses through the designation of activities, and thus services and functions, whether more ecological or more anthropocentric. Program entails the temporal scale as well, e.g., seasonal use of spaces or varied daily or weekly patterns of use.

The first issue that Program addresses is the feasibility of activities, which has already been touched on at the Demand level, and is related to site-specific characteristics so as to offer contextualized solutions. Each location has its own conditions favorable or not for certain activities. The planning of landscape services becomes more tangible through an association with physical activities and a process of listing, selection, hierarchization, and spatial distribution and combination.

Landscape Program primarily aims to promote multifunctionality through a combination of activities, ensuring compatibility between different activities and between activities and context. Multifunctionality, a principal hallmark of landscape, refers to the integration and interaction of different activities on the same piece of land. An example could be a constructed wetland that reduces the quantity of urban runoff, improves its quality,

LANDSCAPE PROGRAM ACTIVITIES PER CATEGORY

ECOLOGICAL ACTIVITY

- Preservation/protection
- Restoration
- Ecological research
- Observation (wilderness areas/ ecological reserves/natural areas/heritage rangelands/ national parks/estuaries/ cultural heritage etc.)

RECREATIONAL ACTIVITY

- General recreation
- Playgrounds
- Athletics (ground or saltwater/ freshwater oriented)
- Extreme sports
- Theme parks
- Events/art
- Recreational transportation
- Ecotourism
- Education

PRODUCTIVE ACTIVITY

- Agriculture (crops/tree plantations for food, energy, fiber, feed)
- Forestry
- Livestock
- Aquaculture
- Horticulture / Plant propagation
- Freshwater
- Quarrying

SUPPORTING ACTIVITY

BIOENGINEERING

- Recharge of groundwater (bioswales etc.)
- Wastewater treatment (constructed wetlands)
- Runoff control (check dams, bioswales etc.)

MITIGATION

- Noise/odor buffer/protective buffer/buffers between conflictive land uses
- Carbon sequestration (carbon forests etc.)

acts as a wildlife habitat and also provides a recreational green space for the community.

"From a cultural perspective landscape hosts multiple human activities such as farming, settlement, and recreation, and from an environmental perspective landscape sustains multiple climatic, hydrologic, edaphic, and ecological processes. With respect to both of these, planners and managers aspire to assure productivity, diversity, stability, and integrity."[20]

In order to ensure compatibility, Program has to define and adjust the number and intensity of activities within a specific place. Depending on the size of spaces, it creates an area of influence and an attractor for users.

It is recommended that planners identify and meticulously enumerate the activities that form a landscape entity's program at the time of planning a landscape system. An activity has to be connected with services, so that when a certain activity is added to or subtracted from a program it is clear what services are gained or lost.

Once the services are translated as specific activities that the system has to include, they can be distributed within the system. This distribution is realized in relation to compatible specific location conditions (site-specific goals). A park, for example, if given different types of activities, should be designed as part of a system of open spaces and according to the system's needs and limitations. It will be assigned the activities needed, not the ones automatically connected with a preconceived idea of park.

MULTIFUNCTIONALITY, FEASIBILITY, AND COMPATIBILITY OF ACTIVITIES

INTEGRATED HIGH-PERFORMANCE LANDSCAPES

LOCATION/LEVEL OF ATTRACTION/SCALE AND PROGRAM OF ACTIVITIES

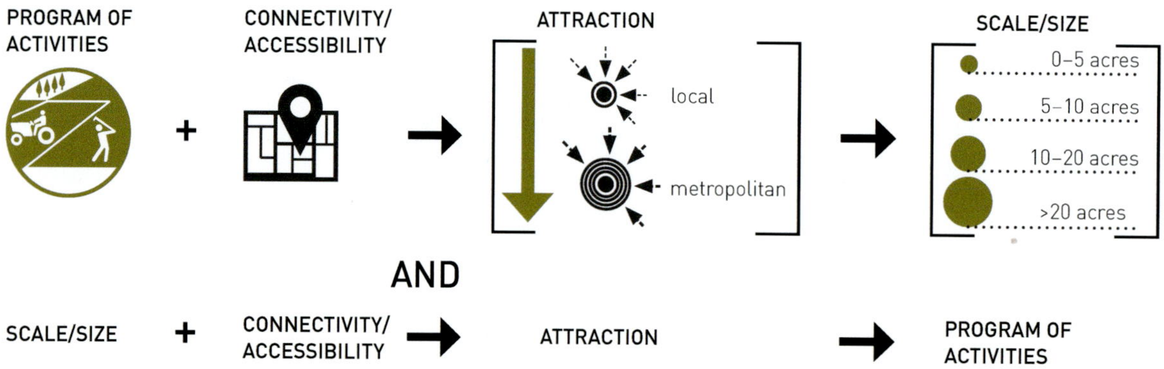

LANDSCAPE PATTERN

Landscape Pattern describes the spatial and functional structure of landscape, and the system's configuration with specific attributes that will enable its provision of services and support specific activities. Understanding how biological and physical systems are structured and how they function is critical to the practice of sustainable design. This has to consider how systems operate across spatial and temporal scales.

Landscape Pattern is a term inaugurated and established by landscape ecology. It is extensively documented and widely accepted, allowing the discussion to shift and finally focus on the functional aspect of landscape. Also known as mosaics, landscape pattern can be defined and configured spatially as a system of corridors and patches that is supplemented by the matrix. At the same time, it can be configured functionally through the six landscape functions: habitat, source, sink, conduit, filter, and barrier. These two levels of landscape configuration, by spatial structure and

LANDSCAPE STRUCTURE SPATIAL PATTERNS

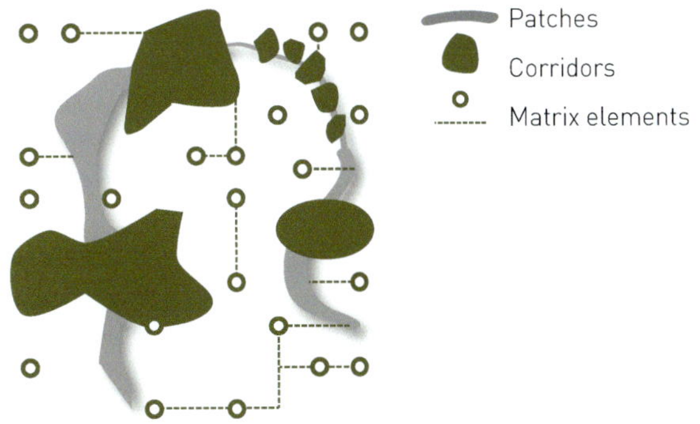

- Patches of various sizes and spacing can be distributed as "stepping stones" supporting transitions and movements
- Corridors can provide uninterrupted connections
- Matrix elements can also be connected, expanding the system

SPATIAL PATTERNS CONFIGURATION

Surfaces configuration

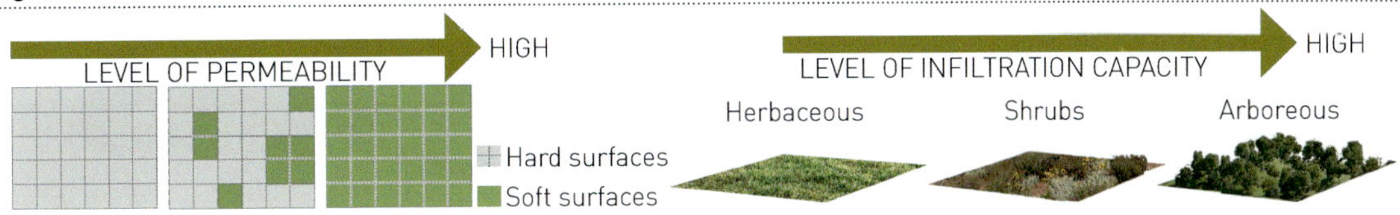

Vegetation density

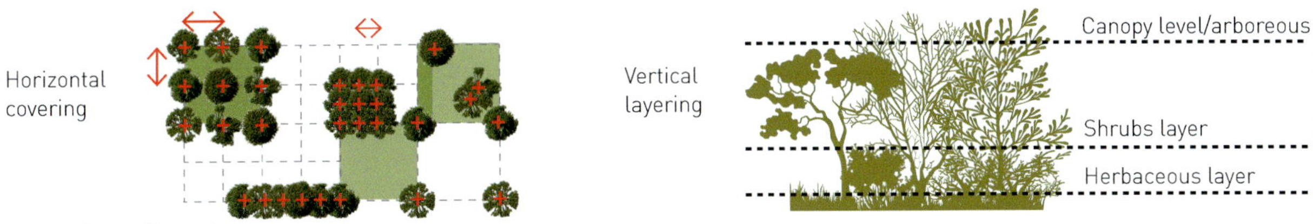

Topography configuration

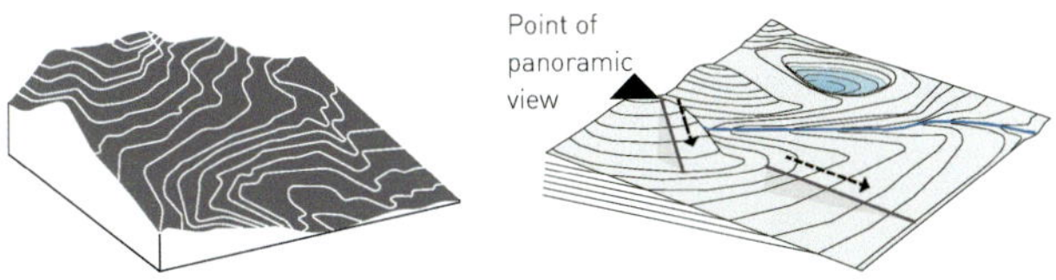

Water accumulation
Water flow
Points of panoramic view
Slope and relation to erosion and pedestrian flows
Orientation

Plant selection

Environmental requirements				Morphological characteristics								Productivity		Functions		
Solar access	Climate	Soil	Water	Form	Height	Canopy	Seasonal pattern	Color	Foliage density	Growth rhythm	Shade	Fruit	Timber	Urban context adaptation	Fauna attraction	Filter

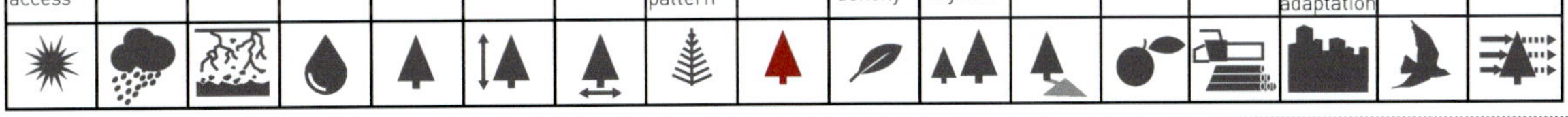

FROM SPATIAL TO FUNCTIONAL PATTERNS

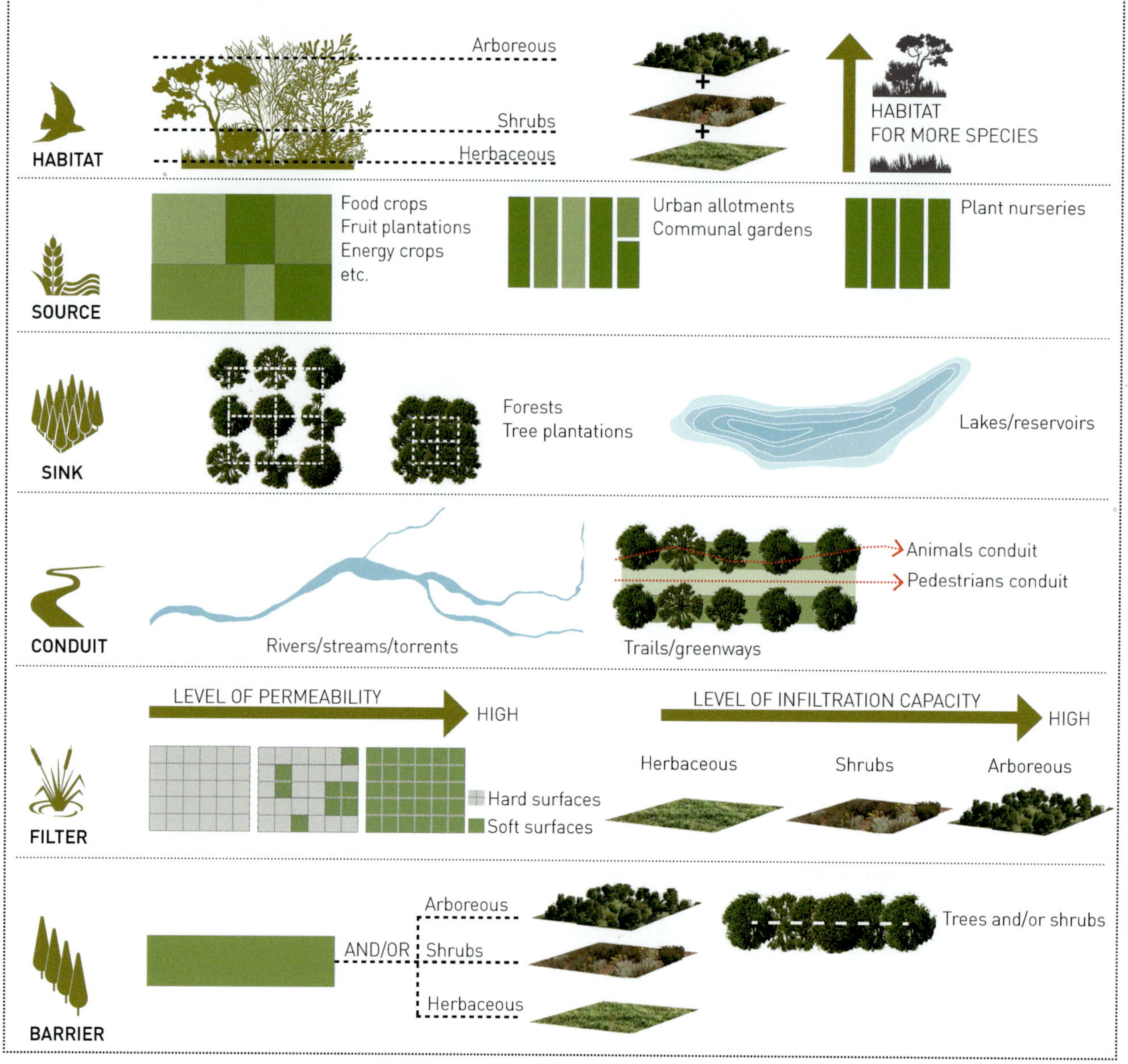

MULTIPLE FUNCTIONS AND PROGRAM

	HABITAT	SOURCE	SINK	CONDUIT	FILTER	BARRIER
ECOLOGICAL	●	●	●	●	●	○
RECREATIONAL	○	○	●	●	●	●
PRODUCTIVE	●	●	●	●	○	○
SUPPORTING	○	○	○	●	●	●

Note:
This is a generic, indicative approach that has to take into consideration the specific design of spaces to be more concrete

● Potential performed function
○ Not performed

function (operation), are highly interconnected.

Regarding spatial patterns, the main defined components already mentioned are corridors and patches. Corridor is "a linear landscape element that provides for movement between habitat patches" that plays "a key role in connectivity" as linkage.[21] "The patches are defined as regions that are more or less internally homogeneous with respect to a measured variable (a set of spatially proximate homogeneous units)."[22] They are configured and described by their area, shape, type, edge, as well as their distribution and density.[23]

The matrix is a component of the landscape altered from its original state by human land use and the context of corridors and patches. It may vary in cover from human-dominated to seminatural. Thus urban development constitutes a matrix. The matrix includes elements that are not considered landscape infrastructure but perform complementarily to it.[24]

Regarding functional patterns, the present approach proposes the characterization of the previously mentioned landscape functions as principal (designated) and potential.

The principal function of each landscape entity is defined using criteria of program, spatial pattern, as well as location and size. This process is essential for sustainable landscape infrastructure planning, given that it will lead the design of the entities and give them specific attributes that will ensure the provision of multiple services.

Patches and corridors can be configured to perform one or more of the six functions. Their performance is determined by the configuration of elements in relation with their material configuration (permeability of surfaces, differentiation between softscapes and hardscapes) and the vegetation configuration (vegetation density, plant selection, plant characteristics). A multitude of characteristics must be taken into consideration in order for plants to serve the functions of the space and facilitate maintenance (horizontal covering, vertical layering).

Topography is central to the function of spaces. It can block, direct, or concentrate multiple elements such as water, wind, solar radiation, noise, soil, etc., as well as movement within spaces and visual perception. Large changes in topography can have multiplying and lasting effects on microclimate, flora, fauna, etc. and are usually to be avoided.

The principal function is one of the highly systemic interpretations of landscape infrastructure. It is the one that distributes roles and purposes (here defined as functions) to system entities. It should lead to an analysis of functions to help an overview of system purpose. The potential function is the high-performance interpretation of the system, the simultaneous multiple purpose of entities. After the definition of the principal function of each landscape entity, other simultaneous functions can be assigned to the same landscape entity for a landscape system with multiple benefits. Multiple functions can potentially coexist within a landscape entity, depending on criteria such as the entity's program, the spatial pattern, its location, etc. Ideally, new landscape elements once established could eventually perform all functions.

The matrix is a very useful concept for urban planning. It forms the background of the landscape infrastructure and can perform complementarily to it if its role is enhanced through planning. It enters the territory of private buildings and streetscape and includes "private and semiprivate gardens, shared (communal) spaces around apartment buildings, backyards, balconies, roof gardens, green roofs, and living walls. ... This category also includes unused land reserved for future use, ... large institutional and manufacturing sites,"[25] as well as transportation nodes.

Matrix expands the system to provide stepping stones for alternative linkages between open spaces to enhance connectivity and establish ecological functions beyond landscape infrastructure limits.

LANDSCAPE MAINTENANCE

The Landscape Maintenance system level represents the way to ensure the maintenance of the cycle of functions and resources of the landscape. It investigates ways to perpetuate landscape's cyclical functions and the regeneration of natural resources. The issues raised deal with the optimization of practices, aiming to minimize the number and intensity of maintenance practices so as to reduce the quantity of resources consumed and the associated cost. Ideally, a landscape once established can become self-sufficient. Maintenance practices are "all actions which have the objective of retaining or restoring an item in or to a state in which it can perform its required function. The actions include the combination of all technical and corresponding administrative, managerial and supervision actions."[26]

A sustainable landscape is "a healthy and resilient landscape that will endure over the long term without the need for high input of scarce resources such as water. The natural functions and processes of the landscape are able to maintain it into the future."[27]

This level puts emphasis on the main attribute of landscape as a nonstatic system of living components that keep evolving beyond the phase of their implementation. Once more it highlights that it is critical to understand the conditions, cycles, and processes that form landscape in order to know how to maintain it. The intensity of practices needed to sustain entities in a desired state and allow their evolution has to be considered during planning. The intensity of maintenance depends upon a place's spatial configuration and its composition of hard and/or soft surfaces. Some of the main maintenance activities are listed here, such as irrigation, using either recycled wastewater or collected rainwater, pruning, and seasonal trimming of plants to support their proper growth or support young plants till established. Hard surfaces and structures have their own needs in order to perform their assigned function.

The program of activities of a landscape entity, its location and consequently its role as attractor, local or metropolitan, will define the intensity of its use and will also determine the need for maintenance. Another characteristic that defines maintenance is the status of an entity; if it is maintained in its original condition, it already includes established processes, but if it is of new configuration, the new processes need to be supported in order to become established. Strategic options in terms of spatial design can minimize the maintenance services needed. In particular, for the maintenance of vegetation, the proper selection of climate-adapted native plants and plants with low water requirements ensures the maintenance cycle of landscapes without intensive services.

A key concept at the Maintenance level is that of mimicking, the process of replicating the biological adaptations to changing conditions and the cyclical patterns inherent in natural systems, which support sustainability and regeneration over time.[28]

The Maintenance level aims to address program activities and configuration of landscape entities in terms of persistence, effort, and financial cost. "Ecosystem degradation is a persistent reduction in the capacity to provide ecosystem services."[29] This definition can be expanded to point out that an entity's degradation leads to the inevitable invalidation of its function.

SOFT SURFACES MAINTENANCE

- Effective irrigation techniques adapted to plant needs
- Use of recycled wastewater for irrigation
- Integration of rainwater collection/use for irrigation
- Seasonal trimming of plants to support their proper growth
- Vegetation clearing/removal of dead or problematic trees
- Recycling of green manure and trimming residues for compost
- Supporting of young plants till established
- Revegetation of spaces
- Soil improvement
- Production-related works

HARD SURFACES AND STRUCTURES' MAINTENANCE

Resistant materials help to ensure low maintenance costs

- Maintenance of materials and structures
- Cleaning of spaces

MAINTENANCE INTENSITY IN RELATION WITH LOCATION AND PROGRAM ACTIVITIES

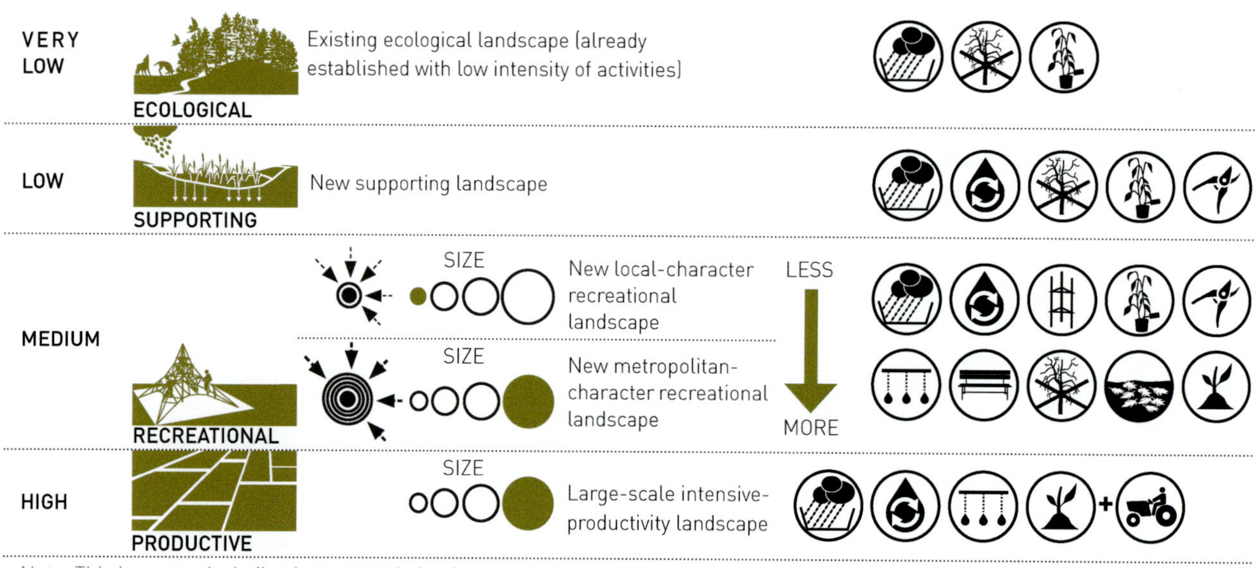

Note: This is a generic, indicative approach that has to take into consideration the specific design of spaces to be more concrete

4.4 | SYNERGIES BETWEEN LANDSCAPE AND WATER INFRASTRUCTURE

Landscape infrastructure stands in the middle of the complex, uncertain, and multifaceted interrelationships between city and nature. Providing a wide range of goods and services, it overlaps with other systems' services in many cases, especially when natural phenomena and processes are involved."Ecosystem services are the benefits provided to humans through the transformations of resources (or environmental assets, including land, water, vegetation and atmosphere) into a flow of essential goods and services e.g. clean air, water, and food."[30]

Given that Landscape consists of both terrestrial and aquatic systems, it overlaps with Water infrastructure in terms of water supply, with groundwater or surface water as sources, regulation or control of flows through waterways, and filtration and storage through wetland processes and land cover permeability. In other words, both in terms of structural components and processes, Landscape is tightly related with Water infrastructure. The relevant guidelines presented at the end of this chapter highlight the recurrent reference to the Landscape system as a way to resolve, treat, and regulate water issues. The connection is representative of the active synergies between different systems, and is worth further analysis. It serves as an example of how the Demand for certain services can be resolved through the Landscape system with Program activities, spatial and functional Patterns, and cycle Maintenance.

Process can be defined as the "flow of materials and energy through the patterns of landscape structure. These flows represent process and relate to the impacts associated with the movement of water, wind, plants, wildlife, and people."[31] It is these flows that give Landscape its regenerative potential. "A regenerative system provides for continuous replacement, through its own functional processes, of the energy and materials used in its operation."[32]

Natural cycles, inherent to and inseparable from Landscape, are cyclical regenerative processes, among which are hydrological structures and processes. The water cycle or "hydrological cycle recycles and purifies water. It is driven by energy and gravity; however, plants play a crucial role in terms of modifying rainfall inflows, water movement into and through the soil, and water outflows through evapotranspiration, surface runoff and subsurface drainage."[33] Development and urbanization can dramatically modify these natural cycles and their ability to produce services, especially through the construction of extensive impervious surfaces and engineered drainage systems.

"The timing and magnitude of runoff, flooding, and aquifer recharge can be strongly influenced by changes in land cover, including, in particular, alterations that change the water storage potential of the system, such as the conversion of wetlands or the replacement of forests with croplands or croplands with urban areas."[34]

A sustainable approach should attempt to integrate

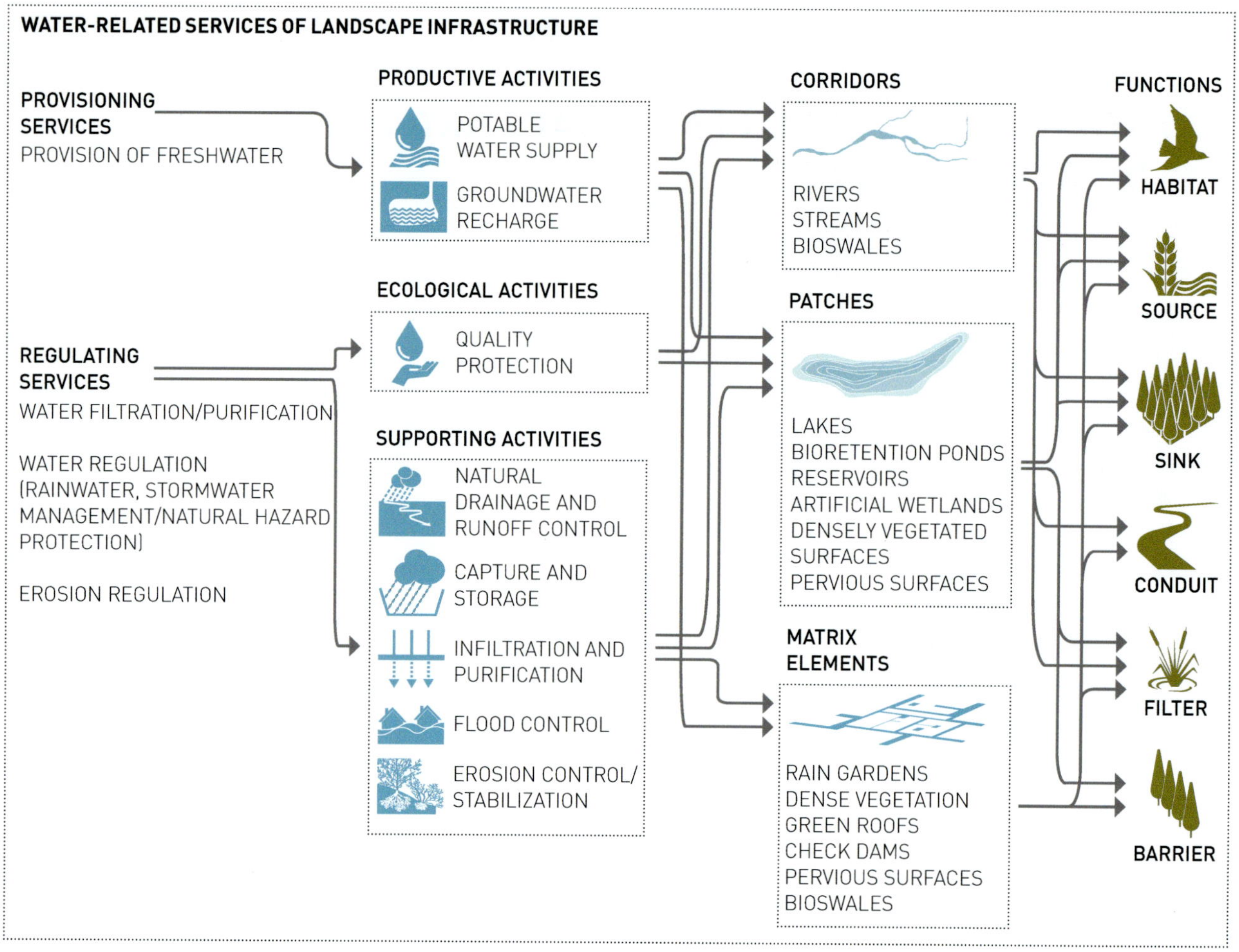

these natural cycles into the urban environment. Water-sensitive planning aims to restore or replicate the natural water cycle within the urban context, treating urban stormwater runoff as a resource. In this direction urban stormwater management could be met exclusively by Landscape infrastructure, as natural hazard protection is among the main regulating services that landscape supports. The inherent functions of landscape make it a key planning strategy for resilience against flooding, whether it is through coastal ecosystems that absorb the impact of extreme phenomena or along inland water bodies.

Integrating stormwater management into the Landscape system offers multiple benefits, such as water quality treatment, wildlife habitat support, opportunities for recreation, and open public spaces, due to the system's inherent multifunctional character. Landscape can provide retention of stormwater rather than rapid conveyance, reducing peak flows and runoff from the urban environment while simultaneously offering infiltration and groundwater recharge. Instead of a drainage system expansion that simply captures, Landscape entities can reuse, store, and infiltrate stormwater through vegetation and soil. These services are translated into supporting Program activities, such as runoff control, groundwater recharge, bioretention, rainwater harvesting, infiltration, erosion control, etc., and attributed to specific spatial and functional patterns (conduit, source, sink, filter, barrier). Topography configuration creates water sinks (natural lakes or artificial reservoirs) for water supply or barriers for flooding control. It canalizes rainwater flow for flood protection (rivers/streams) or slows runoff velocity to recharge groundwater. It infiltrates runoff water to avoid contamination of groundwater through bioretention ponds, bioswales, and constructed wetlands. Hydrological system components when integrated within Landscape may also provide habitat and enhance the condition of public open spaces. Their spatial configuration can support the performance of multiple functions simultaneously.

Vegetation itself cannot be seen separately from water, whether it is water requirements for plants' proper growth, maintenance, and establishment or as a water regulation medium, as in the case of the water-filtering capacity in relation with structural diversity of vegetation cover.

"Green spaces and water systems ... intersperse, connect and provide vital life support for humans and other species within our urban environments."[35]

4.5 | OBJECTIVES AND GUIDELINES FOR A SUSTAINABLE LANDSCAPE SYSTEM

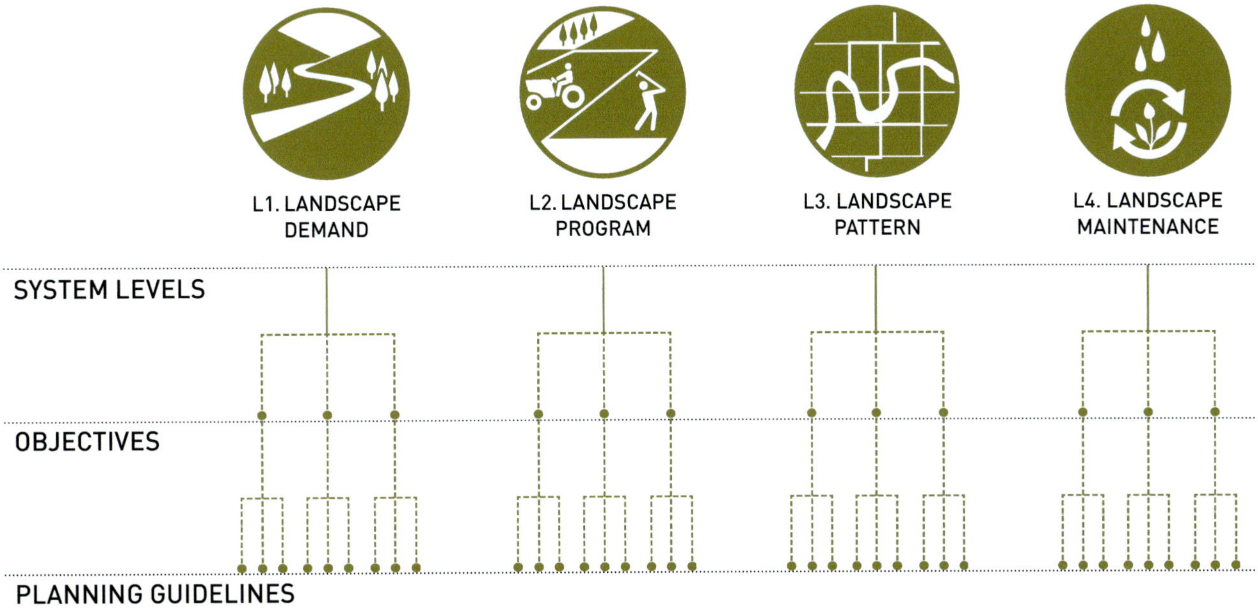

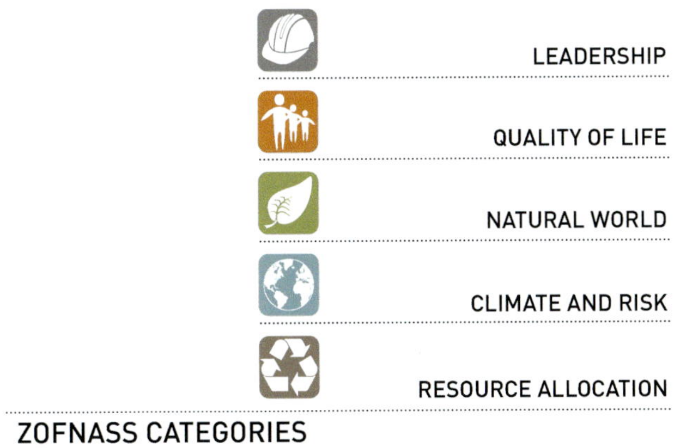

Following the decoding of the Landscape system at four system levels, planning objectives are defined for each level. They represent the main sustainability challenges and strategies toward a sustainable landscape system, expressing what should be the overall final outcome of landscape planning.

The objectives are associated with the five focus categories of the Zofnass Program, namely Leadership, Quality of Life, Natural World, Resource Allocation, and Climate and Risk. The Sustainable Planning Guidelines aim to lead toward these objectives, highlighting synergies and connections with the other six infrastructure categories of Energy, Water, Transportation, Solid Waste, Information, and Food.

OBJECTIVES AND GUIDELINES SYNOPSIS

All levels' objectives and guidelines are related directly or indirectly to the system's overall purpose of multifunctionality, the main characteristic of landscape as infrastructure. Whether it is expressed through multiple services, activities, or functions (and the maintenance and regeneration of these over time), multifunctionality traverses all four system levels to define a targeted sustainable system management.

The objectives and guidelines of the Landscape Demand level are primarily associated with the Leadership category and should be approached through extensive documentation and analysis. The main goal is to identify the opportunities, vulnerabilities, and limitations of a region in order to define the need for services that landscape is asked to perform.

This is the first reading of Landscape infrastructure in connection with the local environmental, social, and economic context. Objectives highlight the dependence on each place's characteristics. They aim to encourage the listing of specific provisioning, regulating, and cultural services that the landscape system has to deliver. The synergistic character of landscape is evident, as mitigation, compensation, regulation, but also provisioning services connect it with all the other infrastructure systems.

One particularity of this infrastructure system is that its services are not as obvious as in other systems. The formulation of ecosystem services is proposed to lead the decision-making and planning process on this system level. These services are further classified as potential or required, to point out the fact that not all places require or are capable of providing the same quantity and quality of services.

While in other infrastructure systems the main objective is to reduce the system's initial demand, in Landscape infrastructure the goal is to support and deliver the maximum services possible without overexploiting the corresponding available resources.

On the Program level, objectives are related to the Leadership, Quality of Life, and Natural World categories. The main objective of this level is to promote the multifunctionality of landscape entities while ensuring the compatibility of different simultaneous activities and their feasibility and integration at certain locations. The system has to select and promote activities in relation to the services needed and the context that supports them. The program has to explore and reconsider the potential of activities to deliver specific services. Therefore this level is supplementary to the Demand level. Its objectives and guidelines aim to analyze services in terms of uses of space and users. It does not encourage direct and obvious connections (one service corresponding to one activity), as this would lead once again to monofunctionality and would miss the optimum gains for environmental, social, and economic development. The guidelines of this level lead to this direction, proposing a process of planning.

On the Pattern level, objectives cover all five of the Zofnass Program focus categories, while connections are noted with all other infrastructure systems. Some of the most important synergies found are associated with Transportation infrastructure. The main objectives concern the optimized spatial distribution of the main structural components of a landscape system and the assignment of the proper ecological function to these components. The

L1. LANDSCAPE DEMAND

- L1.1. Understand urban planning opportunities, vulnerabilities, and limitations through the analysis and documentation of local context and processes
- L1.2. Define potential and required Landscape services preserving the ecological balance of natural communities, habitats, and environmentally critical lands
- L1.3. Define potential and required Landscape services regulating city areas' response to natural phenomena
- L1.4. Define potential and required Landscape services regulating and improving conditions within the urban fabric
- L1.5. Define potential and required Landscape provisioning services
- L1.6. Define potential and required Landscape cultural services
- L1.7. Define potential and required Landscape services regulating the function and performance of other infrastructure systems

L2. LANDSCAPE PROGRAM

- L2.1. Develop an integrated Landscape infrastructure strategy
- L2.2. Select the optimum activities to form the Program of Landscape infrastructure
- L2.3. Make Landscape entities as multifunctional as possible
- L2.4. Define the optimum distribution of Landscape activities across the city
- L2.5. Ensure compatibility of Landscape activities with locations

structure and the function will lead the design and final configuration of spaces, which makes it essential for the planning process.

The Pattern level is the one where the repartition of specific roles within the system takes place, supported through design that preserves or creates new conditions within each landscape entity (goal-specific and site-specific design).

The proper ecological (ecosystem) function for a given entity is defined by its location and by the services and activities it has been assigned to provide. The establishment of functions requires focusing on and designing the site-specific conditions that will support each function. The guidelines of the corresponding objectives propose a process of connection of spatial and functional configuration of entities. As already mentioned, the spatial configuration of an entity can and ought to be translated into function. The topography (or microtopography) of a given place can establish it as a conduit, a sink, or a barrier. The selection of surface cover materials can support the function of filter, if permeability and vegetation cover are selected. A targeted vegetation configuration in terms of density, vertical layering, and species selection can establish the functions of sink, habitat, filter, and barrier. An entity can function as a source in order to create new resources, through topography, plant species selection, etc., or as a result of spatial design. Guidelines thus are proposed to help the spatial design achieve its functional attributes. In order to facilitate the designing process, the guidelines make reference to a certain place's principal function before addressing the multiplicity of simultaneous functions.

L3. LANDSCAPE PATTERN

- L3.1. Identify main Landscape system entities to support identified services and selected activities
- L3.2. Delimit main Landscape system entities
- L3.3. Assign and establish the appropriate ecological functions to Landscape system entities according to identified services, location, etc.
- L3.4. Expand the system of Landscape corridors and patches through a supporting matrix
- L3.5. Integrate Landscape with other infrastructure systems' entities and facilities
- L3.6. Enhance quality of place
- L3.7. Increase connectivity between Landscape entities
- L3.8. Preserve and enhance biodiversity in Landscape corridors and patches
- L3.9. Prepare the Landscape system for short-term hazards and projected future threats
- L3.10. Take advantage of locally available resources for development of Landscape patches and corridors

Landscape Pattern stresses the importance of landscape forming networks that have to be continuous. Landscape connectivity is essential for the continuous flow and persistence of animal populations, goods, and services. Fragmentation and isolation of components make them unstable and more vulnerable to loss of their animal and plant communities. Objectives of increasing connectivity between landscape entities and preserving and enhancing biodiversity are interrelated and crucial for providing the basis for essential functions. The main synergies involve landscape ecology's notion of matrix as the background of corridors and patches. Matrix can expand the landscape system and supplement its performance. Within the urban fabric, for example, the transportation system due to its spatial presence is a potential and very useful matrix element.

Maintenance objectives are more related with Resource Allocation, given that maintenance practices require large quantities of resources. The objectives direct planning and management in procurement for sustainable use of resources, and optimized practices and technologies for water and energy use. They also highlight the strong connection with Solid Waste infrastructure.

The Maintenance level's main objective is to ensure that landscape entities, whether existing or newly designed, will persist in time and will further evolve. This is a precondition for landscape's delivery of potential quantity and quality of services. Ideally, every space once established has to approach a state as close to self-sufficiency as possible, in accordance with natural processes' innate potential. Therefore maintenance should be limited to the minimum. Leadership category objectives point to the need during the planning process to procure with a goal of minimum maintenance. Proper plant and material selection that adapts to climate conditions and intensity of use are indicative ways of avoiding excessive resource consumption and costs. Objectives aim to make clear that, though maintenance is part of the operation phase of a given entity, it needs to be placed at the core of the planning and designing process. Enhancing natural processes is set as a main objective for the Natural World category. The guidelines illustrate ways to approximate this goal through support of nutrients cycles, soil properties, and control of invasive species. To protect natural processes, it is also necessary to control negative flows from landscape entities.

The objectives under the Resource Allocation category focus on the availability of resources and the creation of new or alternative resources. They promote water conservation

L4. LANDSCAPE MAINTENANCE

- L4.1. Minimize intensity of required Landscape Maintenance services, aim for self-sufficiency
- L4.2. Develop effective and flexible Landscape Maintenance mechanisms
- L4.3. Promote community well-being
- L4.4. Enhance natural processes
- L4.5. Reduce greenhouse gas emissions from Landscape maintenance
- L4.6. Prepare Landscape system for projected climate threats
- L4.7. Optimize material use in Landscape Maintenance
- L4.8. Optimize Water use in Landscape Maintenance
- L4.9. Optimize Energy use in Landscape Maintenance
- L4.10. Contribute to Solid Waste reduction and resource recovery through Landscape Maintenance

methods, primarily through the selection of plant species, efficient irrigation techniques, and on-site harvesting of rainwater. The synergy with Water and Solid Waste infrastructure points out the importance of integrating the three systems to exploit landscape's ecological attributes and capacity to offer certain services such as water filtration, remediation, and waste decomposition.

"A useful way to consider the sustainable use of resources, such as water, materials, and energy, is to consider designed landscapes as functioning systems with inputs, outputs and internal cycling. Unsustainable systems tend to be 'open' with high resource inputs, minimal internal recycling, and high outputs of waste and energy. More sustainable systems will be more 'closed', with reduced inputs of materials and energy, a high level of internal recycling, and reduced waste outputs."[36]

GUIDELINES

OBJECTIVES		GUIDELINES	SYNERGIES
L1 LANDSCAPE DEMAND			▲ TO + WITH ▼ FROM
L1.1. Understand urban planning opportunities, vulnerabilities, and limitations through the analysis and documentation of local context and processes	L1.1.1.	Identify large-scale ecological flows and geographic patterns expanding beyond the city's boundaries	
	L1.1.2.	Collate local and regional environmental data	+
	L1.1.3.	Understand the local state of biodiversity and set relevant goals	
	L1.1.4.	Collate local and regional social/economic data	+
	L1.1.5.	Collate local and regional data on cultural aspects	+
L1.2. Define potential and required Landscape services preserving the ecological balance of natural communities, habitats, and environmentally critical lands	L1.2.1.	Define needs for protection and restoration of habitats from loss, fragmentation, and degradation through Landscape infrastructure	
	L1.2.2.	Define needs for protection and restoration of ecological flows and wildlife migration routes through Landscape infrastructure	
	L1.2.3.	Define needs for protection and restoration of carbon sinks through Landscape infrastructure	
	L1.2.4.	Define needs for restoration of contaminated areas through Landscape infrastructure	▲ ▲
	L1.2.5.	Define needs for regulating city growth and expansion patterns through Landscape infrastructure	▲
	L1.2.6.	Define needs for water infiltration/purification through Landscape infrastructure	▲
L1.3. Define potential and required Landscape services regulating city areas' response to natural phenomena	L1.3.1.	Define needs for rainwater/stormwater management through Landscape infrastructure	▲
	L1.3.2.	Define needs for soil control through Landscape infrastructure	
	L1.3.3.	Define needs for city protection against natural hazards and extreme events through Landscape infrastructure	
L1.4. Define potential and required Landscape services regulating and improving conditions within the urban fabric	L1.4.1.	Define needs for climate adaptation and microclimate improvement through Landscape infrastructure	▲
	L1.4.2.	Define needs for air quality improvement through Landscape infrastructure	
	L1.4.3.	Define needs for mitigation of conflicts between land uses through Landscape infrastructure	
L1.5. Define potential and required Landscape provisioning services	L1.5.1.	Identify opportunities for the provisioning of food, fiber, and feedstock provided by the natural context	▲ ▲
	L1.5.2.	Define strategies to enhance (benefit from) the productivity of landscape through Landscape infrastructure	▲ ▲
L1.6. Define potential and required Landscape cultural services	L1.6.1.	Identify recreational/social services opportunities provided by the natural context	
	L1.6.2.	Define strategies to offer recreational/social services required by demographics, population distribution, and culture through Landscape infrastructure	

OBJECTIVES	GUIDELINES		SYNERGIES
● L1.7. Define potential and required Landscape services regulating the function and performance of other infrastructure systems	L1.7.1.	Define needs for mitigating and compensating for other infrastructure systems' impact on the natural world through Landscape infrastructure	▲ ▲ ▲ ▲ ▲ ▲
	L1.7.2.	Define needs for mitigating and compensating for other infrastructure systems' impact on community life through Landscape infrastructure	▲ ▲ ▲ ▲ ▲ ▲
	L1.7.3.	Define needs for protection of infrastructure systems against projected climate change and extreme events through Landscape infrastructure	▲ ▲ ▲ ▲ ▲ ▲

L2 LANDSCAPE PROGRAM

▲ TO
+ WITH
▼ FROM

OBJECTIVES	GUIDELINES		SYNERGIES
● L2.1. Develop an integrated Landscape infrastructure strategy	L2.1.1.	Hierarchize Demand for Landscape services based on city goals for sustainable development	
	L2.1.2.	Trace correlation of alternative Landscape infrastructure strategies with city competitiveness and economic sectors	
● L2.2. Select the optimum activities to form the Program of Landscape infrastructure	L2.2.1.	Translate identified Landscape services to specific activities	▲
	L2.2.2.	Integrate long-term city growth and development goals with Landscape activities Program	
	L2.2.3.	Secure long-term financial viability of Landscape activities Program	
	L2.2.4.	Regularly update Landscape activities Program planning according to community growth and development	
● L2.3. Make Landscape entities as multifunctional as possible	L2.3.1.	Define possible combinations of compatible Landscape activities to create social, financial, and ecological catalysts	
	L2.3.2.	Combine Landscape activities taking into consideration seasonal patterns	
	L2.3.3.	Combine Landscape activities to enhance daily patterns for safe, lively environments	
● L2.4. Define the optimum distribution of Landscape activities across the city	L2.4.1.	Adapt the scale and distribution of Landscape activities to the size of communities and areas they intend to serve	
● L2.5. Ensure compatibility of Landscape activities with locations	L2.5.1.	Ensure the protection of areas with special natural and cultural features through the regulation of Landscape activities within them and their adjacent areas	
	L2.5.2.	Adjust the intensity of Landscape activities according to scale and context	
	L2.5.3.	Integrate Landscape activities with existing natural properties and processes where possible	

OBJECTIVES	GUIDELINES		SYNERGIES

L3 LANDSCAPE PATTERN

▲ TO
✚ WITH
▼ FROM

Objectives		Guidelines
L3.1. Identify main Landscape system entities to support identified services and selected activities	L3.1.1.	Identify and place main patches in connection to land use planning, demographics, and natural context
	L3.1.2.	Identify and place main corridors in connection to land use planning, demographics, and natural context
	L3.1.3.	Integrate corridors and patches to function as an overall network of open spaces
	L3.1.4.	Form connections of Landscape entities with peri-urban areas
L3.2. Delimit main Landscape system entities	L3.2.1.	Define shape and area of main patches for better integration with the natural processes and the urban fabric
	L3.2.2.	Define width of main corridors for better integration with the natural processes within the urban fabric
	L3.2.3.	Avoid fragmentation of large patches
L3.3. Assign and establish the appropriate ecological functions to Landscape system entities according to identified services, location, etc.	L3.3.1.	Support/establish the functions of conduit, sink, and barrier through topography/microtopography of Landscape entities
	L3.3.2.	Support/establish the function of filter through the selection of surface cover materials in Landscape entities
	L3.3.3.	Support/establish the functions of sink, habitat, filter, and barrier with adequate vegetation density, patterns, vertical vegetation layering, and plant selection in Landscape entities
	L3.3.4.	Support/establish the function of source for the creation of new resources through Landscape entities
	L3.3.5.	Add multiple secondary ecological functions to Landscape entities
	L3.3.6.	Optimize Landscape asset management through adequate administrative structures
L3.4. Expand the system of Landscape corridors and patches through a supporting matrix	L3.4.1.	Exploit opportunities for additional Landscape supportive patches and corridors in the private space
	L3.4.2.	Exploit opportunities for additional Landscape supportive patches and corridors given by Transportation entities
L3.5. Integrate Landscape with other infrastructure systems' entities and facilities	L3.5.1.	Configure Landscape entities' vegetation and paving to provide for other infrastructure networks' needs and facilitate access and maintenance
	L3.5.2.	Investigate needs of Landscape attractors' integration with Transportation Nodes
L3.6. Enhance quality of place	L3.6.1.	Use the configuration of Landscape entities to strengthen each location's sense of place
	L3.6.2.	Ensure openness and easy access to Landscape entities for all members of the community

OBJECTIVES	GUIDELINES		SYNERGIES
	L3.6.3.	Cater to the needs of different community groups	
L3.7. Increase connectivity between Landscape entities	L3.7.1.	Avoid or compensate for the interruption of flows, wildlife corridors, and greenways caused by Landscape entities	
	L3.7.2.	Distribute smaller patches to function as stepping stones for wildlife movement	
L3.8. Preserve and enhance biodiversity in Landscape corridors and patches	L3.8.1.	Preserve the natural topography	
	L3.8.2.	Preserve biodiversity, avoiding alien monocultures in Landscape entities	
	L3.8.3.	Enhance biodiversity, mimicking natural vegetation patterns in Landscape entities	
	L3.8.4.	Provide food and shelter for different urban wildlife populations in Landscape entities	
L3.9. Prepare the Landscape system for short-term hazards and projected future threats	L3.9.1.	Assess and manage future risks from natural and man-made phenomena affecting Landscape entities	
L3.10. Take advantage of locally available resources for development of Landscape patches and corridors	L3.10.1.	Reduce excavated material taken off site for the construction of Landscape patches and corridors / Balance cut and fill	
	L3.10.2.	Use materials that are readily and locally sourced for the construction of Landscape patches and corridors	▼

L4 LANDSCAPE MAINTENANCE

▲ TO
+ WITH
▼ FROM

OBJECTIVES	GUIDELINES		SYNERGIES
L4.1. Minimize intensity of required Landscape Maintenance services, aim for self-sufficiency	L4.1.1.	Promote use of plants and materials adequate to local climate, soil, and exposure	
L4.2. Develop effective and flexible Landscape Maintenance mechanisms	L4.2.1.	Identify and assess threats and problems in Landscape Maintenance through a robust and integrated monitoring system	+
	L4.2.2.	Implement diversity in Landscape management methods and techniques	
	L4.2.3.	Develop Landscape Maintenance plans based on voluntarism	
L4.3. Promote community well-being	L4.3.1.	Assess and resolve potential health hazards from Landscape Maintenance activities and substances used	
	L4.3.2.	Assess and resolve noise pollution issues caused by Landscape Maintenance activities	
	L4.3.3.	Mitigate dust issues caused by Landscape Maintenance activities	
	L4.3.4.	Assess and resolve odor issues caused by Landscape Maintenance activities	

OBJECTIVES	GUIDELINES		SYNERGIES
L4.4. Enhance natural processes	L4.4.1.	Support Landscape entities' self-regeneration over time	
	L4.4.2.	Support nutrients cycle during Landscape Maintenance	
	L4.4.3.	Avoid use of persistent pesticides in Landscape Maintenance	
	L4.4.4.	Protect soil properties and functions during Landscape Maintenance	
	L4.4.5.	Control invasive species as part of Landscape Maintenance	
	L4.4.6.	Control flows of fertilizer- and pesticide-laden runoff water from Landscape elements	
L4.5. Reduce greenhouse gas emissions from Landscape Maintenance	L4.5.1.	Minimize use of carbon-fueled equipment for Landscape Maintenance	
L4.6. Prepare Landscape system for projected climate threats	L4.6.1.	Prepare Landscape Maintenance action plans for periods of extreme climatic conditions	
L4.7. Optimize material use in Landscape Maintenance	L4.7.1.	Use locally available resources (soil, plants, raw materials) for Landscape Maintenance	
L4.8. Optimize Water use in Landscape Maintenance	L4.8.1.	Set long-term goals and standards for reducing Water Consumption in Landscape Maintenance	
	L4.8.2.	Use native plants of low water requirements	
	L4.8.3.	Construct and operate efficient Landscape irrigation systems	
	L4.8.4.	Harvest rainwater for Landscape Maintenance where possible	
	L4.8.5.	Use gray or reclaimed water for cleaning and irrigation in Landscape entities	
L4.9. Optimize Energy use in Landscape Maintenance	L4.9.1.	Use energy-efficient equipment for Landscape Maintenance	
L4.10. Contribute to Solid Waste reduction and resource recovery through Landscape Maintenance	L4.10.1.	Conform waste management within Landscape entities to the requirements of Solid Waste Collection and Transfer planning	
	L4.10.2.	Use residues of Landscape Maintenance practices as supply resources for other infrastructure systems	
	L4.10.3.	Incorporate Solid Waste and wastewater Treatment residues and by-products in regular Landscape Maintenance if available	

RELEVANT GUIDELINES

L1 LANDSCAPE DEMAND

FROM → TO + WITH

GUIDELINES		RELEVANT GUIDELINES	
L1.1.2.	Collate local and regional environmental data	I1.2.1.	Identify opportunities to monitor and protect the natural context through information infrastructure and applications
L1.1.4.	Collate local and regional social/economic data	I1.1.3.	Identify opportunities to improve city administration and social amenities through information infrastructure and applications
L1.1.5.	Collate local and regional data on cultural aspects	I1.1.3.	Identify opportunities to improve city administration and social amenities through information infrastructure and applications
L1.2.4.	Define needs for restoration of contaminated areas through Landscape infrastructure	W2.3.3.	Identify and manage basic sources of contaminants within the watershed
		F2.5.5.	Identify and manage basic sources of contaminants within the foodshed
L1.2.5.	Define needs for regulating city growth and expansion patterns through Landscape infrastructure	F2.5.3.	Protect foodshed from urban expansion
L1.2.6.	Define needs for water infiltration/purification through Landscape infrastructure	W2.3.4.	Control contaminants in runoff water through bioretention and bioinfiltration methods
L1.3.1.	Define needs for rainwater/ stormwater management through Landscape infrastructure	W2.3.6.	Provide for rainwater and stormwater management needs with natural drainage, infiltration, and floodplain management
L1.4.1.	Define needs for climate adaptation and microclimate improvement through Landscape infrastructure	E1.1.4.	Improve the urban microclimate
L1.5.1.	Identify opportunities for the provisioning of food, fiber, and feedstock provided by the natural context	F2.5.1.	Map and monitor the city's foodshed and highly productive farmland
		E2.1.1.	Identify all potential Energy sources
L1.5.2.	Define strategies to enhance (benefit from) the productivity of landscape through Landscape infrastructure	W2.1.4.	Promote use of Water resources available within project boundaries
		F2.3.4.	Provide know-how and assistance to local farmers on more productive methods and techniques
L1.7.1.	Define needs for mitigating and compensating for other infrastructure systems' impact on the natural world through Landscape infrastructure	SW3.8.4.	Protect water elements and groundwater from contaminants found in Solid Waste Treatment and Disposal facilities
		SW3.8.3.	Protect soil from contaminants found in Solid Waste Treatment and Disposal facilities
		SW3.8.2.	Protect natural habitats and wildlife corridors and compensate for their fragmentation or disruption by Solid Waste Treatment and Disposal facilities
		SW4.6.2.	Protect soil from contaminants during Solid Waste Collection and Transfer

GUIDELINES	RELEVANT GUIDELINES	
	SW4.6.3.	Protect water elements and groundwater from contaminants during Solid Waste Collection and Transfer
	SW4.6.1.	Protect natural habitats and wildlife corridors and compensate for their fragmentation or disruption by Solid Waste Collection and Transfer processes
	SW3.10.2.	Compensate for Solid Waste Treatment and Disposal emissions through carbon offset and sequestration practices
	E3.6.2.	Protect natural habitats and wildlife corridors and compensate for their fragmentation or disruption by Energy Generation facilities
	E3.6.3.	Protect soil from contaminants found in Energy Generation facilities
	E3.6.4.	Protect Water elements and groundwater from contaminants found in Energy Generation facilities
	E4.7.2.	Protect natural habitats and wildlife corridors and compensate for their fragmentation or disruption by Energy Distribution and Transmission
	E3.7.2.	Compensate for Energy Generation emissions through carbon offset and sequestration practices
	W3.1.6.	Prepare for Water Treatment failures or overflow events
	W3.6.3.	Protect natural habitats and wildlife corridors and compensate for their fragmentation or disruption by Water Treatment facilities
	W3.6.4.	Protect soil from contaminants found in Water Treatment facilities
	W4.6.1.	Protect habitats and wildlife corridors and compensate for their fragmentation or disruption by Water Networks
	W3.8.2.	Compensate for emissions incurred in Water Treatment through carbon offset and sequestration practices
	T3.7.2.	Protect natural habitats and wildlife corridors and compensate for their fragmentation or disruption by Transportation Nodes
	T3.7.6.	Protect soil from contaminants, erosion, and compacting caused by Transportation Nodes
	T4.8.4.	Protect natural habitats and wildlife corridors and compensate for their fragmentation or disruption by Transportation Networks
	T4.8.11.	Protect soil from contaminants, erosion, and compacting caused by Transportation Networks
	I4.7.2.	Protect natural habitats and wildlife corridors and compensate for their fragmentation or disruption by Information Networks

GUIDELINES		RELEVANT GUIDELINES	
		F2.7.4.	Protect Water elements and groundwater from contaminants used in Food Supply and Production activities
L1.7.2.	Define needs for mitigating and compensating for other infrastructure systems' impact on community life through Landscape infrastructure	SW3.6.4.	Assess and resolve potential noise pollution issues caused by Solid Waste Treatment and Disposal facilities
		SW3.6.5	Assess and resolve potential odor issues caused by Solid Waste Treatment and Disposal facilities
		SW3.7.1.	Visually integrate Solid Waste Treatment and Disposal facilities to the local context
		SW4.4.2.	Assess and resolve potential noise pollution issues caused by Solid Waste Collection and Transfer
		SW4.5.1.	Visually integrate Solid Waste Collection and Transfer facilities and equipment to the local context
		E3.4.3.	Assess and resolve noise pollution issues caused by Energy Generation facilities
		E3.5.1.	Visually integrate Energy Generation facilities to the local context
		E4.6.1.	Visually integrate Energy Distribution and Transmission facilities to the local context
		W3.4.2.	Assess and resolve potential odor issues caused by Water Treatment facilities
		W3.4.3.	Assess and resolve potential noise pollution issues caused by Water Treatment facilities
		W3.5.1.	Visually integrate Water Treatment facilities to the local context
		W4.4.2.	Assess and minimize potential odor issues caused by Water collection and transfer facilities and processes
		W4.5.1.	Visually integrate Water Networks to the local context
		I3.6.1.	Visually integrate Information Nodes to the local context
		T3.4.2.	Visually integrate Transportation Nodes to the local context
		T4.5.2.	Visually integrate Transportation Networks to the local context
		T3.6.1.	Assess and resolve noise pollution issues caused by Transportation Nodes
		T4.7.3.	Assess and resolve noise pollution issues caused by Transportation Networks
L1.7.3.	Define needs for protection of infrastructure systems against projected climate change and extreme events through Landscape infrastructure	SW3.9.1.	Assess and manage future threats from natural and man-made phenomena affecting Solid Waste Treatment and Disposal facilities
		SW4.7.1.	Assess and manage future threats from natural and man-made phenomena affecting Solid Waste Collection and Transfer facilities

GUIDELINES		RELEVANT GUIDELINES	
		E3.9.1.	Assess and manage threats from natural and man-made phenomena affecting Energy Generation facilities
		E4.8.1.	Assess and manage future threats from natural and man-made phenomena affecting Energy Distribution and Transmission
		W3.7.2.	Prepare action plans for periods of extreme water scarcity or high precipitation
		W4.8.1.	Assess and manage threats from natural and man-made phenomena affecting Water Networks
		I3.8.1.	Assess and manage risks from natural and man-made phenomena affecting Information Nodes
		I4.8.1.	Assess and manage future threats from natural and man-made phenomena affecting Information Networks
		F2.3.2.	Prepare plans for short-term and long-term hazards affecting local Food Production
		T3.8.1.	Assess and manage future threats from natural and man-made phenomena affecting Transportation Nodes
		T4.9.1.	Assess and manage future threats from natural and man-made phenomena affecting Transportation Networks

L2 LANDSCAPE PROGRAM

FROM ▸ ▸ TO + WITH

L2.2.1.	Translate identified Landscape services to specific activities	F2.4.2.	Provide space for communal gardens or allotments

L3 LANDSCAPE PATTERN

FROM ▸ ▸ TO + WITH

L3.4.2.	Exploit opportunities for additional Landscape supportive patches and corridors given by Transportation entities	T3.7.3.	Develop permeable paving and vegetation configurations for Transportation Nodes to provide natural drainage
		T3.7.4.	Develop permeable paving and vegetation configurations for Transportation Nodes to filter runoff
		T3.7.5.	Develop vegetation configurations for Transportation Nodes to provide habitats
		T4.8.5.	Develop permeable paving and vegetation configurations for Transportation Networks to provide natural drainage
		T4.8.6.	Develop permeable paving and vegetation configurations for Transportation Networks to filter runoff

GUIDELINES		RELEVANT GUIDELINES	
		T4.8.7.	Develop vegetation configurations for Transportation Networks to provide habitats
L3.5.1.	Configure Landscape entities' vegetation and paving to provide for other infrastructure networks' needs and facilitate access and maintenance	W4.3.4.	Integrate Water Networks with Landscape entities
		E4.4.3.	Integrate Energy Distribution and Transmission networks with Landscape entities
		I4.4.2.	Integrate Information Networks with Landscape entities
L3.5.2.	Investigate needs of Landscape attractors' integration with Transportation Nodes	T3.3.3.	Provide required traffic connections to large Landscape infrastructure entities
L3.10.2.	Use materials that are readily and locally sourced for the construction of Landscape patches and corridors	T3.7.7.	Save, preserve, and reuse topsoil before the construction of Transportation Nodes where possible
		T4.8.12.	Save, preserve, and reuse topsoil before the construction of Transportation Networks where possible
		T3.9.4.	Plan for the reuse of elements or recycling of materials after Transportation Nodes decommission
		T4.10.5.	Plan for the reuse of elements or recycling of materials after Transportation Networks decommission

L4 LANDSCAPE MAINTENANCE

FROM ▸ ▸ TO + WITH

L4.2.1.	Identify and assess threats and problems in Landscape Maintenance through a robust and integrated monitoring system	I1.3.2.	Identify other infrastructure systems' processes to be managed through Information infrastructure
		I1.3.1.	Identify infrastructure systems' processes to be monitored/documented through Information infrastructure
L4.4.6.	Control flows of fertilizer- and pesticide-laden runoff water from Landscape elements	W2.3.4.	Control contaminants in runoff water through bioretention and bioinfiltration methods
		F2.5.5.	Identify and manage basic sources of contaminants within the foodshed
L4.8.1.	Set long-term goals and standards for reducing Water Consumption in Landscape Maintenance	W1.3.1.	Reduce water required for Landscape Maintenance
L4.8.2.	Use native plants of low water requirements	W1.3.1.	Reduce water required for Landscape Maintenance
L4.8.3.	Construct and operate efficient Landscape irrigation systems	W1.3.1.	Reduce water required for Landscape Maintenance
		I1.3.2.	Identify other infrastructure systems' processes to be managed through Information infrastructure
		I1.3.1.	Identify infrastructure systems' processes to be monitored/documented through Information infrastructure
L4.8.4.	Harvest rainwater for Landscape Maintenance where possible	W1.3.1.	Reduce water required for Landscape Maintenance

GUIDELINES		RELEVANT GUIDELINES	
L4.8.5.	Use gray or reclaimed water for cleaning and irrigation in Landscape entities	W1.6.3.	Identify opportunities for use of different nonpotable water qualities
L4.10.1.	Conform waste management within Landscape entities to the requirements of Solid Waste Collection and Transfer planning	SW4.3.6.	Solid Waste and Landscape systems to work in synergy for the Collection and Transfer of Landscape Maintenance residues
L4.10.2.	Use residues of Landscape maintenance practices as supply resources for other infrastructure systems	E2.1.1.	Identify all potential Energy sources
L4.10.3.	Incorporate Solid Waste and wastewater Treatment residues and by-products in regular Landscape Maintenance if available	SW2.3.3.	Match Solid Waste Diversion Model residues and by-products with resource and feedstock needs of other infrastructure systems
		F4.9.2.	Identify potential for using Food Processing and Distribution residues and by-products as feedstock for other infrastructure systems
		W3.9.2.	Investigate using Water Treatment residues for material recovery

NOTES

1. Pierre Bélanger, "Landscape Infrastructure: Urbanism beyond Engineering," in Spiro Pollalis, Andreas Georgoulias, Stephen Ramos, and Daniel Schodek, eds., *Infrastructure Sustainability and Design* (New York: Routledge, 2012), 310.
2. K. L. Wolf, "Ergonomics of the City: Green Infrastructure and Social Benefits," in *Engineering Green: Proceedings of the 2003 National Urban Forest Conference* (Washington, DC, 2003).
3. Sandra Naumann, McKenna Davis, et al., "Design, Implementation and Cost Elements of Green Infrastructure Projects," Final report to the European Commission, Brussels, 2011.
4. Nancy Rottle and Ken Yocom, *Basics Landscape Architecture 02: Ecological Design* (AVA Publishing SA, 2010), 48.
5. Richard T.T. Forman, *Landscape Mosaics: The Ecology of Landscapes and Regions* (New York: Cambridge University Press, 1995); J. F. Thorne, "Landscape Ecology: A Foundation for Greenway Design," in D. S. Smith and P. Cawood Hellmund, eds., *Ecology of Greenways: Design and Function of Linear Conservation Areas* (Minneapolis: University of Minnesota Press, 1993).
6. "Sniffer Project UEUB01: Urban Networks for People and Biodiversity—Form and Function," May 2008, http://www.sniffer.org.uk/search?q=Multifunctional+green+networks+.
7. "Ecosystem Services," Beyond the Line, accessed December 05, 2014, http://www.beyond-theline.org/KeyIdeas/EcosystemServices.aspx.
8. Steve Carpenter, Brian Walker, J. Marty Anderies, and Nick Abel, "From Metaphor to Measurement: Resilience of What to What?," Ecosystems (2001) 4: 765-781.
9. "What's the Problem?," Convention on Biological Diversity, accessed December 17, 2014, http://www.cbd.int/agro/whatstheproblem.shtml.
10. P. Selman, "Planning for Landscape Multifunctionality," *Sustainability: Science, Practice, & Policy* 5(2):45-52. Published online Nov 30, 2009. http://www.google.gr/archives/vol5iss2/communityessay.pselman.html.
11. Millennium Ecosystem Assessment, *Ecosystems and Human Well-being: Health Synthesis* (Washington, DC: World Resources Institute, 2005).
12. "2 Green Infrastructure: Concepts and Definitions," Green Infrastructure Evidence Base, accessed August 10, 2014, http://gievidencebase.botanicgardens.sa.gov.au/contents/green-infrastructure-concepts-and-definitions.
13. Millennium Ecosystem Assessment, *Ecosystems and Human Well-being: A Framework for Assessment* (Washington, DC: Island Press, 2003).
14. Robert Costanza, Ralph d'Arge, Rudolf de Groot, Stephen Farber, Monica Grasso, Bruce Hannon, Karin Limburg, Shahid Naeem, Robert V. O'Neill, Jose Paruelo, Robert G. Raskin, Paul Sutton, and Marjan van den Belt, "The Value of the World's Ecosystem Services and Natural Capital," *Nature* 387 (15 May 1997) : 253–260.
15. Carl Folke, Steve Carpenter, Brian Walker, Marten Scheffer, Thomas Elmqvist, Lance Gunderson, C.S. Holling, "Regime Shifts, Resilience, and Biodiversity in Ecosystem Management," Annual Review of Ecology, Evolution, and Systematics, Vol. 35(December 2004): 557-581.
16. S. Baby, "Assessing Morphological Landscape Carrying Capacity for Coastal Areas in Kuwait," accessed December 05, 2014, http://www.niscair.res.in/jinfo/IJMS/IJMS-Forthcoming-Articles/IJMS-PR-Aug%202014/MS%202162%20Edited.pdf.
17. "Ecosystems and Human Well-Being: A Framework for Assessment," Report conducted by Millennium Ecosystem Assessment (MA), 63, http://www.unep.org/maweb/documents/document.300.aspx.pdf.
18. "Mapping and Assessment of Ecosystems and their Services," European Commission, accessed April 14, 2014, http://ec.europa.eu/environment/nature/knowledge/ecosystem_assessment/pdf/MAESWorkingPaper2013.pdf.
19. The CICES classification is being used: "For the purposes of CICES, ecosystem services are defined as the contributions that ecosystems make to human well-being. They are seen as arising from living organisms (biota) or the interaction of biotic and abiotic processes, and refer specifically to the 'final' outputs or products from ecological systems. That is, the things directly consumed, used or enjoyed by people. Following common usage, the classification recognizes these outputs to be provisioning, regulating and cultural services, but it does not cover the so-called 'supporting services' originally defined in the MA. The supporting services are treated as part of the ecosystem processes and ecosystem functions that characterize ecosystems. Since they are only indirectly consumed or used, and may simultaneously facilitate the output of many 'final outputs', it was considered that they were best dealt within environmental accounts, in other ways." "Mapping and Assessment of Ecosystems and their Services," European Commission.
20. Selman, "Planning for Landscape Multifunctionality."
21. Anthony B. Anderson and Clinton N. Jenkins, *Applying Nature's Design: Corridors as a Strategy for Biodiversity Conservation* (New York: Columbia University Press, 2006), 3.
22. Forman, *Landscape Mosaics.*
23. Forman, *Landscape Mosaics.*
24. Katie Meiklejohn, Rob Ament, and Gary Tabor, "Habitat Corridors & Landscape Connectivity: Clarifying the Terminology," Center for Large Landscape Conservation, http://largelandscapes.org/media/publications/Habitat-corridors-and-landscape-connectivity1.pdf.
25. "2 Green Infrastructure: Concepts and Definitions," Green Infrastructure Evidence Base.
26. "Optimize Operational and Maintenance Practices," Whole Building Design Guide, accessed December 03, 2014, http://www.wbdg.org/design/optimize_om.php.
27. "Sustainable Landscapes Project," State of South Australia, accessed June 05, 2014, http://www.environment.sa.gov.au/botanicgardens/Learn/Green_Infrastructure/Sustainable_Landscapes_Project.
28. Rottle and Yocom, *Basics Landscape Architecture 02*, 120.
29. "Glossary of Terms," Biodiversity Information System for Europe, accessed June 01, 2015, http://biodiversity.europa.eu/maes/glossary-of-terms.
30. Costanza et al., "The Value of the World's Ecosystem Services and Natural Capital."
31. Rottle and Yocom, *Basics Landscape Architecture 02*, 64.
32. John Tillman Lyle, *Regenerative Design for Sustainable Development* (New York: Wiley, 1994), 10.

33. "2 Green Infrastructure: Concepts and Definitions," Green Infrastructure Evidence Base.
34. Definition of water regulation in Millennium Ecosystem Assessment, *Ecosystems and Human Well-Being: Synthesis.*
35. Australian Institute of Landscape Architects (AILA), "Green Infrastructure Report," 2012, http://www.greeninfrastructure.org.au/.
36. Nigel Dunnett and Andy Clayden, *Resources: The Raw Materials of Landscape* (London: Spon Press, 2000).

IMAGE CREDITS

P. 30: Photograph by Ales Krivec, accessed on October 05, 2015, https://unsplash.com/photos/XW4C3wPqyJs. **P. 38:** Photograph by Murashkame, accessed on October 01, 2015, https://unsplash.com/photos/cBw5FisOsw8. **P. 39:** Adapted from J. F. Thorne, "Landscape Ecology: A Foundation for Greenway Design," in D. S. Smith and P. Cawood Hellmund, eds., Ecology of Greenways: Design and Function of Linear Conservation Areas (Minneapolis: University of Minnesota Press, 1993). **P. 42:** (Background image) Photograph by Gabriel Garcia Marengo, accessed on October 01, 2015, https://unsplash.com/photos/Dm6Ns3TjQm4. **P. 43:** (Top left) "Forest," Photograph by LEEROY, accessed on July 20, 2015, http://www.lifeofpix.com/gallery/forest-2/; (Center) "Habitat Fragmentation," accessed on April 14, 2014, https://gardenerscoach.files.wordpress.com/2013/03/habitat-fragmentation1.jpg; (On the right) "Logs," Photograph by LEEROY, accessed on July 20, 2015, http://www.lifeofpix.com/gallery/logs/. **P. 46:** (Top) Ecosystem services list from European Commission, "Mapping and Assessment of Ecosystems and Their Services," April 2013, pp. 30–35, accessed on April 17, 2015, http://ec.europa.eu/environment/nature/knowledge/ecosystem_assessment/pdf/MAESWorkingPaper2013.pdf. **P. 53:** Photograph by Dani Corona, accessed on October 01, 2015, https://unsplash.com/photos/NxzovEuz878.

Sustainable Infrastructure Planning Guidelines

Transportation
Infrastructure

Chapter 5

Transportation

serves the movement of people, animals, and products from one location to another: the mobility of passengers and freight.

5.1 | INDUSTRY PERSPECTIVES

Smart Cities and Green Transportation: Creating a Model of Sustainable Urbanism

The guidelines in this chapter speak to a fundamental balance that underlies the essence of community sustainability at all scales, from the smallest villages to the broadest urban regions. A working, durable, inspiring definition of smart growth lives in the balance between transportation infrastructure and land use planning.

This balance can be expressed as two poles of an active, ongoing, generative conversation. One pole represents the policies we advocate and the plans we develop to reduce the absolute demands our settlements place on the natural environment—in short, how we act to minimize our environmental footprint. The other pole seeks to support the demands that remain in the most efficient, appropriate manner with as little impact on the environment as possible.

In the realm of urban transportation, these poles are generally represented by two interlocked principles: walkability and multimodality. Walkability is an urban planning strategy, a commitment to urban form and land use deployment that reduces the overall need to travel. Multimodality, a transportation strategy, asserts that all travel modes from airplane to walking have appropriate roles in a thoughtful, fully connected urban system.

Taken together, walkability and multimodality conspire to reduce the consumed energy of travel and the embodied energy of excessive infrastructure development. The guidelines in this chapter focus on how to achieve, optimize, and sustain this benefit.

The Transportation guidelines are grouped into four system dimensions: Trip Generation, Modes, Nodes, and Networks. Within each level, specific objectives and supporting guidelines address one or more of the Zofnass Program's five areas of focus: Leadership, Quality of Life, Natural World, Climate and Risk, and Resource Allocation.

Reduction of travel demand (including the "travel" of energy, water, food, waste, and commodities) is elegantly expressed in the guidelines for Trip Generation. Enabling people to live within walking distance of all that cities have to offer is the key to reducing the number and length of trips, thus easing transportation's impact on the environment.

The benefits of multimodality lie at the core of Modes and Networks. These guidelines encourage leaders to create and sustain interconnected systems of multiple forms of transportation at all scales of urban geography, from local centers to vast regions.

This strategy extends the benefits of walkable urbanism far beyond the local walk (or other form of human or active transportation) from house to train. Whole cities and urban regions become accessible. People move with ease across multimodal networks, each mode calibrated to the length and type of trip it serves along the journey.

The guidelines for Nodes take multimodality to the streets, confronting what it takes to create and nurture those special places in the urban construct where people and transportation come together. Nodes are everywhere in the multimodal city, from the nearest bus stop to the local village center where you board your train, to downtown were you work and shop, to the airport where you set out for destinations beyond your own walkable region. Trips start and end at nodes, and nodes are where you transfer between modes on a multimodal journey.

The guidelines address Nodes at three levels:

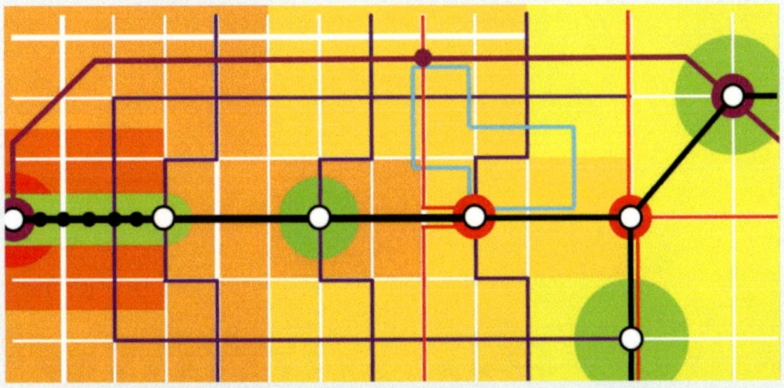

Smart growth transect:
A transit corridor emanates from a downtown core, intersecting other modes and branching out to serve new transit villages in the suburbs

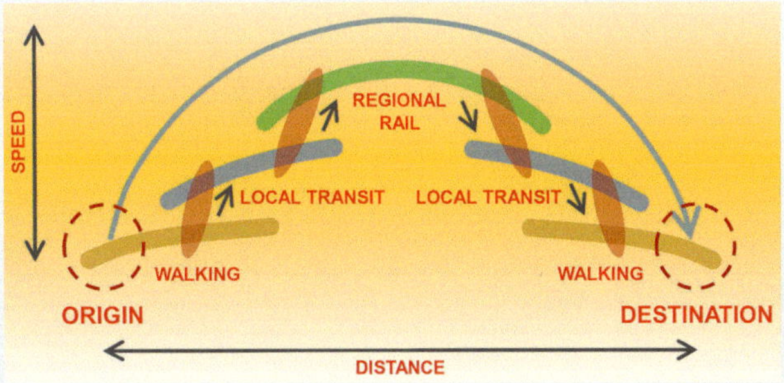

Arc of a trip:
Every urban journey, however complex, begins and ends on foot (or by other form of human-powered transportation)

- Functionality—how nodes work as transportation elements,
- Placemaking—how to fully express nodes as essential and vital urban places,
- Environment—how we minimize and mitigate the impacts of what we build.

It is essential that planners, engineers, and public officials get this right. The blend of workability and quality of life—each supporting the other—is what draws people to take on the practice of sustainable urbanism: to live, work, and play in cities that sit lightly on the earth.

A specific objective of the Zofnass Planning Guidelines is to illustrate synergy, the essential interactions among infrastructural systems that make up the city. The Transportation guidelines cross-reference the infrastructures of Energy, Water, Food, Solid Waste, Landscape, and Information, enriching and lending nuance to the strategies of balancing transportation and land use. These connections show that transportation serves both things and people, and that modes include wires, pipelines, and the electromagnetic spectrum. In all of these realms, the same rules apply: First, reduce the levels of absolute demand. Then, serve remaining demand in the most sustainable fashion.

The guideline "Place Transportation Nodes in correlation to Trip attractors" provides an elegant example. A driving force behind the "locavore" movement sweeping American cities is the recognition that transporting food over great distances consumes major amounts of transportation energy. Placing a transportation node—perhaps a transit intermodal center—adjacent to the trip attractor of a local farmer's market is exactly what this guideline would advocate.

The Zofnass commitment to true multimodality is particularly evident in guidelines that recognize the role of highways within the multimodal spectrum. Technology is revolutionizing the efficiency and effectiveness of motor vehicles, from the continued development of energy-efficient propulsion to the advent of smart highways and self-driving cars. Addressing both vehicle and path, the guidelines embrace this revolution by seeking to optimize the effectiveness of highways while minimizing their effects on the natural and urban environment.

The guidelines create the foundation for smart highways by advocating for energy-efficient and alternative-energy vehicles within a context of long-term emission reduction across all modes of transportation. The emerging technology of self-driving cars is but one example of the innovations these guidelines empower.

The promises of this technology—to dramatically increase the carrying capacity of highway infrastructure, while reducing inefficient and polluting stop-and-go traffic—extend the useful life of existing highway networks while reducing the embodied energy of future highway development and the consumed energy of travel. In a world of political and financial challenges, these benefits support a key Zofnass objective: the long-term financial viability of transportation investments.

In the smart growth model, even as land use guides transportation, transportation creates new urban opportunities. As the unfolding vision of interconnected multimodal transportation transforms the urban landscape, people increasingly seek to live, work, and play at the nodes where transportation modes come together. The possibilities extend well beyond the familiar model of the transit village around a local train station. One example, evident worldwide, is the growth of virtual cities around major international airports.

With extensive mixed-use development and excellent regional intermodal connections, "airport cities" are themselves significant urban destinations catering to the needs of arriving, departing, and transferring airline passengers. Historically, these cities have evolved over time with minimal long-range planning, yet they have the potential to be models of transit-oriented sustainable urbanism.

Transit node:
Trains, buses, bikes, and pedestrians connect at the Downtown Transit Center in Tempe, Arizona

Smart cars:
Vehicle connectivity and automation will transform transportation systems

The Zofnass guidelines speak directly to the possibilities of airport cities—and all cities—by promoting appropriate mixed-use development, compact urban form, reliable public transit, and walkable public realms of refinement and beauty.

At the core, these guidelines sound a clarion call for interdisciplinary collaboration in generating and stewarding broad-based visions of a sustainable urban future. Crafting the delicate balance between transportation and land use requires the contributions of everyone in the city: the leadership of politicians and public officials; the technical prowess of planners, engineers, and architects; the energy of builders and developers; and the collective voice of the city's inhabitants, the most important stakeholders of all.

In their highest expression, cities hold the promise of being the solution to sustainable human settlement, not the problem. Taken together, these guidelines show us the path forward.

Memphis: America's Aerotropolis™

Memphis has earned distinction as "America's premier Aerotropolis" in part because of its location in the heart of the country, with easy access to road, rail, runway, and river.
Image courtesy of Greater Memphis Chamber.

Rick Phillips
Director of Urban Design, Northern California, HNTB Corporation
Acknowledgements: Jim Grant, Jim Barbaresso, Julie Bartels Smith

5.2 | SUSTAINABLE TRANSPORTATION PLANNING

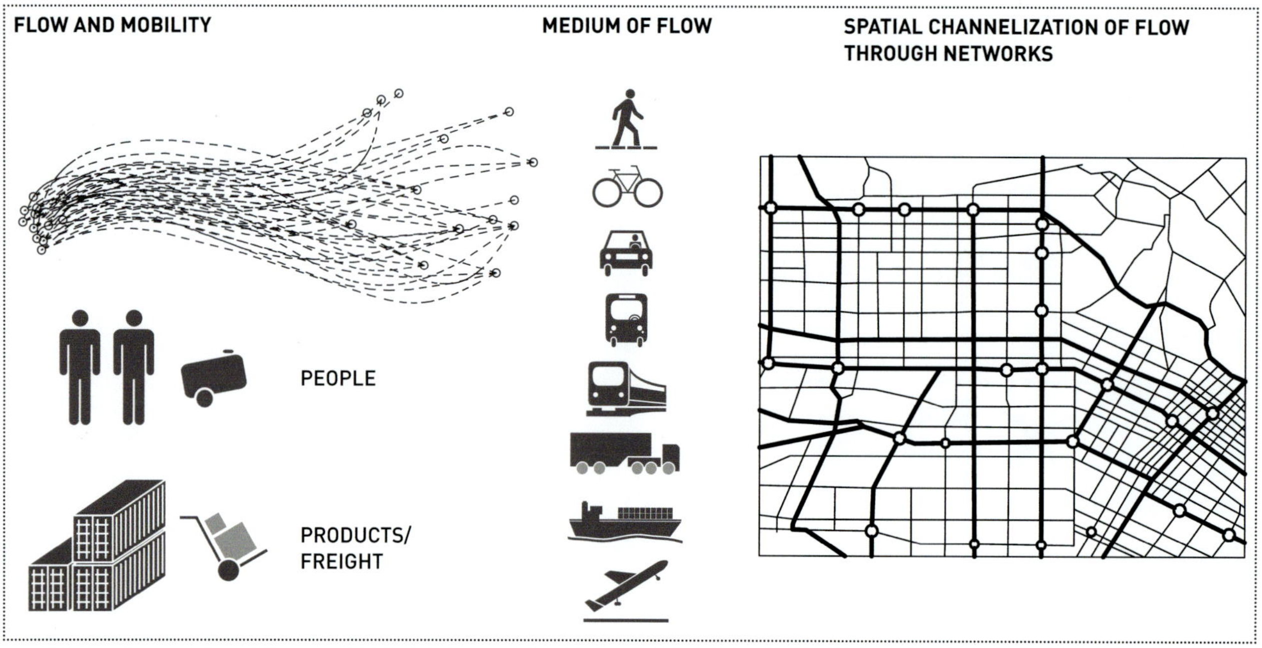

Transportation serves the movement of people, animals, and products from one location to another: the mobility of passengers and freight.

Transportation infrastructure forms linkages between locations, generating routes, intersections, and nodes. It is responsible for the spatial channelization of traffic flows through fixed installations on defined networks of roadways, waterways, airways, terminals, stations, and related facilities, but also includes the vehicles and operations that facilitate conveyance. It has expanded over time to meet the ever-evolving travel demand and behavior dynamics of people, businesses, and communities seeking to interchange, connect, trade, and communicate. Transportation is closely interrelated with urban and rural economies, acting as a catalyst for development. A node can be the organizing and attracting force for human activity, and a route can be the axis along which a city primarily expands.

Transportation infrastructure is both an incentive and a medium that can transform the concepts of proximity and physical distance. It allows the flows of people to expand and reconcentrate in the form of immigration, international student flows, business travel, migration, tourism, etc., affecting economic and cultural dynamics. Similarly, it increases the speed and volume of the transfer and distribution of goods, allowing for the creation of large metropolitan complexes. Based on transportation potential, human settlements can exceed the resource capacity of their immediate region and depend on global hinterlands. In a way, transportation is a prerequisite for the urban condition, not only in providing external connections but also in serving the needs of interconnection and distribution within cities as they grow.

There are substantially different means or modes by which one may perform transport. Each mode has its own infrastructure, vehicles, and operations. Operations deal with the way vehicles are managed, users are engaged, and the procedures set in place for this purpose, including financing, administration, and policies. Operations can be part of both the public and private sectors, depending on the location and mode.

In terms of performance, travel can be configured, quantified, and thus evaluated through data that are directly connected to the selection of modes. Examples of such data include the number of required vehicles, potential volume of passengers or cargo per vehicle, distance/ length of travel, speed, frequency, and daily or seasonal travel patterns. Other data express the direct or indirect impacts of transportation due to operation and vehicles or due to its strong physical presence constantly interacting with the urban fabric and natural land mosaics. Energy intensity, fuel consumption, emissions per trip, volume and quality of runoff, life cycle analysis of materials, road density, and average total road length per unit area serve as measures to assess the potential impact of transportation on local environments and highlight the challenges toward sustainability.

ENVIRONMENTAL IMPACTS OF TRANSPORTATION

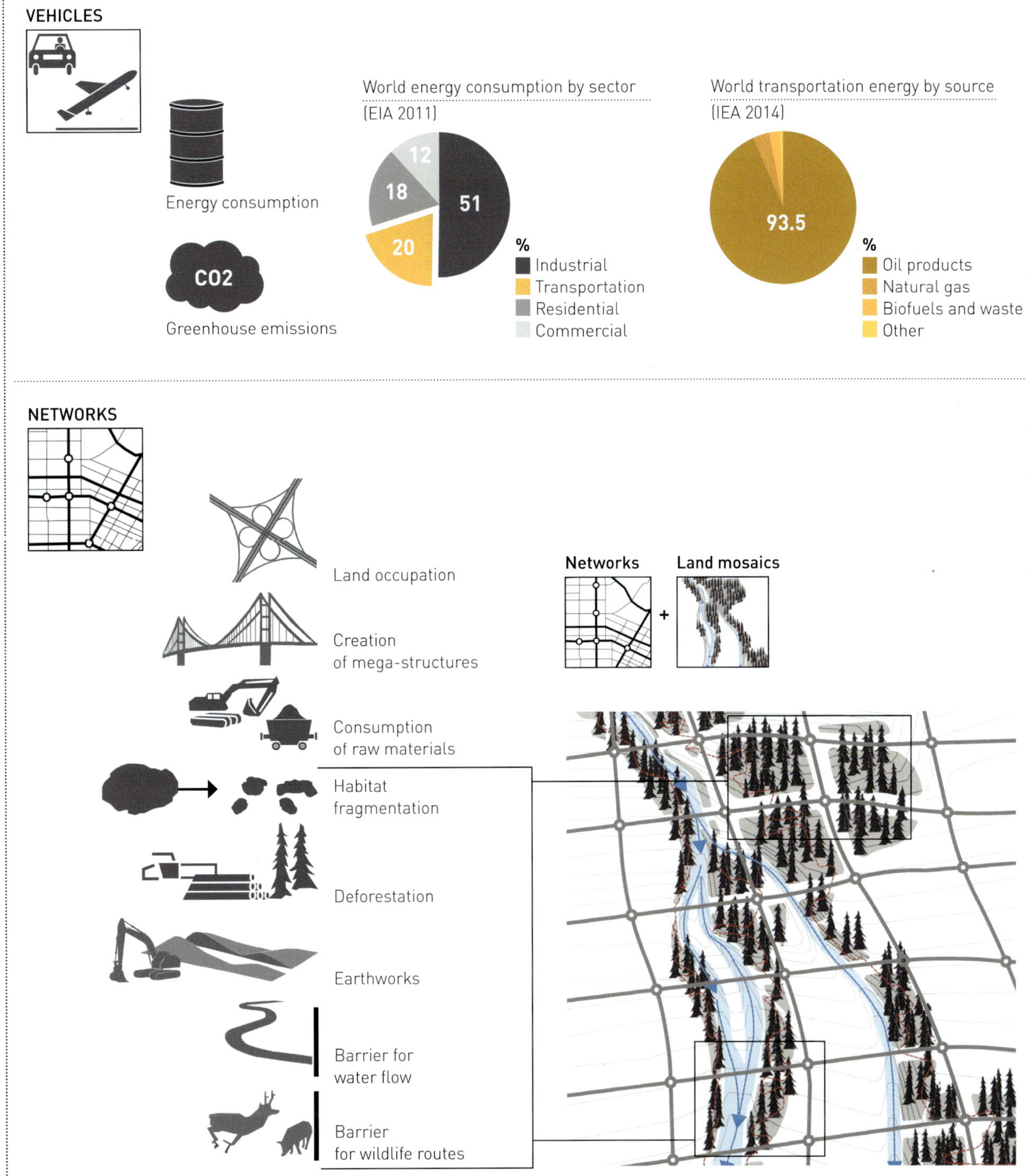

CHALLENGES TOWARD A SUSTAINABLE TRANSPORTATION SYSTEM

The inherent properties of transportation infrastructure pose significant challenges when aiming for environmental, financial, and social sustainability. Transportation infrastructure stands out because of its strong spatial dimension and extensiveness. Its networks and nodes occupy significant areas of land and require vast amounts of materials and resources to be constructed. Sustainable transportation infrastructure should aim to avoid, minimize, and compensate for its negative impacts on individuals, communities, and ecosystems in the process of its inevitable expansion to meet and manage increased user demand. It is important to stress that sustainable transportation cannot be seen independently of sustainable urban development.

"Nature ... the patterns and processes beyond human control ... daily nurturing and periodically perturbing us. We in turn create an extensive infrastructure ... the basic framework and facilities servicing people on land ... with linear features reaching almost everywhere. Nature affecting infrastructure, infrastructure affecting nature ... titanic interactions not to be missed."[1]

"Two giants in uneasy embrace, road system and land"[2] stands as an especially useful metaphor to illustrate the relation between transportation and land. The exposure of transportation structures to the elements brings up the issues of both their impact on natural systems and their resilience against them. Adverse geology, solar radiation, water, and wind are some of the natural forces that impose significant stress on infrastructure.

Aiming for resiliency, roads, canals, ports, vehicles, and vessels have become a dominant force that shapes landscape. Roads in particular, in expanding the potential area of human interaction and activities, apply extended rectilinear networks or grids upon the natural mosaics, enclosing parts of land into isolated polygons. Natural patterns and the processes underlying them are radically transformed. For example, the construction and operation of transportation infrastructure is often responsible for alterations to water routes. Riverbeds can be relocated or confined by retaining walls. As extensive paved surfaces limit the natural infiltration of rainwater, runoff is often collected and disposed of in large quantities in existing water bodies, affecting their natural balance. Topsoil that is not covered with paving materials can still be compacted or eroded. Light, noise, and chemical pollution and changes in moisture, temperature, and wind patterns can further alter the biotic environment. The consequences include the loss and fragmentation of habitats, the obstruction of wildlife corridors, and the degradation of freshwater sources and productive lands. These diverse effects of construction and traffic on plants and animals have increased rates of wildlife mortality and are directly linked with loss of biodiversity.

But the impact of transportation infrastructure is not limited to its adjacent systems and environments. The supply of construction materials, such as gravel and sand, may alter the landscape at extraction locations. Further, the processing and transfer of these materials requires significant amounts of energy, leading to emissions of greenhouse and other pollutant gases.

Sustainable transportation planning is thus called on to address the challenges of transport's inherent controversial nature in micro and macro contexts. Networks should be integrated with local land mosaics, avoiding, minimizing, or compensating for disruption of water flows and wildlife routes. Extensive cut-and-fill operations causing alteration of topography should be avoided. The configuration of paved areas should minimize impervious surfaces and identify opportunities to improve urban microclimate. The use of raw materials can be optimized through adequate sizing of structures or the incorporation of reclaimed materials. Further strategies to mitigate climate change include the promotion of low-energy and low-carbon practices such as innovative vehicle technologies and use of renewable fuels.

"Roads facilitate the movement of people and goods, play a central role in urban and economic development, and enhance social interactions. Without roads or other forms of transport, farm production would be limited to areas adjacent to consumers, and workers would be rooted near their jobs."[3]

From a socioeconomic point of view, transportation is a system that will continue to evolve and grow along with societies in order to meet evolving travel demands. It is the supporting framework for the distribution of goods and services and a driver for development and prosperity in many sectors of the economy. It safeguards the diffusion of resources at a regional scale, eliminating physical barriers to competitiveness for communities, businesses, and individuals. It can form the connective tissue of the various assets and financial clusters of an area. It allows for faster supply chains and extends the physical ground of financial interaction from local to international scale, increasing employment opportunities.

Due to its strong spatial dimension and presence, transportation operates visibly in the public realm. It shapes the public space and has a direct connection with the compactness and coherence of the urban fabric. Its performance and configuration affect substantially a community's everyday life, creating or limiting opportunities for engagement and interaction. While ensuring equal access is the principal objective, infrastructure components, serving as conduits or hubs establishing connections, can simultaneously act as barriers. A highway connecting two parts of a city can disrupt the continuity of the urban fabric at a local scale, negatively affecting quality of space, community coherence, and interactions. A congested network is the cause of lost productive hours. Roads and vehicles challenging distance and facilitating people's lives can pose significant threats to public health and safety through accidents and through air, light, and noise pollution.

Sustainable transportation planning should aim to create a pedestrian-friendly environment and shift passengers' preference from private vehicles to public transport. Among motorized modes, road and rail public transport have the lowest energy intensity and carbon emissions. Efficient service, improved punctuality, affordability, and reliability provide the main incentives that will increase their share in the modal split.

But for transportation planning to achieve its sustainability objectives and effectively contribute to quality of life, it must be closely associated with urban planning elements like land use zoning and urban density. Collaborative planning efforts should aim for compact and mixed-use development that provides a choice of close-by destinations and multiple transit options. This model of development allows for shorter trips to everyday amenities than can be covered easily on foot. At the same time it creates lively environments that enhance the sense of safety for pedestrians. Compactness is also crucial for the financial viability of the system as each infrastructure component is able to serve a larger number of passengers.

Planning for a sustainable transportation system includes multiple objectives with strong overlaps and linkages, initiated by diverse economic, environmental, or social drivers. The major challenge lies in achieving these goals through the integration of initiatives and apparently conflicting strategies. Transportation should exist within the natural world without interfering with natural processes, should create linkages and not barriers, and should support existing economic activities without conflicting with long-term goals. The shift to a sustainable system and development therefore implies organizational changes and behavioral shifts for both transportation users and service providers.

ENVIRONMENTAL

Challenges for environmental sustainability include:

- Reducing land use and footprint
- Avoiding significant land alteration
- Protecting and restoring habitats and wildlife corridors
- Protecting water cycle processes and water routes
- Reducing greenhouse gas and air pollutant emissions
- Resolving soil and water contamination from polluted runoff
- Minimizing use of nonrenewable energy and material resources

SOCIAL

Challenges for social sustainability include:

- Ensuring accessibility and affordability for all social groups
- Protecting public health and safety
- Maintaining quality of open public spaces
- Adapting to and preserving cultural resources
- Resolving traffic congestion

ECONOMIC

Challenges for economic sustainability include:

- Adapting to economic sectors' needs by efficiently supporting different supply chains
- Securing long-term economic viability
- Maintaining competitive costs
- Resolving traffic congestion

STRATEGIES TOWARD A SUSTAINABLE TRANSPORTATION SYSTEM

SYSTEM PURPOSE: RELIABLE AND EQUAL ACCESSIBILITY	**UNNECESSARY TRIP REDUCTION:** Reduce length and occurrence of passenger and freight trips
	MODAL SHIFT: Prioritize nonmotorized and low-energy-intensity modes
	SYSTEM EFFICIENCY: Improve functionality of infrastructure, users' behaviors, and efficiency of vehicles
	MINIMIZATION OF FOSSIL FUEL USE: Link transport fuel supply with renewable energy sources

The primary purpose and service of Transportation as a system is to provide equal opportunities of access. The structures and operations put in place create several sustainability challenges, as discussed in the previous section. The strategies listed above suggest basic steps that could be included in the planning of a transportation system. They do not aim to outline all the requirements for a sustainable system but instead highlight approaches that if implemented could resolve multiple issues simultaneously.

Unnecessary trip reduction is the first strategy listed. The main target of each transportation system should be to increase connectivity and encourage mobility; however, a large number of trips could be avoided without negatively affecting social interactions or productivity. Eliminating unnecessary trips, reducing their length, or regulating their occurrence within the day can work as preventive measures. Benefits include reductions in traffic congestion, accidents, air pollution, and fossil fuel consumption, contributing significantly to quality of life and environmental protection. Less traffic translates to lower requirements for transportation system capacity and thus smaller needs for materials and resources for construction and operation. The actions used to implement this strategy may come from a wide range of sectors not directly connected with infrastructure, indicating that integrated and interdisciplinary approaches are necessary for sustainable planning.

Advanced information technologies, online services, and teleconference applications can play a key role in reducing trip numbers both for goods and passengers. Improved logistics will optimize freight transport trips, and awareness-raising activities help shift consumer preferences to locally produced and shipped goods. At the same time, many daily transactions could be performed through online services without need for actual travel. Examples of such services include internet banking or online shopping. Recent studies have revealed that online purchases are among the greenest ways to shop. A full van making multiple deliveries in the same neighborhood can produce 50 percent less carbon dioxide emissions than if the same households made individual visits to a supermarket.[4] Similarly, information technologies can provide many professionals with the option to work from home. Other approaches to optimizing travel could include flexible working hours or a compressed work week.

Urban planning has a defining role in trip generation numbers and trends. Promotion of mixed-use development and designing of compact urban areas are key in reducing length of trips and allowing for combined trips. Providing the basic amenities in every residential area and across the city ensures that daily needs can be met with minimum need for travel.

Modal split describes how travel demand is covered by the different available means of transportation. The distribution should shift toward less energy-intensive and polluting means of transport. For short-distance passenger travel this equals prioritizing nonmotorized modes and mass transit or, for longer distances, high-speed rail or sea transport over aviation. Similarly for freight, shipping and rail transport should be preferred in place of air cargo and road transport. At city scale specifically, the shift should be from individual motorized transport toward public transport or cycling and walking. The need for seamless urban planning and infrastructure development is very evident when aiming for a more sustainable modal split. Strategies for pedestrian traffic and public transport should be supported through urban space compactness, density, and land use mix and distribution.

For each selected transportation mode, sustainability performance can be improved via optimized operations. The overall efficiency of the system can increase through the functionality of infrastructure facilities and entities, selection of vehicles, and engagement of users.

This involves optimizing the placement of nodes and routes, the length of networks, their width, section, and variability, and the hierarchy of corridors, as well as pursuing integration between different modes. Advanced technology applications and use of information infrastructure can further increase efficiency. For example, advance traveler information systems (ATIS) and variable message signs (VMS) can be used to support traffic management.

In addition to infrastructure configuration, selection and promotion of more efficient and hybrid vehicles (and regular maintenance of them) are key for the performance of the transportation system. Integrating new technologies, such as robotics or vehicle-to-vehicle communication, has the potential to notably increase transport efficiency. Finally, as with every system, the participation of users is crucial for effective operation. In the case of transportation, shifting users' behaviors toward habits like eco-driving or active decisions such as car sharing should complement the design of infrastructure.

Finally, one of the basic causes for the environmental impact of the transportation sector is its demand for large amounts of energy, mostly derived from fossil fuels. This leads to air pollution, greenhouse gas emissions, and overdependence on fossil fuel sources. Reducing travel needs, promoting nonmotorized or low-energy-intensity modes, and improving overall system efficiency already minimizes the need for energy and thus fuel. Shifting to electric-, hydrogen-, and biofuel-powered vehicles, whose energy supply may be produced by renewable energy sources, would be another significant step in reducing fossil fuel consumption and the consequent environmental impacts.

MODAL SHIFT

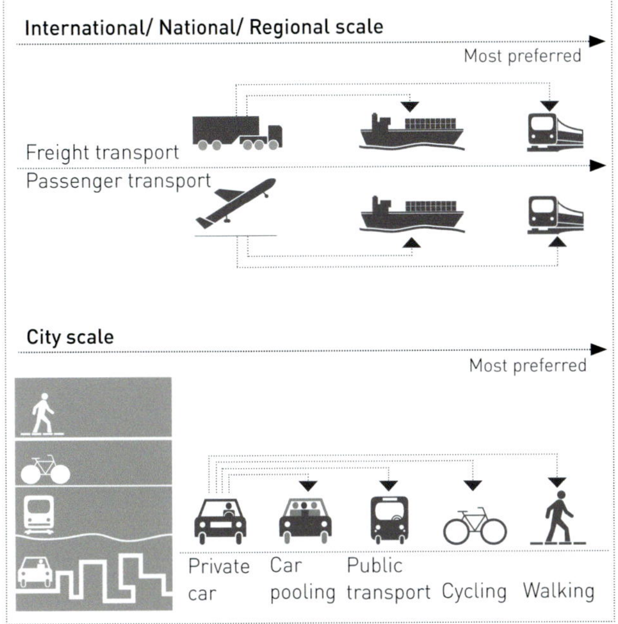

RENEWABLE ENERGY SOURCES FOR TRANSPORTATION SYSTEM

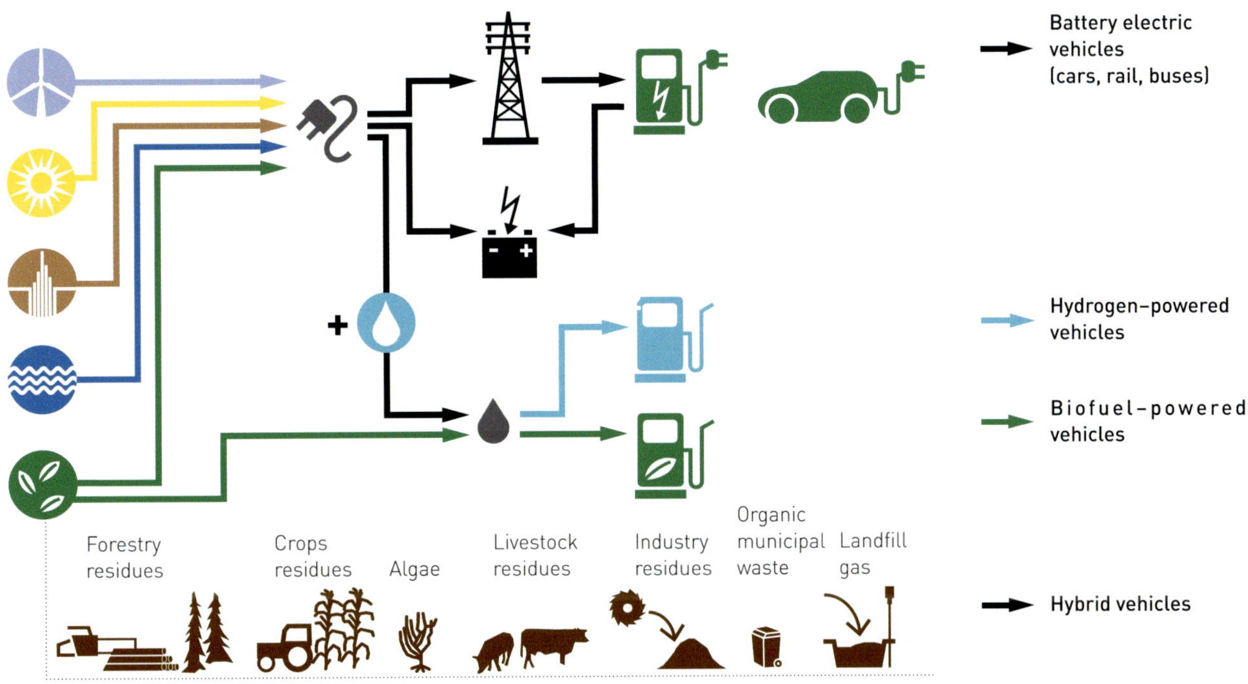

5.3 | TRANSPORTATION SYSTEM DECODING

In line with the general framework of the guidelines, the Transportation infrastructure system is decoded on four system levels: Trip Generation and Transportation Modes, Nodes, and Networks.

The first level, Trip Generation, examines the demand to be served by the infrastructure system. Trip Generation aims to promote the understanding and optimization of the needs for transportation. It addresses the number of trips generated, their length, distribution in time, and their classification in connection with land use trip generators and attractors.

The second level, Transportation Modes, represents the basic strategic decisions concerning the selection of modes, technologies, and resources that will be used to meet the demand for transportation. The modes promoted or introduced will determine to a large extent the organization and planning of the entire transportation system and its impact on community, economy, and natural systems. The benefits, limitations, and environmental impacts in the specific context have to be examined and resolved prior to planning the individual components of the system. The percentages of trips performed by all available transportation modes comprise the city's modal split, an important indicator of sustainability performance in urban areas.

The third and fourth levels, Transportation Nodes and Networks respectively, represent the main infrastructure entities that form the system. Nodes are the points of the system where different transportation modes interact and major flows originate or terminate. The spatial organization and efficient function of transportation nodes determine whether or not the entire system will operate properly, and most importantly how accessible it will be for users. The fourth level includes the most extensive physical infrastructure present in modern cities, transportation networks. Networks ensure connectivity, a fundamental factor for the well-being and growth of modern societies. Dense and widespread transportation networks may improve traffic flows to some extent, but at the same time act as barriers with negative impacts on community life and natural-world processes. Transportation networks also form strong synergies with other utility and infrastructure networks, as they commonly host a wide range of pipes, cables, etc.

TRIP GENERATION

The Trip Generation level examines the number and type of trips taking place in the area of study. The trips generated depend on urban configuration, such as land use distribution, and on the overall socioeconomic, cultural, and environmental context, which varies significantly among different urban areas, cities, and countries. Land-use-related variables that affect travel include population, employment, and economic activity.[5] For the purpose of trip generation analysis, a geographic area can be divided into zones. The land use and urban fabric properties of these zones can then be translated into trips produced and attracted for purposes like work, commerce, manufacturing, and recreation.[6] Every trip has an origin, referred to as the trip production point, and a destination that constitutes the trip attraction point. Passenger trips are usually divided into three main categories based on the points of trip production and trip attraction, as the graph below shows. Trips can also be classified by land use, with residential, industrial, commercial, and recreational land use categories most commonly used. In addition to spatial distribution, another important characteristic to be taken into consideration during planning is the trips' distribution in time, throughout the day, seasons, or year.

TRIP ATTRACTORS AND GENERATORS

PASSENGER TRIPS

Attractors and generators:

Type of trips:

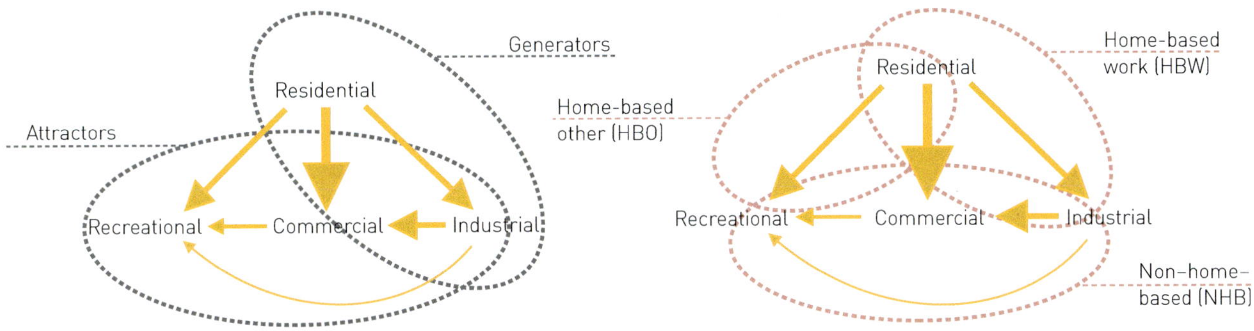

FREIGHT TRIPS

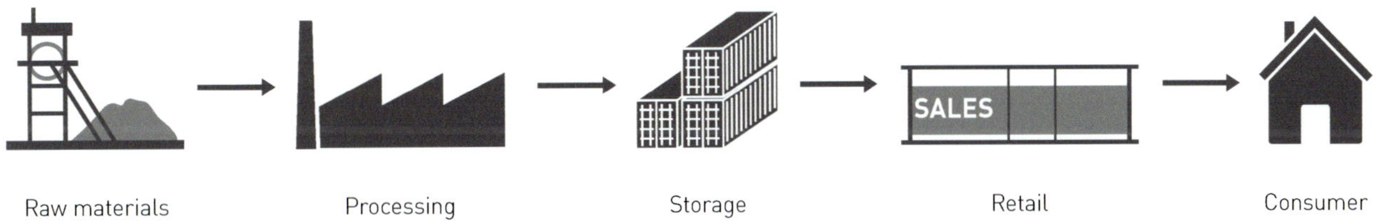

TRANSPORTATION MODES

The transportation modes that will constitute or dominate the system are major components of transportation planning. A first fundamental distinction should be made between modes used for intercity and for city-scale travel. Intercity transportation modes include aviation, land transportation vehicles (rail, road, off-road), shipping transportation, and pipelines that transfer oil or natural gas. At city scale, the most commonly used modes are walking, cycling, public transport (bus, tram, rail), and private motorized vehicles (cars, motorcycles, trucks, etc.). Less popular but still an option for many cities around the world are ferries and cable transport systems.

City-scale modes differ significantly in terms of benefits, limitations, and environmental impact. Walking is commonly accepted as the most sustainable mode. It has no negative environmental impact, has significant health benefits for the population, and makes for more vibrant and engaged communities. However, it is usually limited to short journeys and is more suitable for compact urban environments. Cycling is a flexible and similarly sustainable solution. It offers zero greenhouse emissions, health benefits, and minimum disruption to the communities' functions and quality of life. Limitations include the requirement of relatively good levels of fitness and difficulties stemming from adverse weather conditions or city terrain. Safety concerns may arise when cycling is used alongside motorized traffic. Public transport is the most environmentally friendly mode of motorized transportation, with low emissions per passenger mile. It is a low-cost and efficient mode of transport, suitable for short or long journeys. Prioritizing dense and well-organized public transport networks reduces congestion and guides cities to develop more compactly. While public transport networks may cause some disruption to a city's quality of life, this can be minimized through efficient planning and suitable solutions. Use of private motorized vehicles is considered the most unsustainable mode for city transport. Motor vehicles contribute significantly to congestion, greenhouse gas emissions, and air quality problems and occupy a significant amount of the city's land surface. The continuous expansion of road networks, in order to accommodate the needs of private vehicles, creates some of the most challenging sustainability issues affecting today's cities.

As already mentioned, the portion of daily journeys performed with each transportation mode constitutes the modal split of the city. Setting modal split targets is usually

INTERCITY MAIN MEANS OF TRANSPORT

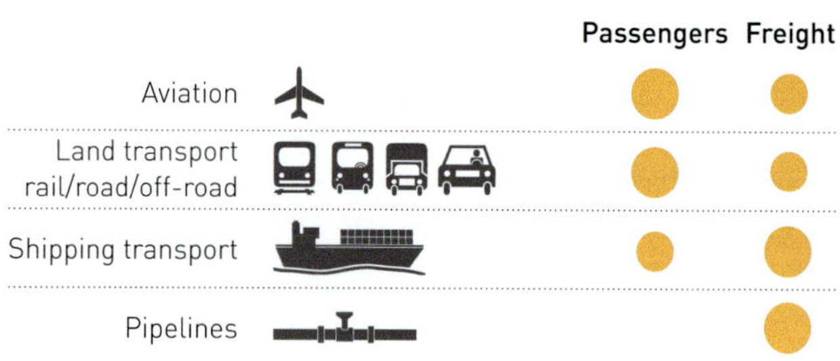

CITY MAIN MEANS OF TRANSPORT

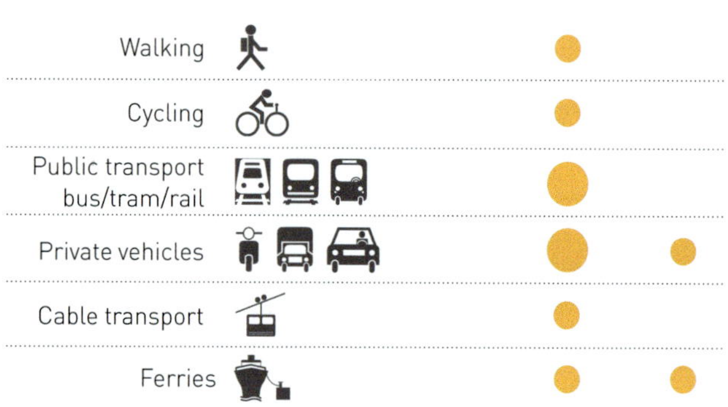

CO_2 EMISSIONS PER PASSENGER KILOMETER IN CITY AND INTERCITY SCALE (UK Department for Transport, 2011)

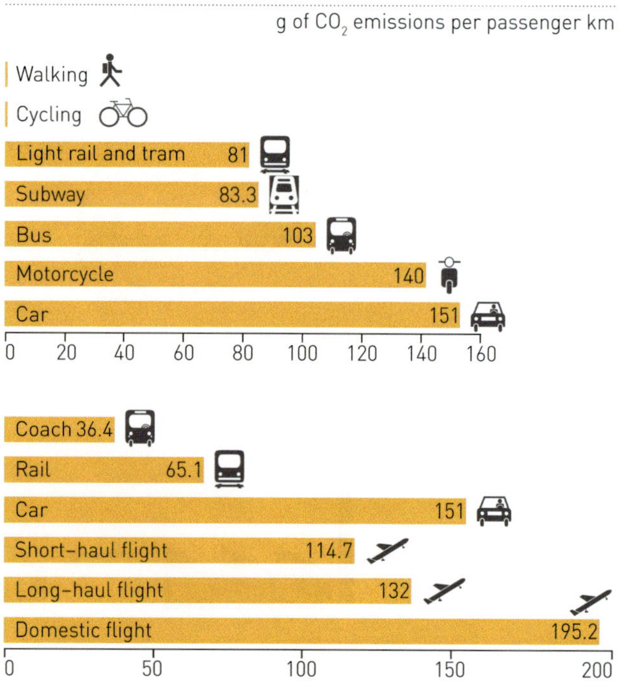

BENEFITS AND LIMITATIONS OF TRANSPORTATION MODES

BENEFITS	LIMITATIONS
Walking: The most common green transport option	
• Minimum environmental impact • Improved health and well-being • Zero emissions – no noise or air pollution • More vibrant and safe cities • Support of local shops and businesses	• Short journeys • Exposure of users to weather phenomena • Limitation by extreme topography
Cycling: Environmentally friendly way of transport	
• Insignificant environmental impact • Quick access – reduced traffic congestion • Improved health and well-being • No air or noise pollution • Reduced financial cost for the user and society • More vibrant and safe cities • Support of local shops and businesses • Access to both roads and paths • Minimum parking and storage needs	• Safety concerns when sharing road infrastructure • Exposure of users to weather phenomena • Not suitable for very long or intercity trips • Requirement for basic physical fitness • Increased exposure to air pollution in urban areas • May be limited due to city terrain
Public Transport: Energy-efficient way to travel	
• Reduced air pollution – low emissions per passenger-mile • Quick access across local and more distant destinations • Enhanced social equity ensuring all members of society are able to travel • Reduced traffic congestion • Low cost for users • Increased safety levels • Use of environmentally friendly fuel alternatives • Facilitates increased urban population densities, allowing cities to be built more compactly	• Noise pollution • Urban fragmentation mainly by rail networks • Reduced flexibility – depends on networks density and routes frequency • Large upfront costs requiring large investments • Limited access to remote destinations
Private Vehicles: The dominant form of transportation in most of the world	
• Higher flexibility for users – on-demand transportation • Ability to reach remote destinations	• Dependence on fossil fuels • Air and noise pollution • Traffic congestion • Urban and habitat fragmentation • High accident rates • Significant land occupation for circulation and parking • Encouragement of urban sprawl • Car dependence leading to inactivity and obesity • High costs for users

MAIN MEANS OF TRANSPORT WORLDWIDE (Land Transport Authority Singapore, 2011)

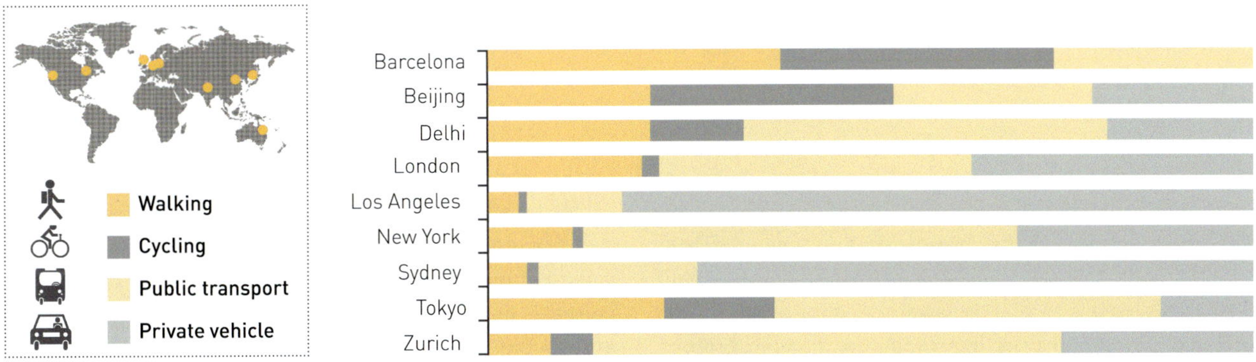

an important part of a city's sustainable transportation plan. The preceding diagram summarizes the modal splits of nine big cities, illustrating how urban transportation can vary around the globe. Los Angeles and Sydney highly depend on private motor vehicles, a fact that is reflected in the typology of their urban fabric. In Delhi, New York, Tokyo, and Zurich, on the other hand, public transportation predominates. In Beijing cycling is the most popular means of transport, while in Barcelona the greatest number of daily journeys are done on foot.

TRANSPORTATION NODES

Nodes include all the neuralgic points of the transportation system. These include airports, ports, train and bus stations that usually function as intercity nodes, as well as subway, tram, and suburban rail stations, ferry slips, parking lots, depots, etc. that function as city nodes facilitating urban transportation. The spatial distribution and organization of nodes are critical for the efficiency and successful operation of the transportation network. Nodes should be located in relation to major trip attraction points and areas. It should be mentioned, though, that large nodes acting as hubs of the system can further increase the trip attraction of surrounding areas. In addition to their spatial distribution, nodes should be easily accessible by all users of the system and capable of serving the peaks in traffic flow. Finally, it is of particular importance that nodes of different modes be combined and interconnected, allowing for fast intermodal transfer and offering alternative routes, adding to the efficiency, flexibility, and resilience of the system as a whole.

NODE ENTITIES

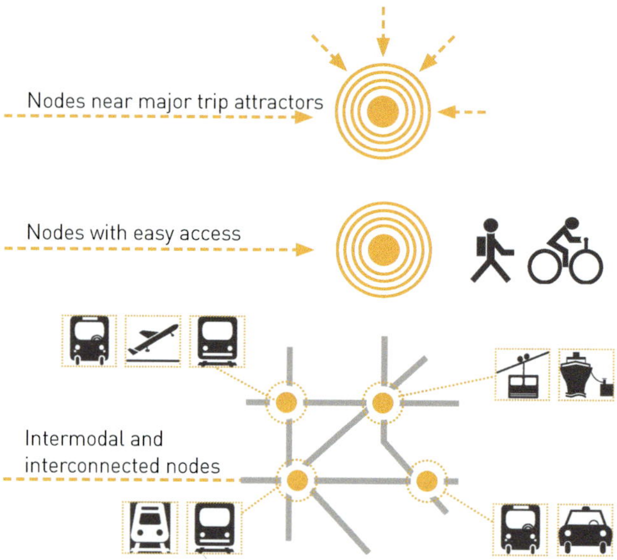

TRANSPORTATION NETWORKS

Transportation networks can be separated into two main categories, those that need a spatial installation and those that don't. Road, rail, pipeline, and cable transport networks require extensive physical infrastructure, while aviation, shipping, and off-road vehicles do not need any permanent infrastructure besides their terminal nodes. The main entities of the Transportation Networks level include roads, railways, cycle and pedestrian paths, canals, waterways, pipelines, and cableways. A range of subentities are also part of networks, including interchanges, bridges, tunnels, culverts, sidewalks, traveler information systems, toll stations, service reservation corridors, etc.

Road networks are the most extensive and common infrastructure worldwide, used daily by the vast majority of the global population. They demand the extraction of large amounts of materials for their construction and maintenance and are often responsible for habitat fragmentation. They contribute to heat island effects and increase noise, light, air, and water pollution in adjacent areas. Overdependence on road networks may lead to urban sprawl and continuously increasing congestion. Among their main impacts are habitat fragmentation and noise pollution. Pedestrian and cycle networks usually share the same infrastructure with road networks. With the appropriate selection of permeable and light-colored paving materials, their negative impact can be reduced. Rail networks exist in more than 151 countries worldwide and have significantly less environmental impact. Finally, canals, once used mainly for freight transport, are now in most cases decommissioned or used only for recreational purposes. Canals can disrupt natural water flows and affect or pollute wetland ecosystems.

NETWORK ENTITIES BY TRANSPORTATION MODE

NETWORKS		USERS
Roads	→	Cars, motorbikes, buses, trucks
Railways	→	Trains, trams
Cycle paths	→	Bicycles
Pedestrian paths	→	Pedestrians
Canals, waterways	→	Ships, boats, ferries
Cableways	→	Cable cars, chairlifts
Pipelines	→	Other infrastructure networks

NETWORKS' MAIN CHARACTERISTICS AND IMPACTS

NETWORKS	MAIN CHARACTERISTICS (CIA 2013)	MAIN IMPACTS
Roads	Road networks are the most common and extensive infrastructure, with more than 64,285,009 km (2013) worldwide Materials: sand, gravel, geosynthetics, concrete, asphalt, etc.	Noise pollution, road dust, water pollution, light pollution, habitat fragmentation, heat island effect, urban sprawl, deterioration of local air quality
Railways	Rail networks exist in 151 countries woldwide with a total length of 1,148,186 km (2013) Materials: steel, timber, concrete, ballast, etc.	Noise pollution, habitat fragmentation
Pedestrian and cycle paths	Pedestrian and cycle networks usually share the same infrustructure with roads Segregated cycle facilities increase cycle usage and cyclists' safety	No significant environmental impacts
Canals, waterways	Early industrial man-made canals were mainly used for freight transport but have ceased to carry significant amounts of trade; many have been abandoned to navigation, used only for recreation.	Canals can disrupt natural water circulation, and alter or pollute wetlands ecosystems.

ROADS HIERARCHY

Freeways: Provide uninterrupted traffic, high speed limit
Arterials: Carry large volumes of traffic
Collectors: Collect traffic from local roads and distribute to arterials
Local roads: Low speed limit, carry low volumes of traffic, direct access to properties

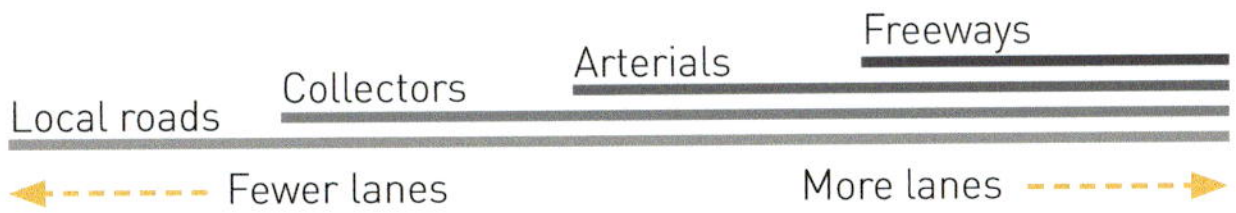

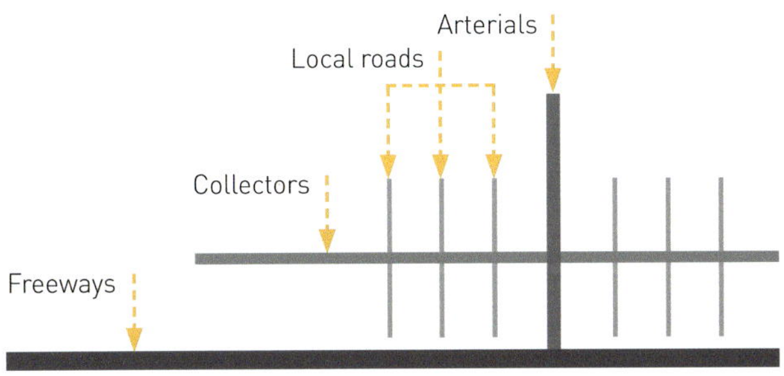

5.4 | OBJECTIVES AND GUIDELINES FOR A SUSTAINABLE TRANSPORTATION SYSTEM

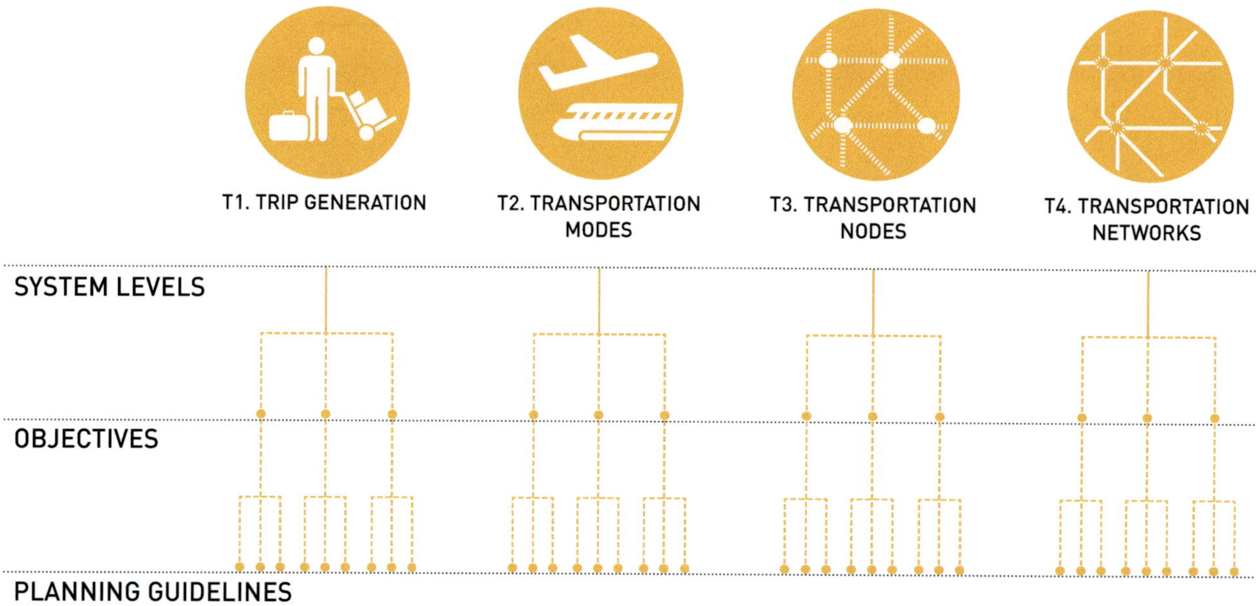

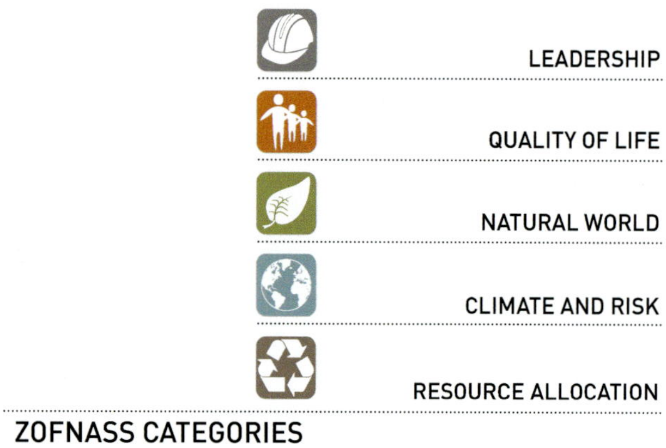

Following the decoding of the Transportation system on four system levels, planning objectives are defined for each level. They represent both challenges and strategies toward a sustainable transportation system, expressing what should be the overall final outcome of transportation planning.

The objectives are associated with the five focus categories of the Zofnass Program, namely Leadership, Quality of Life, Natural World, Resource Allocation, and Climate and Risk. The Sustainable Planning Guidelines aim to lead toward these objectives, highlighting synergies and connections with the other six infrastructure categories of Energy, Water, Solid Waste, Landscape, Information, and Food.

The objectives and guidelines of the Trip Generation level are primarily associated with the Leadership category. They do not refer directly to sustainable infrastructure planning, but instead aim to create a solid base through user engagement and monitoring mechanisms. Trip reduction is a central strategy at this level. Objectives call for the reduction of trip numbers and length, not by discouraging mobility but through information applications and urban planning strategies. These include improved logistics, remote access services, new telecommunication technologies, as well as the promotion of mixed-use development and urban compactness. Other objectives require the provision of monitoring for trip generation trends and regulation of their rates and patterns. Corresponding guidelines include identifying the main trip attractors in the area of study and reducing peaks in traffic. A thorough understanding of the demand for transportation services allows tailoring of infrastructure to local needs, to anticipate and prevent system failures but also to have a strong base of collected data upon which to develop future action plans. The Trip Generation level features strong synergies with the Information infrastructure system. Telecommunications or Information infrastructure as a subsystem of Transportation can be used to optimize, monitor, and regulate trip generation.

T1. TRIP GENERATION

- T1.1. Reduce number of unnecessary Trips
- T1.2. Reduce length of Trips
- T1.3. Monitor Trip Generation trends
- T1.4. Regulate Trip Generation rates and patterns

On the Transportation Modes level, the guidelines aim at strategic planning of the city's modal split. Objectives are related to the Leadership, Quality of Life, and Climate and Risk categories. As discussed earlier, the selection of modes directly affects the life of communities and the emissions generated by the transportation sector. The main objective of this level is to ensure connectivity and provide alternative connections and means of transport for all system users. Since transportation accounts for a large percentage of global greenhouse and air pollutant emissions, the guidelines emphasize the promotion of nonmotorized, low- or clean-energy modes. Resolving traffic congestion is another challenge to be confronted from the very early stages of the decision-making process. It is an issue for most modern cities, affecting quality of life, reducing productivity, and further increasing greenhouse gas emissions. Guidelines here call for strong connection of planning to the city's economic sectors and to projections for long-term development. Specific benchmarks should be adopted in order to monitor the system's performance and the implementation of planning. The selection of fuels to be used by vehicles and the steps required to ensure their adequate operation and maintenance are also addressed here. An important synergy is created with Energy infrastructure, as the decisions taken within this Transportation system level will have a significant impact on the energy demand of the city.

T2. TRANSPORTATION MODES

- T2.1. Secure long-term city connectivity at local and supraregional level
- T2.2. Define clear goals to be achieved through the transportation modal split
- T2.3. Reduce traffic congestion
- T2.4. Reduce greenhouse gas and air pollutant emissions

On the Nodes level, objectives cover all five of the Zofnass Program focus categories. In terms of Leadership, the main objectives concern high-level decisions for optimized spatial distribution of nodes and their correlation with trip attraction points, as well as ensuring adequate coverage and equal access for all citizens. Nodes have a significant role in facilitating traffic flows. They should be designed with a capacity compatible with long-term city needs, supported by adequate administrative structures and operated based on real-time information. Nodes of different transportation modes should be combined and integrated in order to create hubs and support efficient intermodal transfer. Quality of Life is directly connected to quality of place and public spaces. Transportation nodes can have a very strong spatial presence and in many cases function as dominant public reference points. They should be planned with respect to the local context, without creating barriers along pedestrian routes or negative visual impact on local character. The well-being of communities is further connected with the elimination or mitigation of Transportation infrastructure's negative impacts, such as noise pollution. In the Climate and Risk and Natural World categories, objectives concern the resilience of the system against climate change risks and the protection of natural processes, respectively. The guidelines focus on assessing and managing potential risks from natural and man-made phenomena, addressing response and recovery times in case of emergency, and mitigating the heat island effect. Transportation nodes often feature extensive hardscaped surfaces that can affect the microclimate in adjacent areas. The placement and configuration of transportation nodes should aim to maintain the integrity of natural-world elements and systems. Guidelines toward this end include avoiding construction of transportation nodes on undeveloped land, providing for natural drainage, and protecting soil quality. The objectives for Resource Allocation address the need for optimized material use and solid waste management. They are expressed through guidelines promoting the use of reclaimed and recycled materials in transportation nodes construction and the management of waste according to Solid Waste infrastructure requirements, such as separation at source. At this system level synergies are noted with all other infrastructure systems. Indicatively, transportation planning should take into consideration the transfer and connection needs of solid waste and wastewater processes, such as the transportation of waste and sludge. Information infrastructure as a subsystem can support the real-time monitoring and management of nodes. Some of the most important synergies, however, link with Landscape infrastructure. Landscape can be used both to mitigate the negative impacts of transportation nodes, such as noise pollution or contaminated runoff, and to protect nodes themselves from extreme weather phenomena through the creation of barriers, buffers, etc.

T3. TRANSPORTATION NODES

- T3.1. Place and distribute Transportation Nodes for optimized Transportation system performance
- T3.2. Facilitate traffic flows
- T3.3. Integrate Transportation Nodes with other infrastructure systems
- T3.4. Enhance quality of place
- T3.5. Enhance public space
- T3.6. Promote community well-being
- T3.7. Maintain the integrity of natural-world elements and systems
- T3.8. Prepare for short-term hazards and projected future threats
- T3.9. Optimize material use in Transportation Nodes
- T3.10. Contribute to Solid Waste reduction and resource recovery

The Networks system level repeats many of the Nodes objectives, as it similarly refers to the more physical dimension of the system. Objectives relate to all five focus categories of the Zofnass Program. Because of the large scale of transportation networks' physical infrastructure, its impact can be very significant. In the Leadership category the main objectives concern extending the transportation networks for optimized system performance, facilitating flows, and integrating with other infrastructure networks. Required considerations include providing adequate coverage and length of networks for all transportation modes implemented, combining and integrating networks of different modes, and clearly hierarchizing networks to serve different levels of flow intensity. Another aspect of Leadership is to establish adequate administrative structures and the financial viability of networks throughout construction and operation. In the Quality of Life category, objectives focus again on improving public space and minimizing impacts such as urban fragmentation and light or air pollution, while ensuring safety for all users of the networks. An objective that stands out is the creation of a pedestrian-friendly environment. It aims to stress the importance not only of planning pedestrian routes but of incorporating supporting measures that will attract users through improved safety and comfort. These measures are directly connected with urban planning, such as concentrating activities along pedestrian routes that enhance daily patterns for safe, lively environments. In Climate and Risk, providing resilience against natural and man-made phenomena is again key for securing the continuous and smooth operation of the system. Potential risks should be assessed and response and recovery times addressed. Taking measures to mitigate the heat island effect is particularly important due to the large ratio of city area occupied by roads and the dark color of asphalt paving. In Natural World, the objectives point to minimizing disruptions of natural systems by avoiding landscape alterations, maintaining the water cycle, and protecting habitats and wildlife corridors. Again because of the significant area taken up by transportation networks, developing impermeable paving and vegetation configurations to provide natural drainage and filtration of runoff can have tremendous positive impact. As vegetated corridors are often created along transportation networks, the guidelines advise adapting plant selection to local context, controlling invasive species, providing habitats, and implementing maintenance practices that avoid the use of pesticides. The Resource Allocation category has considerable importance, including objectives for materials, energy, and water consumption. It is important to source materials locally and promote the use of reclaimed and recycled materials and energy- and water-saving practices at a high level. Synergies with all other infrastructure systems are traced. Road networks in particular commonly host elements of other systems, including energy, water, and information networks and waste collection equipment. Achieving integration between the different infrastructure networks is central for a successful systemic approach to the city. Synergies with Information and Landscape infrastructures can assist the Transportation system to achieve sustainability goals. Information applications can again be used for optimized traffic management and provision of real-time traffic data, and landscape can help in mitigating negative impacts such as chemical, noise, and light pollution. Conversely, transportation planning can support the function of the other systems. Controlling the contaminants in runoff contributes to basic objectives of protecting the watershed and foodshed. Transportation networks have the potential to utilize by-products of other infrastructure processes. Materials recovered through solid waste management and reclaimed water can be used for construction and maintenance.

The Objectives with the respective Guidelines as well as the relevant Guidelines of other infrastructure systems are presented in the following tables. (TO) and (FROM) indicate that the guideline can assist or can be assisted by the other infrastructure system, where the arrow is placed. (WITH) indicates that the guideline might require planning for an Information or Landscape subsystem and should follow multiple Guidelines from these two systems.

T4. TRANSPORTATION NETWORKS

- T4.1. Develop Transportation Networks for optimized Transportation system performance
- T4.2. Facilitate traffic flows
- T4.3. Integrate Transportation Networks with other infrastructure systems
- T4.4. Enhance public space
- T4.5. Enhance quality of place
- T4.6. Create a pedestrian-friendly environment
- T4.7. Promote community well-being
- T4.8. Maintain the integrity of natural-world elements and systems
- T4.9. Prepare for short-term hazards and projected future threats
- T4.10. Optimize material use in Transportation Networks
- T4.11. Optimize Energy consumption in Transportation Networks
- T4.12. Optimize Water consumption in Transportation Networks

GUIDELINES

OBJECTIVES		GUIDELINES	SYNERGIES
T1 TRIP GENERATION			▲ TO + WITH ▼ FROM
T1.1. Reduce number of unnecessary Trips	T1.1.1.	Provide and promote remote access and e-services	+
T1.2. Reduce length of Trips	T1.2.1.	Promote mixed-use development	
	T1.2.2.	Include basic amenities within each residential area	▼
	T1.2.3.	Promote compactness in urban fabric	▲
	T1.2.4.	Optimize placement of infrastructure facilities that generate significant traffic	▼ ▼
T1.3. Monitor Trip Generation trends	T1.3.1.	Identify main city Trip attractors	
	T1.3.2.	Collect and organize data on Transportation needs of goods and passengers	+
	T1.3.3.	Monitor daily and seasonal Trip Generation fluctuations across the city	+
T1.4. Regulate Trip Generation rates and patterns	T1.4.1.	Reduce peaks in Trip Generation	
	T1.4.2.	Optimize itineraries within other infrastructure systems	▼ ▼
T2 TRANSPORTATION MODES			▲ TO + WITH ▼ FROM
T2.1. Secure long-term city connectivity at local and supraregional level	T2.1.1.	Ensure coverage of passenger and freight needs through multiple and alternative Transportation Modes	
	T2.1.2.	Trace correlation of different Transportation Modes with city competitiveness and economic sectors	
	T2.1.3.	Secure long-term financial viability of Transportation system	
T2.2. Define clear goals to be achieved through the transportation modal split	T2.2.1.	Set goals for the Transportation modal split	
	T2.2.2.	Adopt quantitative indicators to benchmark and monitor Transportation system performance	
	T2.2.3.	Integrate long-term city growth and development goals with Transportation Modes planning	
	T2.2.4.	Regularly update Transportation Modes planning, according to community development and Transportation system performance	
T2.3. Reduce traffic congestion	T2.3.1.	Provide and promote reliable public and mass transit	▲
	T2.3.2.	Provide car sharing and bike sharing options	▲
T2.4. Reduce greenhouse gas and air pollutant emissions	T2.4.1.	Set long-term goals and standards for reducing emissions in the Transportation sector	
	T2.4.2.	Promote nonmotorized Transportation Modes	▲
	T2.4.3.	Use efficient and/or alternative-fuel vehicles	▲ ▼

OBJECTIVES		GUIDELINES	SYNERGIES
	T2.4.4.	Put in place mechanisms to ensure regular maintenance of vehicles	▲
	T2.4.5.	Educate users on the efficient operation of vehicles	▲

T3 TRANSPORTATION NODES

▲ TO
+ WITH
▼ FROM

OBJECTIVES		GUIDELINES	SYNERGIES
T3.1. Place and distribute Transportation Nodes for optimized Transportation system performance	T3.1.1.	Place Transportation Nodes in correlation to Trip attractors	
	T3.1.2.	Ensure equal connection opportunities for all areas of the city	
	T3.1.3.	Combine and integrate Nodes of different Transportation modes for efficient intermodal transfer / Create hubs	
	T3.1.4.	Reserve spaces for Transportation Nodes facilities taking into consideration long-term city development	
	T3.1.5.	Optimize placement of Transportation Nodes with regard to Network routes	
T3.2. Facilitate traffic flows	T3.2.1.	Provide adequate Transportation Nodes capacity to cover long-term city needs	
	T3.2.2.	Cater to the special needs of different groups / Provide access for the disabled	
	T3.2.3.	Provide Node traffic operators with real-time information on traffic conditions	+
	T3.2.4.	Provide users of Transportation Nodes with real-time information on itineraries and traffic conditions	+
	T3.2.5.	Secure long-term financial viability of Transportation Nodes	
	T3.2.6.	Optimize Transportation Nodes asset management through adequate administrative structures	
T3.3. Integrate Transportation Nodes with other infrastructure systems	T3.3.1.	Investigate needs of Transportation Nodes integration with Solid Waste Treatment and Disposal facilities	▲
	T3.3.2.	Investigate needs of Transportation Nodes integration with Water and wastewater Treatment facilities	▲
	T3.3.3.	Provide required traffic connections to large Landscape infrastructure entities	▲
T3.4. Enhance quality of place	T3.4.1.	Resolve fragmentation of urban fabric and pedestrian routes caused by Transportation Nodes	
	T3.4.2.	Visually integrate Transportation Nodes to the local context	+
T3.5. Enhance public space	T3.5.1.	Design Transportation Nodes and equipment to extend and enhance public space	
T3.6. Promote community well-being	T3.6.1.	Assess and resolve noise pollution issues caused by Transportation Nodes	+

OBJECTIVES	GUIDELINES		SYNERGIES
T3.7. Maintain the integrity of natural-world elements and systems	T3.7.1.	Avoid construction of Transportation Nodes on undeveloped land	▲
	T3.7.2.	Protect natural habitats and wildlife corridors and compensate for their fragmentation or disruption by Transportation Nodes	+
	T3.7.3.	Develop permeable paving and vegetation configurations for Transportation Nodes to provide natural drainage	▲ ▲
	T3.7.4.	Develop permeable paving and vegetation configurations for Transportation Nodes to filter runoff	▲ ▲
	T3.7.5.	Develop vegetation configurations for Transportation Nodes to provide habitats	▲
	T3.7.6.	Protect soil from contaminants, erosion, and compacting caused by Transportation Nodes	+
	T3.7.7.	Save, preserve, and reuse topsoil before the construction of Transportation Nodes where possible	▲ ▲
T3.8. Prepare for short-term hazards and projected future threats	T3.8.1.	Assess and manage future threats from natural and man-made phenomena affecting Transportation Nodes	+
	T3.8.2.	Plan Transportation Nodes response and recovery in case of emergency	
	T3.8.3.	Mitigate heat island effect through Transportation Nodes design and configuration	
T3.9. Optimize material use in Transportation Nodes	T3.9.1.	Source materials required for Transportation Nodes construction and maintenance locally when possible	
	T3.9.2.	Promote use of reclaimed and recycled materials in Transportation Nodes construction	
	T3.9.3.	Incorporate Solid Waste management residue and by-products in Transportation Nodes construction and maintenance	▼
	T3.9.4.	Plan for the reuse of elements or recycling of materials after Transportation Nodes decommission	▲
T3.10. Contribute to Solid Waste reduction and resource recovery	T3.10.1.	Conform waste management practices within Transportation Nodes to requirements of Solid Waste Collection and Transfer planning	▲

T4 TRANSPORTATION NETWORKS

▲ TO
+ WITH
▼ FROM

OBJECTIVES	GUIDELINES		SYNERGIES
T4.1. Develop Transportation Networks for optimized Transportation system performance	T4.1.1.	Provide adequate coverage and length of Networks for all Transportation modes implemented	
	T4.1.2.	Provide adequate coverage and length of routes for public transport modes implemented	
	T4.1.3.	Combine and integrate Networks of different Transportation modes	

OBJECTIVES		GUIDELINES	SYNERGIES
	T4.1.4.	Reserve spaces for Transportation Networks facilities taking into consideration long-term city development	
	T4.1.5.	Locate safe intersections and interconnections at adequate points / Optimize density of intersections	
T4.2. Facilitate traffic flows	T4.2.1.	Define a clear hierarchy in Transportation Networks to serve different levels of flow intensity	
	T4.2.2.	Provide adequate Transportation Networks capacity to cover long-term city needs	
	T4.2.3.	Organize road networks to promote use of high-occupancy vehicles	▲
	T4.2.4.	Cater to the special needs of different groups / Provide access for the disabled	
	T4.2.5.	Provide Network traffic operators with real-time information on traffic conditions	+
	T4.2.6.	Provide users with real-time information on itineraries and traffic conditions	+
	T4.2.7.	Regularly adapt frequency of itineraries for public transport implemented	
	T4.2.8.	Secure long-term financial viability of Transportation Networks	
	T4.2.9.	Optimize Transportation Networks asset management through adequate administrative structures	
T4.3. Integrate Transportation Networks with other infrastructure systems	T4.3.1.	Configure Transportation Networks to provide for needs of Water Networks and facilitate their maintenance	▲
	T4.3.2.	Investigate needs of Water and wastewater Treatment facilities for connection to Transportation Networks	▲
	T4.3.3.	Configure Transportation Networks to provide for needs of Energy Distribution and Transmission networks and facilitate their maintenance	▲
	T4.3.4.	Configure Transportation Networks to provide for needs of Information Networks and facilitate their maintenance	▲
	T4.3.5.	Investigate needs of Solid Waste Treatment and Disposal facilities' connection to Transportation Networks	▲
	T4.3.6.	Investigate needs of Solid Waste Collection and Transfer facilities' connection to Transportation Networks	▲
	T4.3.7.	Provide spaces for Solid Waste Collection equipment along Transportation Networks	▲
T4.4. Enhance public space	T4.4.1.	Design Transportation Networks and equipment to extend and enhance public space	
T4.5. Enhance quality of place	T4.5.1.	Resolve fragmentation of urban fabric and pedestrian routes caused by Transportation Networks	
	T4.5.2.	Visually integrate Transportation Networks to the local context	+

OBJECTIVES	GUIDELINES	SYNERGIES
T4.6. Create a pedestrian-friendly environment	T4.6.1. Concentrate activities along pedestrian routes that enhance daily patterns for safe, lively environments	
	T4.6.2. Size urban blocks adequately for a dense pedestrian grid	
	T4.6.3. Make provisions for pedestrians' protection from harsh climatic conditions	
	T4.6.4. Provide adequate lighting along pedestrian routes	
T4.7. Promote community well-being	T4.7.1. Regulate speed limits	
	T4.7.2. Assess and resolve light pollution issues caused by Transportation Networks	
	T4.7.3. Assess and resolve noise pollution issues caused by Transportation Networks	+
T4.8. Maintain the integrity of natural-world elements and systems	T4.8.1. Avoid the construction of Transportation Networks on undeveloped land	▲
	T4.8.2. Adapt Transportation Network routes to natural surface-water flows	
	T4.8.3. Minimize natural Landscape alteration for the construction of Transportation Networks	
	T4.8.4. Protect natural habitats and wildlife corridors and compensate for their fragmentation or disruption by Transportation Networks	+
	T4.8.5. Develop permeable paving and vegetation configurations for Transportation Networks to provide natural drainage	▲ ▲
	T4.8.6. Develop permeable paving and vegetation configurations for Transportation Networks to filter runoff	▲ ▲
	T4.8.7. Develop vegetation configurations for Transportation Networks to provide habitats	▲
	T4.8.8. Adapt plant selection for Transportation Networks to local context	
	T4.8.9. Control invasive species along Transportation Networks	
	T4.8.10. Avoid use of persistent pesticides in Transportation Networks maintenance practices	
	T4.8.11. Protect soil from contaminants, erosion, and compacting caused by Transportation Networks	+
	T4.8.12. Save, preserve, and reuse topsoil before the construction of Transportation Networks where possible	▲ ▲
T4.9. Prepare for short-term hazards and projected future threats	T4.9.1. Assess and manage future threats from natural and man-made phenomena affecting Transportation Networks	+
	T4.9.2. Plan Transportation Networks response and recovery in case of emergency	
	T4.9.3. Configure Transportation Networks so as to mitigate the urban heat island effect	▲

OBJECTIVES	GUIDELINES		SYNERGIES
T4.10. Optimize material use in Transportation Networks	T4.10.1.	Select low-maintenance materials and structures adequate for their intensity of use	
	T4.10.2.	Source materials required for the construction of Transportation Networks locally when possible	
	T4.10.3.	Promote use of reclaimed and recycled materials in Transportation Networks construction	
	T4.10.4.	Incorporate Solid Waste management residues and by-products in Transportation Networks construction and maintenance	
	T4.10.5.	Plan for the reuse of elements or recycling of materials after Transportation Networks decommission	
T4.11. Optimize Energy consumption in Transportation Networks	T4.11.1	Minimize Energy required for lighting and operation of Transportation Networks	
T4.12. Optimize Water consumption in Transportation Networks	T4.12.1.	Use gray or reclaimed water for cleaning and irrigation in Transportation Networks	

RELEVANT GUIDELINES

GUIDELINES		RELEVANT GUIDELINES	
T1 TRIP GENERATION		FROM ▸ ▸ TO + WITH	
T1.1.1.	Provide and promote remote access and e-services	I1.1.3.	Identify opportunities to improve city administration and social amenities through Information infrastructure
T1.2.2.	Include basic amenities within each residential area	F3.1.1.	Promote the equal and dense distribution of food retail facilities across the city
T1.2.3.	Promote compactness in urban fabric	E1.2.3.	Plan the urban fabric to provide for short trips and public transportation options
T1.2.4.	Optimize placement of infrastructure facilities that generate significant traffic	F4.5.2.	Optimize placement of Food Processing and Distribution facilities to reduce number and length of food distribution trips across the city
		SW4.2.1.	Optimize placement of Solid Waste Transfer stations between Solid Waste Generation locations and Treatment and Disposal facilities
		SW3.3.1.	Optimize placement of Solid Waste Treatment and Disposal facilities so as to minimize distances from Waste Generation areas
T1.3.2.	Collect and organize data on Transportation needs of goods and passengers	I1.3.1.	Identify infrastructure systems' processes to be monitored/documented through Information infrastructure
T1.3.3.	Monitor daily and seasonal Trip Generation fluctuations across the city	I1.3.1.	Identify infrastructure systems' processes to be monitored/documented through Information infrastructure
T1.4.2.	Optimize itineraries within other infrastructure systems	SW4.5.3.	Schedule Solid Waste Collection and Transfer itineraries so as to mitigate traffic congestion
		F4.5.3.	Regulate Food Distribution itineraries so as to mitigate traffic congestion
T2 TRANSPORTATION MODES		FROM ▸ ▸ TO + WITH	
T2.3.1.	Provide and promote reliable public and mass transit	E1.2.1.	Optimize the Transportation modal split for reduced Energy Consumption
T2.3.2.	Provide car sharing and bike sharing options	E1.2.2.	Optimize the selection and operation of Transportation vehicles for reduced Energy Consumption
T2.4.2.	Promote nonmotorized Transportation Modes	E1.2.1.	Optimize the Transportation modal split for reduced Energy Consumption
T2.4.3.	Use efficient and/or alternative-fuel vehicles	E1.2.2.	Optimize the selection and operation of Transportation vehicles for reduced Energy Consumption
		SW2.3.3.	Match Solid Waste Diversion Model residues and by-products with resource and feedstock needs of other infrastructure systems

GUIDELINES			RELEVANT GUIDELINES	
T2.4.4.	Put in place mechanisms to ensure regular maintenance of vehicles		E1.2.2.	Optimize the selection and operation of Transportation vehicles for reduced Energy Consumption
T2.4.5.	Educate users on the efficient operation of vehicles		E1.2.2.	Optimize the selection and operation of Transportation vehicles for reduced Energy Consumption

T3 TRANSPORTATION NODES

FROM ▶ ▶ TO + WITH

T3.2.3.	Provide Node traffic operators with real-time information on traffic conditions	+	I1.3.1.	Identify infrastructure systems' processes to be monitored/documented through Information infrastructure
		+	I1.3.2.	Identify other infrastructure systems' processes to be managed through Information infrastructure
T3.2.4.	Provide users of Transportation Nodes with real-time information on itineraries and traffic conditions	+	I1.3.1.	Identify infrastructure systems' processes to be monitored/documented through Information infrastructure
T3.3.1.	Investigate needs of Transportation Nodes integration with Solid Waste Treatment and Disposal facilities		SW3.5.1.	Provide required traffic connections to all Solid Waste Treatment and Disposal facilities
T3.3.2.	Investigate needs of Transportation Nodes integration with Water and wastewater Treatment facilities		W3.9.4.	Resolve routes and methods for the collection, treatment, and disposal of Water Treatment residues and by-products
T3.3.3.	Provide required traffic connections to large Landscape infrastructure entities		L3.5.2.	Investigate needs of Landscape attractors' integration with Transportation Nodes
T3.4.2.	Visually integrate Transportation Nodes to the local context	+	L1.7.2.	Define needs for mitigating and compensating for other infrastructure systems' impact on community life through Landscape infrastructure
T3.6.1.	Assess and resolve noise pollution issues caused by Transportation Nodes	+	L1.7.2.	Define needs for mitigating and compensating for other infrastructure systems' impact on community life through Landscape infrastructure
T3.7.1.	Avoid construction of Transportation Nodes on undeveloped land		F2.5.3.	Protect foodshed from urban expansion
T3.7.2.	Protect natural habitats and wildlife corridors and compensate for their fragmentation or disruption by Transportation Nodes	+	L1.7.1.	Define needs for mitigating and compensating for other infrastructure systems' impact on the natural world through Landscape infrastructure
T3.7.3.	Develop permeable paving and vegetation configurations for Transportation Nodes to provide natural drainage		L3.4.2.	Exploit opportunities for additional Landscape supportive patches and corridors given by Transportation entities

GUIDELINES		RELEVANT GUIDELINES	
		W2.3.6.	Provide for rainwater and stormwater management needs with natural drainage, infiltration, and floodplain management
T3.7.4.	Develop permeable paving and vegetation configurations for Transportation Nodes to filter runoff	L3.4.2.	Exploit opportunities for additional Landscape supportive patches and corridors given by Transportation entities
		W2.3.4.	Control contaminants in runoff water through bioretention and bioinfiltration methods
T3.7.5.	Develop vegetation configurations for Transportation Nodes to provide habitats	L3.4.2.	Exploit opportunities for additional Landscape supportive patches and corridors given by Transportation entities
T3.7.6.	Protect soil from contaminants, erosion, and compacting caused by Transportation Nodes	L1.7.1.	Define needs for mitigating and compensating for other infrastructure systems' impact on the natural world through Landscape infrastructure
T3.7.7.	Save, preserve, and reuse topsoil before the construction of Transportation Nodes where possible	F2.5.3.	Protect foodshed from urban expansion
		L3.10.2.	Use materials that are readily and locally sourced for the construction of Landscape patches and corridors
T3.8.1.	Assess and manage future threats from natural and man-made phenomena affecting Transportation Nodes	L1.7.3.	Define needs for protection of infrastructure systems against projected climate change and extreme events through Landscape infrastructure
T3.9.3.	Incorporate Solid Waste management residue and by-products in Transportation Nodes construction and maintenance	SW2.3.3.	Match Solid Waste Diversion Model residues and by-products with resource and feedstock needs of other infrastructure systems
T3.9.4.	Plan for the reuse of elements or recycling of materials after Transportation Nodes decommission	L3.10.2.	Use materials that are readily and locally sourced for the construction of Landscape patches and corridors
T3.10.1.	Conform waste management practices within Transportation Nodes to requirements of Solid Waste Collection and Transfer planning	SW4.3.2.	Solid Waste and Transportation systems to work in synergy for the Collection and Transfer of solid waste

T4
TRANSPORTATION NETWORKS

FROM ▸ ▸ TO
+
WITH

T4.2.3.	Organize road networks to promote use of high-occupancy vehicles	E1.2.	Reduce Energy Consumption in the Transportation system
T4.2.5.	Provide Network traffic operators with real-time information on traffic conditions	I1.3.1.	Identify infrastructure systems' processes to be monitored/documented through Information infrastructure

GUIDELINES			RELEVANT GUIDELINES
		I1.3.2.	Identify other infrastructure systems' processes to be managed through Information infrastructure
T4.2.6.	Provide users with real-time information on itineraries and traffic conditions	I1.3.1.	Identify infrastructure systems' processes to be monitored/documented through Information infrastructure
		I1.3.2.	Identify other infrastructure systems' processes to be managed through Information infrastructure
T4.3.1.	Configure Transportation Networks to provide for needs of Water Networks and facilitate their maintenance	W4.3.1.	Integrate Water Networks with Transportation Networks
T4.3.2.	Investigate needs of Water and wastewater Treatment facilities for connection to Transportation Networks	W3.9.4.	Resolve routes and methods for the collection, treatment, and disposal of Water Treatment residues and by-products
T4.3.3.	Configure Transportation Networks to provide for needs of Energy Distribution and Transmission networks and facilitate their maintenance	E4.4.1.	Integrate Energy Distribution and Transmission networks with Transportation Networks
T4.3.4.	Configure Transportation Networks to provide for needs of Information Networks and facilitate their maintenance	I4.4.1.	Integrate Information Networks with Transportation Networks
T4.3.5.	Investigate needs of Solid Waste Treatment and Disposal facilities' connection to Transportation Networks	SW3.5.1.	Provide required traffic connections to all Solid Waste Treatment and Disposal facilities
T4.3.6.	Investigate needs of Solid Waste Collection and Transfer facilities' connection to Transportation Networks	SW4.2.2.	Provide required traffic connections to all Solid Waste Transfer facilities
T4.3.7.	Provide spaces for Solid Waste Collection equipment along Transportation Networks	SW4.3.2.	Solid Waste and Transportation systems to work in synergy for the Collection and Transfer of solid waste
T4.5.2.	Visually integrate Transportation Networks to the local context	L1.7.2.	Define needs for mitigating and compensating for other infrastructure systems' impact on community life through Landscape infrastructure
T4.7.3.	Assess and resolve noise pollution issues caused by Transportation Networks	L1.7.2.	Define needs for mitigating and compensating for other infrastructure systems' impact on community life through Landscape infrastructure
T4.8.1.	Avoid the construction of Transportation Networks on undeveloped land	F2.5.3.	Protect foodshed from urban expansion
T4.8.4.	Protect natural habitats and wildlife corridors and compensate for their fragmentation or disruption by Transportation Networks	L1.7.1.	Define needs for mitigating and compensating for other infrastructure systems' impact on the natural world through Landscape infrastructure

GUIDELINES			RELEVANT GUIDELINES	
T4.8.5.	Develop permeable paving and vegetation configurations for Transportation Networks to provide natural drainage		L3.4.2.	Exploit opportunities for additional Landscape supportive patches and corridors given by Transportation entities
			W2.3.6.	Provide for rainwater and stormwater management needs with natural drainage, infiltration, and floodplain management
T4.8.6.	Develop permeable paving and vegetation configurations for Transportation Networks to filter runoff		L3.4.2.	Exploit opportunities for additional Landscape supportive patches and corridors given by Transportation entities
			W2.3.4.	Control contaminants in runoff water through bioretention and bioinfiltration methods
T4.8.7.	Develop vegetation configurations for Transportation Networks to provide habitats		L3.4.2.	Exploit opportunities for additional Landscape supportive patches and corridors given by Transportation entities
T4.8.11.	Protect soil from contaminants, erosion, and compacting caused by Transportation Networks	+	L1.7.1.	Define needs for mitigating and compensating for other infrastructure systems' impact on the natural world through Landscape infrastructure
T4.8.12.	Save, preserve, and reuse topsoil before the construction of Transportation Networks where possible		L3.10.2.	Use materials that are readily and locally sourced for the construction of Landscape patches and corridors
			F2.5.3.	Protect foodshed from urban expansion
T4.9.1.	Assess and manage future threats from natural and man-made phenomena affecting Transportation Networks	+	L1.7.3.	Define needs for protection of infrastructure systems against projected climate change and extreme events through Landscape infrastructure
T4.9.3.	Configure Transportation Networks so as to mitigate the urban heat island effect		E1.1.4.	Improve the urban microclimate
T4.10.4.	Incorporate Solid Waste management residues and by-products in Transportation Networks construction and maintenance		SW2.3.3.	Match Solid Waste Diversion Model residues and by-products with resource and feedstock needs of other infrastructure systems
T4.10.5.	Plan for the reuse of elements or recycling of materials after Transportation Networks decommission		L3.10.2.	Use materials that are readily and locally sourced for the construction of Landscape patches and corridors
T4.12.1.	Use gray or reclaimed water for cleaning and irrigation in Transportation Networks		W1.6.3.	Identify opportunities for use of different nonpotable water qualities

NOTES

1. Richard T. T. Forman, "Infrastructure and Nature: Reciprocal Effects and Patterns for Our Future," in Spiro Pollalis, Andreas Georgoulias, Stephen Ramos, and Daniel Schodek, eds., *Infrastructure Sustainability and Design* (New York: Routledge, 2012), 35.
2. Richard T. T. Forman et al. *Road Ecology: Science and Solutions* (Washington, DC: Island Press, 2003), xv.
3. Ibid., 25.
4. "Grocery Delivery Service Is Greener than Driving to the Store," University of Washington, last modified April 29, 2013, http://www.washington.edu/news/2013/04/29/grocery-delivery-service-is-greener-than-driving-to-the-store/.
5. B. G. Hutchinson, *Principles of Urban Transportation Systems Planning* (New York: McGraw-Hill, 1974).
6. Bhanu Murali Krishna Yerra, "The Emergence of Hierarchy in Transportation Networks" (M.S. thesis, University of Minnesota, 2003), 20.

IMAGE CREDITS

P. 72: Photograph by Charles Forerunner, accessed on August 31, 2015, https://images.unsplash.com/uploads/1413387158190559d80f7/6108b580?q=80&fm=jpg&s=91d5cbdf1cd789befb96c6eb2f6c6f17. **P. 75:** Rick Phillips, "Smart Growth Transect." **P. 75:** Rick Phillips, "Arc of a Trip." **P. 76:** Rick Phillips, "Transit Node." **P. 76:** Rick Phillips, "Smart Cars." **P. 77:** Image courtesy of Greater Memphis Chamber. **P. 79:** (World energy consumption by sector) Data source: US Energy Information Administration, 2011, accessed on September 14, 2015, http://www.eia.gov/tools/faqs/faq.cfm?id=447&t=1. **P. 79:** (World transportation energy by source) Data source: International Energy Agency, "Key World Energy Statistics," 2014, p. 37, accessed on September 14, 2015, http://www.iea.org/publications/freepublications/publication/keyworld2014.pdf. **P. 86:** (Lower right) Data source: UK Department for Transport, Transport statistics and reports factsheets, accessed on September 14, 2015, http://www.winacc.org.uk/resource/transport-statistics-and-reports-factsheets. **P. 87:** Data source: Journeys Issue 7, November 2011, LTA Academy, Land Transport Authority Singapore, accessed on September 14, 2015, http://www.lta.gov.sg/ltaacademy/doc/JOURNEYS_Nov2011%20Revised.pdf. **P. 89:** (Networks' main characteristics) The World Factbook CIA, accessed September 14, 2015, https://www.cia.gov/library/publications/resources/the-world-factbook/geos/xx.html.

Sustainable Infrastructure Planning Guidelines

Water Infrastructure

Chapter 6

Water

manages and processes a naturally available resource.

6.1 | INDUSTRY PERSPECTIVES

Changing the Water Paradigm in Urban Infrastructure: Managing Water throughout the Water Cycle

Water is a fundamental requirement for the existence of life on the planet. The classic photo of the earth taken by astronauts orbiting the moon shows a sphere that is clearly influenced by water—on the land, in oceans, and in the atmosphere. Water supports a host of functions including ecosystem provisioning and regulation, human and social systems, and economic development. And we have all the water that we will ever have on the planet right now. Water isn't created and it isn't destroyed.

Early civilizations recognized the importance of water. Nomadic tribes ultimately settled at locations where there was a water supply that met their needs. As a community outgrew its water sources, infrastructure—such as the Roman aqueducts—was built to bring water from far distances.

As urban areas continued to develop, establishing an effective approach to managing waterborne wastes produced by the population became critical to public health and the sustainability of communities. Sewers were constructed to carry wastes rapidly from urban areas to more remote waterways. As urban areas became less pervious (e.g., with more buildings and a higher percentage of paved areas), stormwater was collected and conveyed through combined storm and sanitary sewers or with dedicated storm drainage away from the urban center.

This strategy was viable as long as sufficient freshwater from the hydrologic water cycle was available to replenish surface waters and recharge groundwater. But increasing population and its associated agricultural and other economy-based stresses have strained that cycle immensely. Additional stressors to the hydrologic cycle, including (1) competing and increasing demands for available resources, (2) changes in climate that are exacerbating water shortages, increasing flood risk, and raising sea level, and (3) aging infrastructure that has reached the end of its useful life, all suggest that an entirely new approach is necessary for water resource development, management, and oversight.

Going forward, integrated urban water management is needed that merges traditional functions within a watershed and promotes coordinated planning and development. In this approach, water, land, and related resources (e.g., energy) are linked to urban areas in the application of water-sensitive design for the urban environment. Water sources should be diverse, fit-for-purpose, and renewable, using infrastructure that is adaptive and multifunctional and reinforces water-sensitive behaviors.

From a planning perspective, it will be critical to understand the overall water budget from traditional and nontraditional sources that are available within the watershed. If water has been imported in the past and/or if safe yield of existing supplies is likely to be compromised in the future, planners must understand the extent of new resources available to replace those threatened sources, such as stormwater capture and recharge as well as the reuse of less traditional sources. Matching demand and supply options with fit-for-purpose water will be critical, recognizing that different levels of water quality are required for different uses.

Because a great deal of buried infrastructure (e.g., pipes, conduits) is aging, there is an opportunity for planners to rethink location and sizing as a function of demand forecasts for a specific water need and source type. The ability to decentralize treatment and distribution to more locally match demand and available resources can have significant cost efficiencies from pumping and distribution perspectives. The opportunity to cost-effectively decentralize is much greater in new cities, although the need for repair and replacement of aging infrastructure in established urban areas (where buried piping may comprise 70 percent or more of a water utility's assets) provides an opening for new thinking and innovation there as well.

While water management will become increasingly critical for urban sustainability—particularly since more than 50 percent of the world's population now lives in urban centers with aging infrastructure—it is important to understand current water usage and demand for all human and environmental needs. A global water supply-and-demand model suggests that there will be a 40 percent deficit in accessible, reliable, and sustainable supplies by 2030, if water demand only increases 2 percent annually. Agriculture represents two-thirds of the current and 2030 global water demand, and adequate food production is critical to reducing poverty and addressing the United Nations Millennium Development Goals. Focusing attention on smart water use for agriculture will create additional water supply for urban needs.

Another critical resource for sustainable cities—energy—is inextricably linked to water. It takes water to produce energy (either directly through steam generation or indirectly through cooling), and it takes energy to treat and transport water for consumer use. Efficient urban infrastructure needs to recognize and capitalize on this relationship. For example, water conservation strategies are necessarily energy conservation strategies. The ability to manage all water resources within the watershed and minimize import/export across watershed boundaries will significantly reduce energy use for water transport. Data suggest that 7 percent of global energy demand is consumed in water management, whereas up to 15 percent or more of total energy consumption in California is related to water. This difference can be attributed to a measurable extent to the movement of water across vast distances from source to urban locations in California. Again, these data suggest that efficiently managing all available water resources locally—and making them fit-for-purpose—can reduce energy consumption as well.

Finally, our environment provides important water provisioning and regulating services, such as water purification, erosion mitigation, carbon sequestration, and wildlife habitat. Urban centers stress freshwater supplies needed for such environmental services, and consequently our "natural capital" is deteriorating. A recent example of this is in the San Francisco Bay/San Joaquin River Delta, where agricultural and urban water apportionment has been at least temporarily suspended to maintain minimum flows to support a deteriorating Bay-Delta ecosystem. Changing precipitation patterns and more frequent drought in some areas will only exacerbate this sensitive balance. Because human populations depend upon the provisioning and regulating services provided by the environment, urban needs will not be sustainably met without consideration of all human and natural-system water needs.

From an institutional perspective the challenge is to create governance structures that enable sustainable approaches to water supply. In the United States, regulatory frameworks for water were developed according to its place in the water cycle; that is, there are drinking water regulations, wastewater regulations, reuse water regulations, stormwater regulations, and so on. Often these are implemented by different agencies; thus a holistic view of water management is difficult because oversight is fragmented. One of the best examples of integrated water management, in Singapore, was enabled to an important extent because one entity had control over all water regulation, allowing for systemic planning, management, and innovation.

In summary, planning assumptions need to change dramatically to achieve a sustainable water future. A broadened understanding of available water resources, to include fit-for-purpose water reuse and local replenishment of water resources through green infrastructure, is critical to meet changing water demands in resource-constrained urban areas affected by population growth and climate change. Finally, regulatory constructs need to reflect and incentivize sustainable water approaches. In the end, urban areas need to be water-sensitive, using multifunctional infrastructure and urban design that reinforces water-sensitive behaviors.

Douglas M. Owen
PE, BCEE, ENV SP, Executive Vice President
and Chief Technical Officer, ARCADIS

City Planners and Engineers as Stewards of Sustainable Water Infrastructure

City planners and engineers have proven to be resilient and innovative in developing systems to overcome water supply and sanitation challenges throughout history. On a list of 14 engineering challenges for the 21st century released in 2008 by the National Academy of Engineering, three can be directly linked to water security and sustainability: (1) provide access to clean water; (2) manage the nitrogen cycle; and (3) restore and upgrade urban infrastructure. To answer these 21st-century challenges, planners and engineers must truly acknowledge and embrace their role as stewards of the built environment. Their means should be innovation and technology responsiveness to natural processes, and their methods should be sustainable approaches to planning and design. Resilient engineering solutions combined with well-planned operating strategies will be necessary to balance environmental and ecosystem preservation with city water supply and distribution demands. The end result should be improvement of the quality of life for all of those impacted by city water system solutions. Sustainability of water infrastructure means focusing on the environmental, economic, and social well-being of the built and natural environment.

Water crisis is the third highest global risk of concern in the World Economic Forum's Global Risks 2014 report. Access to clean water is a local distribution issue. We know how to collect, treat, and convey water and wastewater, but how do we do it equitably and affordably? Water is not naturally evenly distributed over geographies. Fewer than 10 countries possess 60 percent of the world's available freshwater supply. There are also significant regional variations within these countries. A region with water scarcity is an unstable region. The Greater Tokyo Region provides a model for integrated water management efforts toward regional stability and sustainability. Greater Tokyo encompasses one of the world's largest high-density populations with an industrial base. Much of the region is located within high-risk seasonal floodplains. Through integrated infrastructure, a systematic framework for flood control and water use, a legal system to regulate water use, upstream/downstream community cooperation, and river restoration, government planners and engineers in Japan provide an example of water security stewardship.

Overconsumption or pollution by cities and communities upstream can be destructive for those downstream. Pollutant discharges significantly affect the ecosystem and human health; for instance, agricultural and industrial nitrogen inputs (pollution) to the global nitrogen cycle have been increasingly creating nitrogen nutrient imbalances in the natural environment. In rivers and near-shore coastal systems, increases in nitrogen levels often lead to anoxia (no oxygen) or hypoxia (low oxygen) conditions. The impact to the natural environment is altered biodiversity and general habitat degradation favoring parasites and infectious diseases. Planners and engineers need to provide local and innovative solutions. The Massachusetts Estuaries Project (USA), for example, is exploring approaches to nitrogen reduction such as tidal flushing, stormwater control and treatment, attenuation via wetlands and ponds, and on-site wastewater treatment systems for local communities.

In many cities and communities, the existing water, wastewater, and stormwater infrastructure is ill-equipped to address the growing demands for clean water and management of the nitrogen cycle, as well as climate conditions of increased tropical storms, potential sea level rise, or droughts. It will be necessary to add new infrastructure as well as renew and expand existing infrastructure. Existing urban water infrastructure presents the most difficult renewal challenge, since it is typically difficult to access without major disruption to traffic and services. The Organization for Economic Cooperation and Development (OECD) has estimated that US $1.3 trillion needs to be invested annually to replace and maintain water infrastructure in developed countries and emerging markets alone. Planners and engineers understand that this investment is necessary, but disagree on how to prioritize spending for renewal and expansion of infrastructure. Guidelines for sustainable city infrastructure planning can provide a framework for prioritizing infrastructure investment. They can also provide a roadmap for integrating and applying innovative approaches, practices, and technologies. The answers to water sustainability challenges will need to be a combination of policies and adequate guidelines, along with transformation of professional practice paradigms. It is the agile and forward-thinking planners and engineers who will develop truly sustainable solutions tailored to the local people and environment that will support generations to come.

Melissa M. Carter,
PE, PMP, Director of Project Management, MWH Global

Do It Like You Did It Last Time? Reexamining Water Infrastructure

As the world has developed over the last 40+ years, the planning and design of water and most other infrastructure have become standardized. Infrastructure projects are designed to meet the adopted standards and design criteria of each jurisdiction in which the project exists. Over time, the basis for the historical standards or criteria may no longer apply, at best may not be understood. Unfortunately this has led to a generation of engineers who design projects using a code compliance design approach without addressing whether the criteria make the most sense for the project. As an example, engineers estimate future demands for infrastructural services, but rarely actively attempt to influence those demands or question whether they are designing the "right" project. Conversely, planners and architects have more readily adopted sustainability principles. They are designing great buildings, projects, master-planned communities, even "new towns" all over the world—but the infrastructure that serves those projects tends to be nearly the same as it was 30 years ago.

We are at a point in the state of our engineering practice where there can be a significant disconnect between historical use, design criteria, the needed capacity of infrastructure, and consumption of resources. This same disconnect can artificially inflate the need or capacity of the built infrastructure. Nowhere is that more apparent than in the planning and design of our water infrastructure systems. Conservative assumptions based on historical use that once served us well can now introduce significant and costly design criteria that do not actually estimate future demands. Engineers and planners need to challenge historical standards and provide guidance on how we can influence water demands and drive change to these existing standards so that the design of water infrastructure is integrated, more project-specific, and appropriate for the project context.

Engineering tends to be a very traditional profession divided into individual disciplines—potable water, wastewater, recycled water, and stormwater—without regard for how those systems can synergistically complement each other. In addition, water infrastructure is designed using a historical, "do it like you did it last time" perspective. How much water we used last year guides how much water we need next year. Planners and engineers need to influence future water demands—not just design infrastructure projects to meet demands—and focus on water projects as part of integrated water systems.

We must reexamine our consumption of water whether for domestic or agricultural use. It is essential for engineers to influence the use of water as it relates both to quantity and quality. Our focus must begin with reducing demand, and then design projects to meet demand requirements with water of appropriate quality from the best available sources, considering environmental as well as urban and agricultural water needs.

The availability of water will continue to be a constraint on new development as well as on the densification of our existing cities worldwide. Through their overarching design of water systems, engineers must focus on every opportunity to reduce use (conservation) and increase efficiency (water fixtures that use less water). Reducing both indoor and outdoor use can have an order-of-magnitude impact on future water demand (affecting water supply, treatment, and distribution) and on wastewater generation (reduced collection and treatment needs) and the production of recycled water. We need to plan and design all of our water systems using an integrated strategy (which should also include stormwater runoff as a potential source) rather than our traditional silo approach.

Our ability to influence demand and preserve water as a resource can be achieved through several approaches:

New development. All new development should be required to use very-low-flow fixtures and appliances to reduce indoor potable demands. The design of outdoor irrigation systems should be based on a maximum allowable water budget. Outdoor irrigation should use recycled water whenever possible. Plant palettes should reflect native species, emphasizing irrigation only for establishment and discontinuing irrigation after one year. Turf should be used only for active spaces. Irrigation should only be allowed on alternate days using street addresses (odd and even) to reduce maximum day demands, allowing us to optimize the design of our key system facilities (water treatment, pumping, and storage). In the United States we should reduce our current average use from 60+ gallons per person per day to no more than 40 gallons per person per day. This can easily be accomplished with no additional cost for new development or redevelopment.

Existing development. We should require the retrofit of existing indoor fixtures (to very-low-flow fixtures) for homes as each sells. (The state of California has adopted this requirement for homes built prior to 1994, starting in 2017.) This will significantly reduce existing water use and wastewater generation, creating capacity in our existing supply and treatment facilities as well as in our distribution and conveyance systems. This will allow densification using already constructed water and wastewater infrastructure capacity.

Decentralized systems. As we developed and standardized, we also centralized much of our water infrastructure. We need to give greater thought to opportunities to decentralize water and wastewater treatment, taking into greater consideration how that could positively affect reliability, resiliency, and opportunities to provide recycled water closer to where it is generated or where there are demands.

Low-impact development. Strategies for stormwater should be incorporated into landscape wherever possible to slow time of concentration, improve water quality, and prioritize surface conveyance over traditional piped systems. We need to mimic the natural hydrology of our water systems rather than designing flood control projects. We need to influence where it makes sense for development to occur rather than simply agreeing to design projects to try to protect development from the probability of a significant weather event.

Reliability, redundancy, and resiliency. It is not enough just to reduce, reuse, and recycle. We need to continue to consider the reliability and redundancy as well as the resiliency of our potable water systems. Where there are adjacent water supply systems we need to invest in cross connections and interagency agreements, including emergency connections to agricultural, recycled, or stormwater supplies where those sources could be conveyed to potable water treatment facilities.

Water pricing. We need to revise water pricing with strategies to reflect the real cost of water and also to influence its consumption. The cost of basic indoor water use (40 gallons per person per day in the US) should be relatively low. A second tier should be more costly, but still be reasonable up to 60 gallons per person per day. There should be a third tier for a small amount of outdoor use in warmer months based on the local climate, and then a significantly higher tier as outdoor use increases. We should also consider providing credits or reduced cost for non-peak hour use to further reduce the water infrastructure requirements (similar to some strategies for energy pricing).

Time of use. Much of our water infrastructure is sized for maximum day or peak hour demands. For example, if outdoor watering is only allowed at night from 7 pm to 7 am (when temperatures are cooler, making irrigation more efficient), then the utility must provide all of the irrigation demand in a 12-hour period. However, if odd and even addresses are required to irrigate on opposite days, then we can reduce the infrastructure required to meet the total outdoor demand by approximately half.

Water quality. We need to provide the minimum safe water quality for the required use and better incorporate the use of "natural" systems to improve water quality. This can be as simple as directing urbanized stormwater runoff into landscape areas or using recycled water to provide fire flows through an irrigation system rather than a potable water system—something that has been done in only a few places in the United States.

Sustainability as it relates to water infrastructure must become an occasion for a group of engineers, planners, and decision makers to have a conversation about why they might design something differently rather than by the adopted code or historical way of doing things. Every small incremental design decision we make influences the consumption, size, resiliency, and cost of our water infrastructure.

Laura Bonich
PE, LEED AP, Director Business Development, NV5

6.2 | SUSTAINABLE WATER PLANNING

PROJECTED GLOBAL WATER SCARCITY, 2025 (International Water Management Institute)

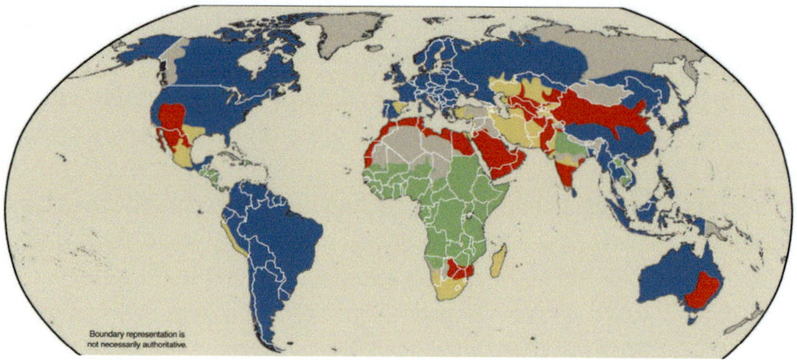

Physical water scarcity: More than 75% of river flows are allocated to agriculture, industries, or domestic purposes. This definition of scarcity relating water availability to water demand implies that dry areas are not necessarily water-scarce.

Approaching physical water scarcity: More than 60% of river flows are allocated. These basins will experience physical water scarcity in the near future.

Economic water scarcity: Water resources are abundant relative to water use, with less than 25% of water from rivers withdrawn for human purposes, but local water scarcity exists.

Little or no water scarcity: Abundant water resources relative to use. Less than 25% of water from rivers is withdrawn for human purposes.

Not estimated

Present-day preparedness has stimulated "infrastructure recalibration" all over the world, alluding to the fundamental urgency of primary resource protection. Freshwater is one such primary resource of vital essence for human life, responsible for natural ecological systems and for the workings of other critical infrastructures of food, energy, transportation, and waste. "Water infrastructure" can be understood as the medium of management and conveyance (natural or man-made) of freshwater (a public resource) to domestic as well as industrial consumers, comprising drinking and wastewater mechanisms. Establishing and maintaining highest standards of water infrastructure for the conveyance of this resource is a main priority of every governing body. Components of water infrastructure systems—source, storage, treatment, network, and distribution of freshwater, and treatment, reuse, and discharge of wastewater—are described in detail in the following section.

Close to 1 billion people do not have access to drinking water of a reliable quality, and as many as 3.5 billion are expected to face water scarcity by 2025.[1] In addition, the number of people with poor access to proper sanitation is estimated at over 2.5 billion.[2] "Right to water" as an International Law was passed by the UN Human Rights Council in September 2010, recognizing access to water and sanitation as a human right.[3] Under this law, governments are expected to take responsibility for ensuring that people can enjoy "sufficient, safe, accessible and affordable water, without discrimination amongst citizens." In reality, implementation of water as a human right at global scale has yet to overcome the challenge of the varying nature of delivery systems of water infrastructure in different parts of the world. Water security is "the availability of an acceptable quantity and quality of water for health, livelihoods, ecosystems and production, coupled with an acceptable level of water-related risks to people, environments and economies."[4] A critical observation is that water infrastructure delivers water essentially to properties, households, businesses, and industries and not to individual human beings; furthermore, integrating different public and private delivery systems within national legislation of several countries poses a complex challenge.

The sociopolitical challenge of water as human right within national legislation is just one part of the complexity of this infrastructure; there is also an important physical dimension in the ways freshwater exists as a resource. It comes in various forms of aquifers, rivers, streams, lakes, and networks of one or more of these systems; these are all geographically identified as part of "watersheds" or "drainage basins." "Watershed" is a land area within which all the water drains to the same place; it can be of any size or shape, can lie across urban and rural areas and state or even national borders. As a result, governance scopes and political boundaries do not always coincide with watershed boundaries, impacting the management and delivery of freshwater resources in significant ways, most often resulting in conflicts at state or even international levels.[5] Worldwide there are about 260 transboundary river basins, lakes, and groundwater aquifers shared between two or more nations, with 40 percent of the world's population relying on these shared freshwater sources.[6,7] Studies indicate that water allocation disputes in recent times are on the increase at local scales, with the participants ranging from residential, tribal, and ethnic groups to local industries and business. Climate-related risks and extreme events will trigger water-related conflicts. With sea level rise, in some areas there will be a greater intrusion of saltwater into freshwater resources, transforming farming and riparian patterns and ecologies. Water management practices will need to respond to unpredicted outcomes of growing droughts, changes in water flow, quantity and its related effects on water quality—the most important characteristic for the next generation of resilient water infrastructure.

As water is a "common-pool resource," the outcome of this resource utilization is inherently collective and not individual. Even though water can be appropriated at the individual consumer level, the effect has a shared impact. Water is thus a public good that needs to be effectively managed in a sustainable way through integrated and holistic approaches. The infrastructure planning paradigm is changing: historically, infrastructure systems responded to the need for a resource; today, the infrastructure planning requisite is to conserve the resource and support system resiliency that is responsive to emergent as well as future needs. This demands true integrity of a system.

FRESHWATER USE BY COUNTRY (WRI 2000)

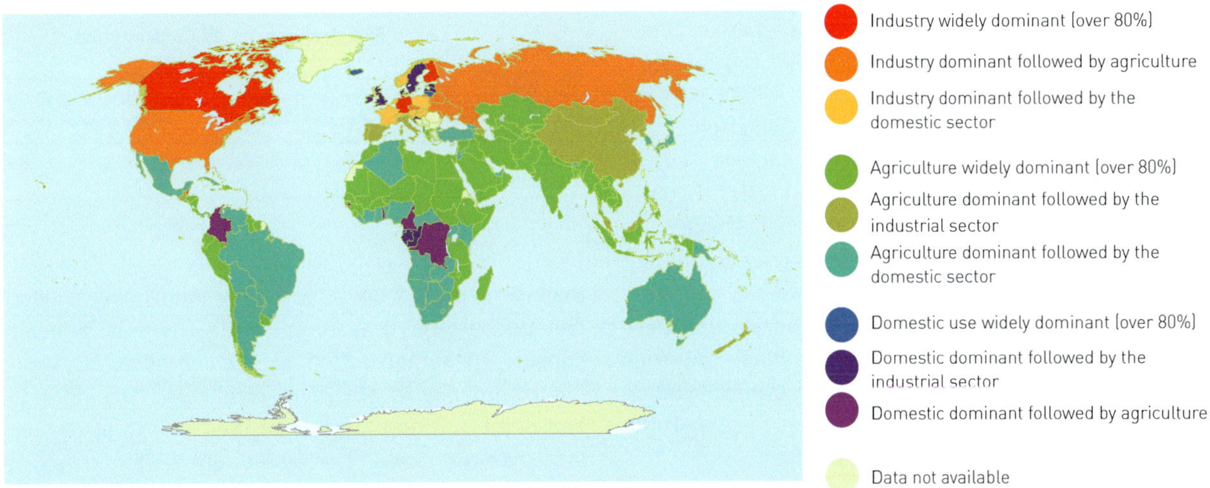

CHALLENGES TOWARD A SUSTAINABLE WATER SYSTEM

Freshwater sources exist naturally in the form of groundwater aquifers and as surface water in the form of reservoirs, lakes, streams, rivers, marshes, and wetlands. Natural freshwater resources are limited and unevenly distributed by region and time, influenced by factors such as climatic trends, seasonal precipitation variations, and geographical characteristics like arid or dry regions. In some dry regions desalination processes are used to obtain potable water from seawater. Along with the limited water quantity, human activities like overexploitation of groundwater resources cause degradation of water quality, collectively amounting to global water scarcity. Declining ecosystems, urban pressures, industrialization, and agricultural processes that discharge untreated effluent into the water stream significantly compromise the quality of water. According to the UN Department of Economic and Social Affairs, almost one-fifth of the world's population (1.2 billion people) live in areas of physical water scarcity, and another 1.6 billion face economic water scarcity due to lack of necessary water delivery infrastructure. The overlap of physical and economic water scarcity reveals the overall burden of the water scarcity challenge. Even more than uneven distribution of available water sources, degrading of water quality and mismanagement of water are influencing factors that lead to water scarcity. The UN Environmental Protection Agency (UNEP) alerts, "Using water the way we have in the past simply will not sustain humanity in future."[8]

Water consumption by sector use can be broadly categorized as domestic, agricultural, and industrial. Globally, 70 percent of water is used for agriculture, 22 for industry, and 8 for domestic needs; these numbers vary regionally. Freshwater use by country profile along with sector can provide a good cumulative understanding of global water consumption trends. Calculations derived from water withdrawal and consumption assessments show a growing increase in water wastage since the 1990s. Also, more water evaporates from reservoirs than is consumed by human use; this is part of the natural hydrological cycle influencing global water quantities. Going forward, climatic trends like temperature rise will increase the rate of evaporation, at the same time as increasing urbanization is also expected to greatly increase water consumption. In the context of global water quantities, where physical water is defined beyond the watershed, geographic, and political boundaries, it is important to include "virtual water," which is essentially the quantity of water required for producing a particular commodity or service. The location of consumption of a commodity is usually different from the location of production; for instance, countries that pay to grow their food in other countries that have more water now have a political stake in those countries growing their food. This may also lead to overexploitation of the producing countries' water resources to support the consuming countries' virtual water footprint.

Groundwater as a freshwater resource is one of the easily accessible forms of water, considered affordable and reliable. However, water overexploitation, or abstraction of groundwater that surpasses the sustainable threshold safe yield of the aquifer, results in several direct and indirect adverse impacts. Besides water quality degradation, which can be more serious than surface water pollution in some cases, these include decrease in recharge of the groundwater aquifer, eventual depletion of water tables, reduction in groundwater flow allowing for seawater intrusion, and land subsidence.

In an ideal scenario, human activity must have minimal impact on the water cycle, enabling self-balance of the system. Regional trends across sectors, however, show a significant imbalance, calling for a stimulus to water conservation. This chapter will also outline reduction in consumption as a first preference in the hierarchy of sustainable strategy for water infrastructure. Planning practices in this regard can prove most influential by tackling this issue from both sides: by implementing progressive planning paradigms improving infrastructure management and delivery, and by influencing consumer behavior at the grassroots level. There is a need to support consumer action at the individual as well as the community level, giving consumers the right tools and demonstrating impacts of their efforts at the larger systems-level scale. The agriculture sector, responsible for 70 percent of global water use, is a default target for water conservation and demand management (UNEP, 2007). High-efficiency irrigation methods and maximizing the use of rainfall for crops are expected to have a positive influence. Farming practices can minimize emission of pesticides and can control and reduce nutrient discharges from fertilizers, reducing contamination of surface and groundwater. Industrial water use for purposes of heat transfer and as a processing medium is being addressed to influence conservation of water quantity and quality. All industries today are encouraged to monitor their water footprint. Reused water means the use of adequately treated wastewater for an appropriate use in accordance with the regulatory requirements.[9] Recycled water means more-than-one-time use of water before returning it to the natural hydrological cycle. Water reuse can be successfully implemented with clear classification of indoor and outdoor water uses and quality standards, to successfully supplement potable as well as nonpotable water uses. The "one water" concept could provide a critical next-generation sustainable practice paradigm, in which any

WATER WITHDRAWAL VERSUS CONSUMPTION (UNEP 2008)

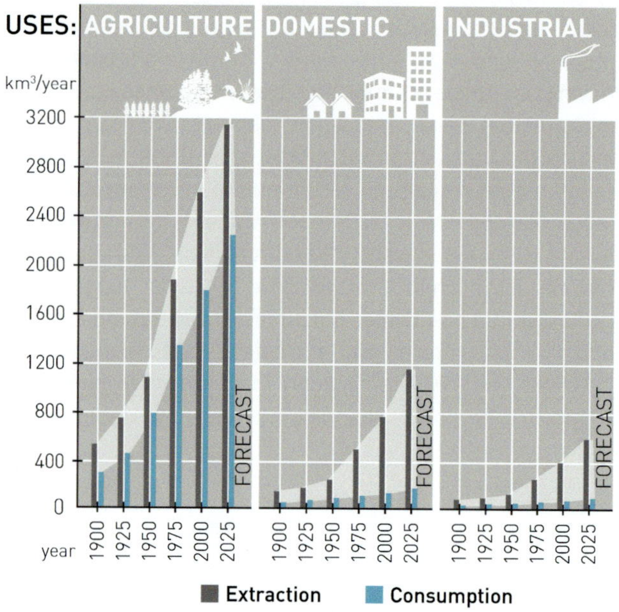

The light gray band represents the difference between the amount of water extracted and that actually consumed. Water may be extracted, used, recycled (or returned to rivers or aquifers), and reused several times over. Consumption is final use of water, after which it can no longer be reused. That extractions have increased at a much faster rate is an indication of how much more intensively we can now exploit water. Only a fraction of water is lost through evaporation.

type of water is basically water, negating the conventional industry terms like drinking water, wastewater, etc. The determinant factor is essentially "quality of water," which needs to be appropriated to its best respective use. This calls for a change in mindset from the infrastructure and institutional structures that are now in place.

Freshwater ecosystems are among the most extensively altered systems on earth, leading to degrading of ecosystem functions due to large-scale surface water and habitat alteration.[10] Of 227 largest rivers, 60 percent are moderately to greatly engineered by dams, diversions, and canals.[11] These may prove beneficial for immediate or short-term purposes, but in the long run they can have negative impacts on natural ecosystem functioning and can compromise effective floodplain management at interregional scales. The World Resources Institute reports that today freshwater and coastal aquatic ecosystems contain more than 500 dead zones, compared to 24 reported dead zones in 1960. Massive land use alterations also disrupt surface water flows, causing erosion. Another challenge is the urban water management that involves nonengineered water outside of the pipe network: for instance, integration of combined sewer overflows (CSO) and on-site treatment of rainwater runoff in infrastructure upgrading, or responsiveness to flood risks from high precipitation, rivers, coastal surges, or winter storm events. Traditional water management systems channeled stormwater runoff away from the city into the watercourses through a combined system of wastewater and stormwater. With the increase in hard surfaces this has resulted in problems of flash flooding and contamination of receiving watercourses; increased intensity of storms and precipitation patterns will only amplify these problems. Sustainable urban water management techniques retain or divert water within the urban areas, where possible replenish groundwater aquifers, integrate water reuse for potable or nonpotable purposes, help slow down runoff to reduce flood risk downstream, and resist coastal surges in urbanized coastal settlements. High energy and cost requirements to treat the declining quality of surface and ground wastewater also challenge the overall sustainability of water infrastructure systems. While this chapter largely addresses the controlled or engineered water system, the Landscape infrastructure chapter addresses the topic of urban water management in more detail. Synergy of ecological elements is the most critical strategy for planning of natural or bioengineered systems in response to urban water challenges.

ENVIRONMENTAL

Challenges for environmental sustainability include:

- Integrating ecology in infrastructure planning and design
- Attention to the carrying capacity of the watershed in floodplain management
- Suitable adaptation of global technological solutions
- Cognizant monitoring of chemical and biological discharge in receiving waters
- Embracing comprehensive soft infrastructure approaches for urban water management

SOCIAL

Challenges for social sustainability include:

- Achieving water security for livelihood as well as ecosystem support
- Protecting public and property from flooding hazards
- Positively influencing consumer behavior to create a stimulus for water conservation
- Overcoming traditional mindsets working around the infrastructure and institutional structures
- Encouraging comprehensive public knowledge of indoor versus outdoor water quality need

ECONOMIC

Challenges for economic sustainability include:

- Acknowledging water first as a social commodity and then as an economic commodity
- Adapting ecological economies for water, as provider of ecosystem services as well as natural capital
- Helping people understand the true cost of water
- Prioritizing financial investment for development of new infrastructure and retrofit or decommissioning of dysfunctional infrastructures
- Generating healthy market competition through inclusive criteria for issuing bonds in water investment

STRATEGIES TOWARD A SUSTAINABLE WATER SYSTEM

SYSTEM PURPOSE: RELIABLE ACCESS TO WATER	**WATERSHED MANAGEMENT:** Preserve long-term renewability and quality of water resources
	WATER SAVING: Reduce water consumption
	WATER RECOVERY: Reuse water with fit-for-purpose treatment
	SYSTEM EFFICIENCY: Minimize leakages

STRATEGIES TOWARD A SUSTAINABLE WATER SYSTEM

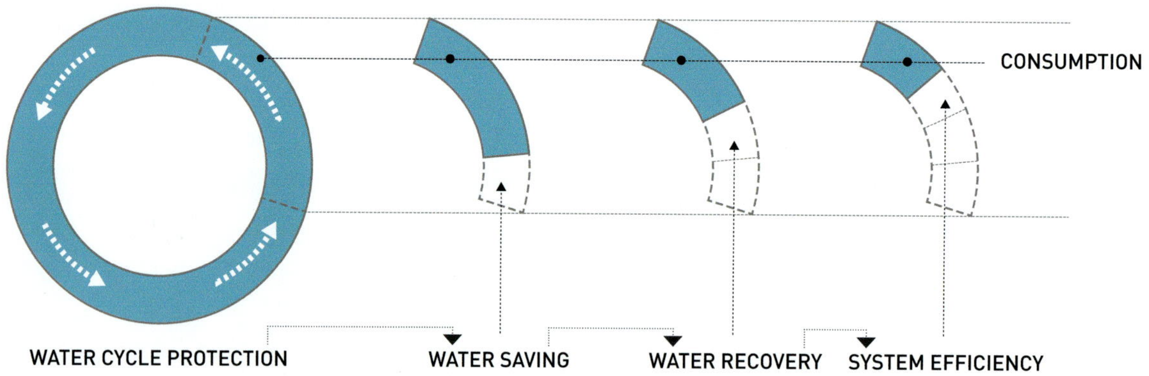

The primary purpose and service of Water infrastructure systems is to provide equal and reliable access to water. The structures and operations put in place create several sustainability challenges, as discussed in the previous section. The strategies listed above suggest basic steps that could be included in the planning of a water infrastructure system. They do not aim to outline all the requirements for a sustainable system but instead highlight approaches that if implemented could resolve multiple issues simultaneously.

Water is a valuable and vitally important resource; consequently the first strategy of a water infrastructure system should be to protect and enhance water resources. This is only possible by setting and implementing a careful watershed management plan that preserves the long-term renewability and quality of water. Water follows a hydrological cycle of evaporation, condensation, precipitation, and percolation or transpiration and uses streams, rivers, lakes, the water table, and the oceans during this process. A successful watershed management strategy aims to minimize any intervention in and disruption of the water cycle and in some cases even to restore its function. It includes measures to protect water resources from contamination, degradation, or depletion, changes in flow, landscape alterations, and any change in biodiversity caused by human activities. Protecting the watershed ensures the long-term availability of high-quality water that requires less infrastructure and consumes fewer resources in water treatment and conveyance. It also reduces flood risk to people, property, and infrastructure.

In order to reduce intervention in the water cycle to a minimum, the resources used to cover water consumption needs should also be kept to a minimum. The more efficient strategy toward that end is water saving, which refers to all those policies, strategies, and technologies used to reduce unnecessary consumption of water. Water saving applies to domestic use by changing consumers' behavior and habits and promoting appliances and fixtures that incorporate water-saving techniques. It also applies

THE NATURAL WATER CYCLE

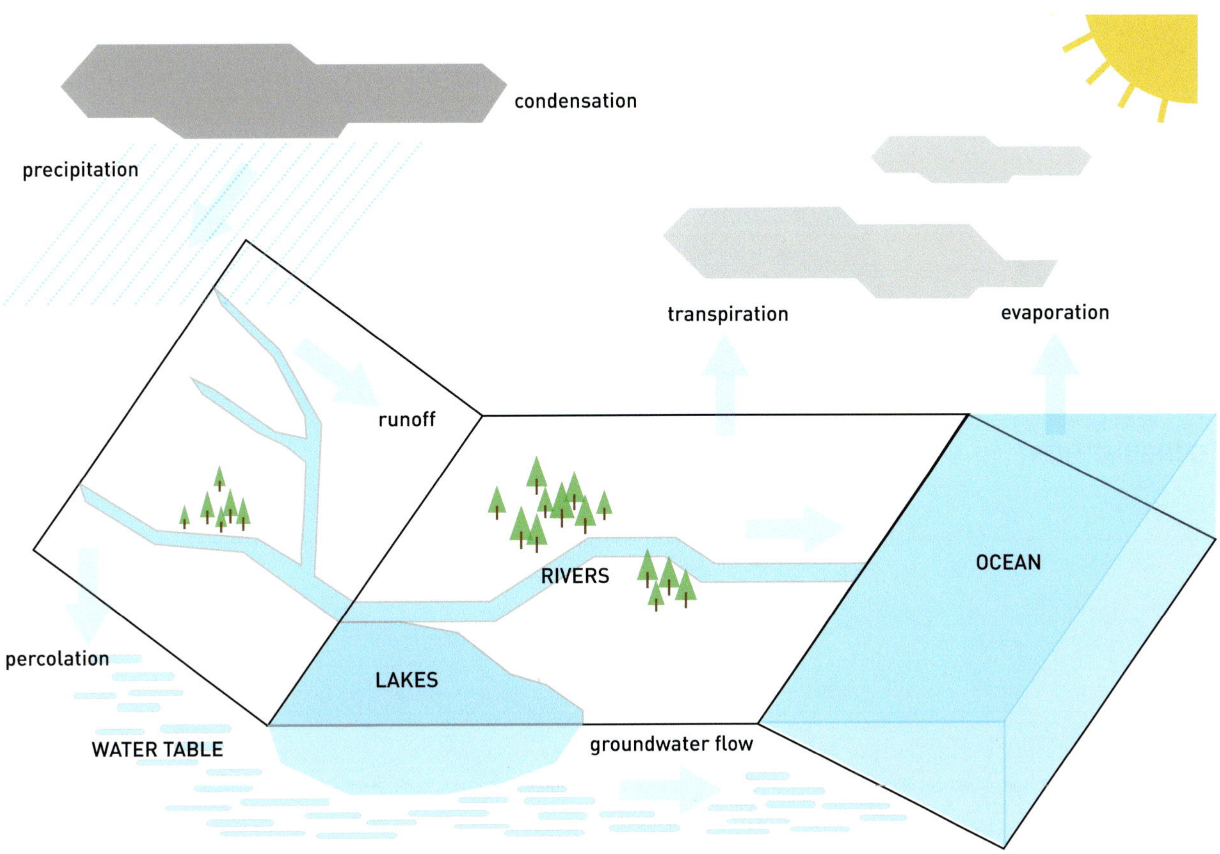

to landscaping and irrigation by introducing water-saving irrigation technologies, like drip irrigation, and promoting the use of climate-adapted plants. Finally, it applies to the energy generation sector, a sector that consumes large amounts of water, by using alternative sources such as wind and solar instead of water-intensive thermal plants, and by reducing energy consumption in general.

Another strategy that significantly reduces freshwater demand is water recovery. By recycling gray water or treated wastewater or by harvesting rainwater, the pressure on freshwater resources is reduced substantially. Recycled water can replace freshwater in several uses like irrigation, street cleaning, firefighting, industrial processes, etc. Furthermore, the technology already exists for direct conversion of wastewater to high-quality potable water. This process currently has limited application due mainly to a high degree of public skepticism. However, increasing water scarcity combined with raised public awareness and familiarization will make direct conversion of wastewater to potable water a more common process in the near future.

Finally, water demand and the consumption of other valuable resources like energy and construction materials can be further reduced by increasing a system's efficiency. Due to leakages, large amounts of water are lost in water networks before even reaching consumers. Leakages are the result of poor design, deteriorating aging infrastructure, and lack of metering. Information infrastructure technologies can play a key role here. Installation of information applications can track consumption in real time through remote meter reading. Real-time monitoring reinforces the detection of leaks and can reduce the time of intervention. Consumption patterns can be identified, providing valuable input and contributing to more efficient resource management. Optimization of network configuration planning combined with efficient maintenance can limit materials usage, extend useful life, and significantly reduce energy consumption in conveyance and treatment.

6.3 | WATER SYSTEM DECODING

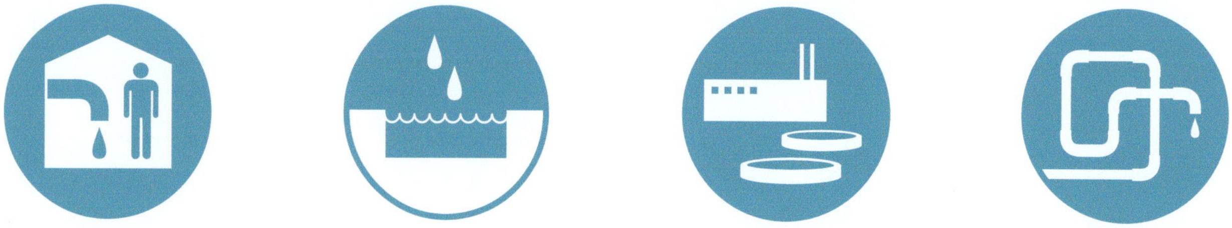

The four system levels used to decode the Water infrastructure system are Consumption, Supply, Treatment, and Networks.

Consumption, the first system level, provides an analysis of the demand to be served by the infrastructure system, examining and monitoring the main factors that affect consumption in conjunction with all available water-saving or water conservation techniques. It categorizes the main water consumption sectors and suggests water conservation methods for each one. Emerging planning and management practices refer to system operators, while engineering and behavioral practices refer to system users, i.e., consumers. The second system level, Supply, represents the basic decisions concerning the main strategies, goals, and standards set for the water supply system. It explores all possible supply sources and the main environmental impacts accompanying their use. In short, it sets the framework for a secure and uninterrupted water supply through a sustainable watershed management. The third system level, Treatment, analyzes the treatment processes and quality standards required for different water uses. It explores the environmental benefits and impacts of each method and deals with issues of by-product handling. Purification of freshwater for potable use, sewage treatment for water reuse or discharge to the natural environment, desalination, and stormwater management through landscape topography for pollutant removal and flood prevention are the main processes covered by this level. Finally, the fourth system level, Networks, deals with various distribution or collection grids and their relevant and usually extensive infrastructure. Distribution and collection grids include potable water networks, recycled nonpotable networks, sewage networks, and stormwater management networks. Resilience, coverage, quality of distributed water, and high efficiency are the most important factors for this level. The configuration of physical infrastructure—the layout, hierarchy, and components of networks—is also explored in this section. Benefits and drawbacks of commonly used distribution methods are analyzed, and the most sustainable, cost-effective, efficient, and reliable solutions are highlighted.

WATER CONSUMPTION

The Consumption level examines the demand for water per person or per urban sector. Water consumption is determined by a number of external factors and varies significantly between different regions and cities. Such factors include economic prosperity and growth, demographic changes, technological developments, and sociocultural values. Urban water consumption may be divided into four main sectors: residential, commercial, industrial, and landscape and agriculture.

The worldwide rapid increase in water demand is putting a strain on the planet's water resources and escalates environmental degradation. Thus water conservation becomes necessary, and water-saving strategies should play a decisive and primary role in the design and implementation of a sustainable water infrastructure system.

Reducing water consumption helps sustain the water cycle and brings other significant benefits, including habitat conservation, reduced need for infrastructure, and reduced energy consumption. To sustain the water cycle, the withdrawal of freshwater from an ecosystem should not exceed its natural replacement rate. Sustaining the water cycle preserves freshwater habitats and local wildlife resources found in them. Moreover, reducing water consumption means that less supply infrastructure like dams or aqueducts will be needed, saving resources and further protecting natural habitats from degradation or fragmentation. Finally, conveying water from the natural environment to cities consumes significant amounts of energy for pumping, delivery, or treatment; therefore any reduction in water demand results in energy conservation.

Water conservation measures should target the two basic groups of water system stakeholders: system users (such as residential users, industries, farmers, etc.) and system operators (such as municipalities, state and local governments, and privately owned suppliers) and can be divided into two main categories based on the type of intervention. The first category includes measures that are based on engineering practices and usually increase the efficiency of the system and minimize losses due to leakages or malfunction. The second category concerns behavioral practices, measures based on educating, increasing awareness, and changing citizens' habits in the direction of responsible and water-saving consumption.[12]

GLOBAL FRESHWATER USE (WBCSD 2005)

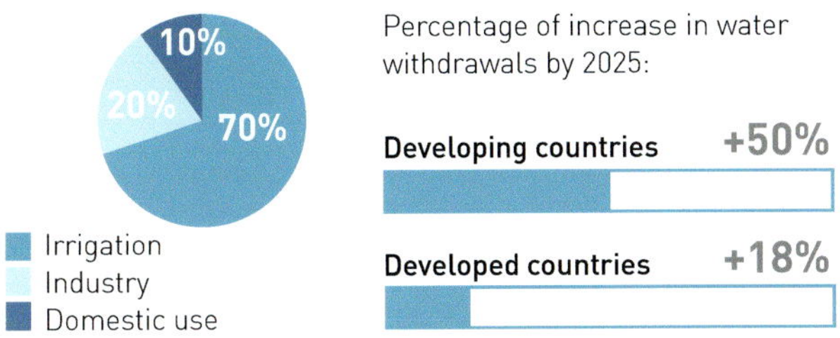

DOMESTIC WATER CONSUMPTION PER CAPITA (WBCSD 2005)

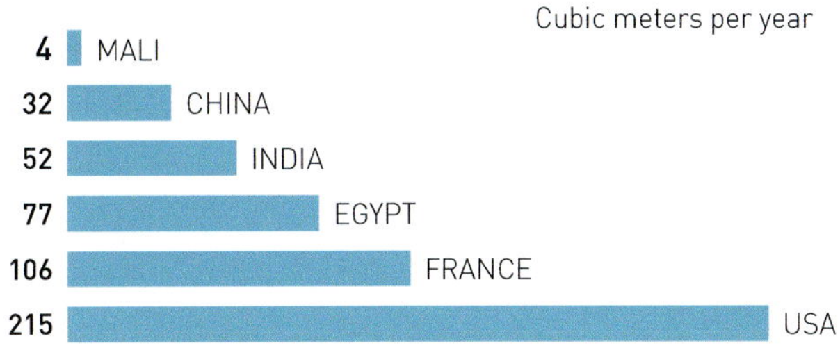

INDICATIVE INDOOR DOMESTIC WATER USE IN USA (WATERRF)

WATER CONSUMPTION SECTORS

MAIN WATER CONSERVATION METHODS

SYSTEM USERS	ENGINEERING PRACTICES	BEHAVIORAL PRACTICES
RESIDENTIAL COMMERCIAL	• **Indoor appliances and fixtures that save water** - Low-flush / Dual flush toilets - Low-flow showerheads - Faucet aerators - Pressure-reducing valves • **Decentralized gray water reuse** • **Decentralized rainwater harvesting**	• **Change water use habits**
INDUSTRIAL	• **Water reuse and recycling in industry** • **Cooling water recirculation**	• **Monitor water systems**
LANDSCAPE AGRICULTURE	• **Reduced landscape irrigation** - Low-water-requirement plants - Native vegetation – appropriate crops - Cycle irrigation - Scheduling lawn irrigation for early morning or evening hours - Low-precipitation-rate sprinklers – drip irrigation - Controlling runoff • **Climate-controlled irrigation** - Automated irrigation using weather data • **Xeriscape landscapes**	

SYSTEM OPERATORS	ENGINEERING PRACTICES	PLANNING AND MANAGEMENT PRACTICES
OPERATORS	• **Metering** (reduces consumption, aids in detecting leaks) • **Leak detection systems** • **Monitoring – Creation of database** • **Promoting water reuse and recycling**	• **Tiered pricing – higher rates on excessive water use** • **Retrofit programs – incentives to replace water-consuming fixtures** • **Public education programs – water use efficiency awareness** • **Planning for resource protection**

WATER SUPPLY

The Supply level has a dual significance. It includes the entire physical infrastructure used to convey water from sources to urban centers as well as the strategic decisions through which these sources are selected, decisions that heavily affect the sustainability of the system. The main purpose of this level is to ensure city water supply by protecting the watershed, exploiting resources within their regenerative capacity, minimizing energy consumed for conveying water, and resolving any issues of waste or by-products connected to the type of supply.

The main potable water supply sources are surface water conveyed from lakes and rivers, groundwater pumped from aquifers, and seawater following a desalination treatment. Alternative water supply sources—usually for nonpotable use—include recycled wastewater and harvested rainwater.

Surface water is the most commonly used source, with the vast majority of cities around the world exploiting adjacent or in many cases distant lakes and rivers to cover their needs for water. Overexploitation and energy consumed to convey water are the main sustainability issues regarding surface water. Overexploitation can result in drainage and then habitat loss, while conveying water from remote locations demands large amounts of energy and extended physical infrastructure and may lead ultimately to reliability/resilience issues. Moreover, dams used to control and collect river water result in fragmentation of river ecosystems, with many environmental and social side effects including disruption of river animals' migratory patterns, increased erosion, resettlement of communities, and a false sense of security from flooding, among others. Groundwater is also a commonly used and often abundant source of water supply. However, overdrawing has a significant adverse effect. Every aquifer has a different recharge rate depending on local climate conditions, vegetation, and soil composition. If the rate of withdrawal exceeds the recharge rate, the level of the water table declines, creating a risk of subsidence or saltwater intrusion. Saltwater intrusion is further aggravated by sea level rise and coastal erosion. As a result, deeper, more expensive and energy-consuming wells need to be drilled.

Surface and groundwater resources are interconnected, recharging and flowing to each other. Consequently, any degradation of one resource affects the other. The same applies to pollution, with surface water pollution eventually permeating into the groundwater aquifers. Saltwater after desalination is a useful source of water in coastal water-stressed areas. Disadvantages include the large amounts of energy required as well as the increased greenhouse gas emissions and the pollution in nearby water bodies from returning brine. It also remains a costly process.

On the other hand, recycled water provides an additional source of water for potable or nonpotable uses, with great environmental benefits. It prevents the depletion of freshwater sources while decreasing wastewater discharges and pollution. Decentralized rainwater harvesting is another sustainable solution for nonpotable-use water supply; it reduces flooding, stormwater runoff, erosion, contamination of surface waters, and energy use. Large-scale rainwater harvesting is not a popular practice, but it could become more popular in the future with the right incentives.

The physical infrastructure used to collect, convey, and store water from remote locations or the aquifers to urban centers includes dams, uncovered ground-level aqueducts, covered tunnels, underground water pipes, reservoirs, bore holes, wells, etc.

ACCESS TO SAFE DRINKING WATER (UNICEF/WHO 2012)

The last few decades have seen a constant increase in global population with access to safe drinking water. Climate change and population growth, though, are predicted to reverse this progress in the very near future.

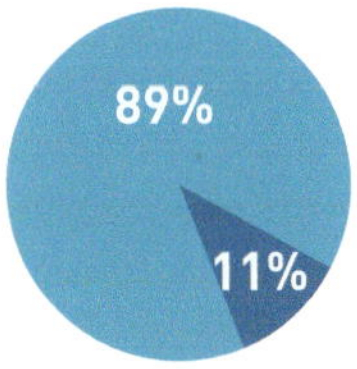

89% of global population has access to piped water through house connections or improved water sources (2010)

11% of global population has to cover their water needs through unprotected sources

WATER SUPPLY SOURCES

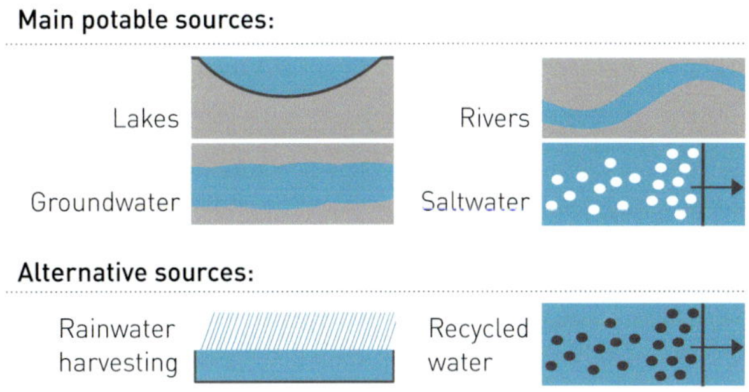

WATER TREATMENT

The Water Treatment level involves infrastructure and processes needed to make water quality acceptable for the desired end use. The main purpose of this system level is to secure reliable and flexible treatment as well as to resolve issues of waste or by-products connected to the type of treatment. It should protect ecosystems from any possible contamination by these by-products and prevent any disruption of natural processes. Equally important is protecting community quality of life from negative impacts such as noise and visual pollution, odors, etc. Finally, a significant target is minimizing resources consumed, especially energy, since many commonly used treatment processes demand large amounts of energy.

Besides the protection of the water source quality in the first place so that the minimum treatment is required, the main sustainability strategy should be the fit-for-purpose treatment. For example, treatment of water to drinking standards is expensive and consumes large amounts of energy, yet only a small fraction of water treated for potable use is actually used for drinking or food preparation. In addition, it is important before a choice is made to research all available options; new alternative treatment techniques based on innovation and high-tech approaches are constantly being developed that can significantly reduce the environmental impact of treatment processes.

The four most commonly used water treatment processes are water purification, desalination, sewage treatment, and stormwater treatment.

Through purification treatment, water becomes suitable for potable use. In urban centers the process takes place in water treatment plants, and the main objectives of potable water treatment are to remove any toxic or health-hazard materials, remove or inactivate disease-producing organisms, provide a water of consistent quality, enhance the aesthetic acceptability of the water, and achieve these goals at reasonable cost.[13] Water purification for drinking purposes usually follows quality standards set by each government and the World Health Organization (WHO). There are many different types of treatment, and the one selected depends on the quality of the water source. Common stages of water purification processes are coagulation, where dirt and other particles suspended in water stick together with the addition of chemicals to form large particles; sedimentation, where these large particles sink under their heavy weight and settle at the bottom; filtration, where water passes through several filters for the removal of any smaller particles; and finally disinfection, for the elimination of any bacteria or microorganisms in the water. Water is then stored in closed tanks or reservoirs and is ready for distribution.

Desalination is the process that converts saline water to potable water. As already mentioned, it can be a valuable and unlimited source of water in arid coastal areas; it is still, however, a costly and energy-intensive process. It takes place at desalination plants, and the most common processes are distillation and reverse osmosis.

Sewage treatment removes contaminants from wastewater, making it suitable for reuse or for return to the natural environment without adverse ecological impact. While conversion of wastewater to potable water is technologically feasible, recycled water is usually used for nonpotable uses due to public skepticism as well as high cost and energy requirements. At some plants, there is indirect use of treated wastewater for potable use: treated wastewater is recycled back into the water system that supplies water treatment facilities, thus short-circuiting the natural water cycle. Sewage can be treated close to where it is created, in individual properties or local neighborhood installations (a decentralized system), or it can be collected and transported through a sewage network to a large treatment plant (a centralized system). Most urban centers use a centralized system in which sewage is treated in industrial-scale sewage treatment plants (STPs). Decentralized systems, though, can provide a more sustainable solution, since they require far less infrastructure and reinforce resilience.

There are numerous techniques that can be used for sewage treatment depending on the type and extent of contamination, including physical, chemical, and biological processes. Four main stages can usually be identified: pretreatment for screening and grit removal, primary treatment for sedimentation and removal of sludge or floatables, secondary treatment for degradation of biological content, and tertiary treatment for any additional requirements such as filtration, nutrient removal, disinfection, etc. All commonly used sewage treatment processes produce a residual sludge usually treated through the waste infrastructure system. For large-scale operations anaerobic digestion is advisable, while for midsize operations aerobic digestion and for small-scale operations composting. Incineration is also a possible but not a common option, while disposal of sludge in landfill is the least preferred and less sustainable option.

The purpose of stormwater treatment is to prevent

MAIN WATER TREATMENT PROCESSES

SUPPLY	TREATMENT AND TREATMENT ENTITIES	COMMON USE
Freshwater	**Water purification** (water treatment plant)	Potable water
Seawater	**Desalination** (desalination plant)	Potable water
Wastewater	**Sewage treatment** (sewage treatment plant)	Nonpotable water/ returning to natural environment
Stormwater	**Remove of pollutants** (retention ponds and artificial wetlands)	Returning to natural environment

COMMON STAGES OF DRINKING WATER PURIFICATION

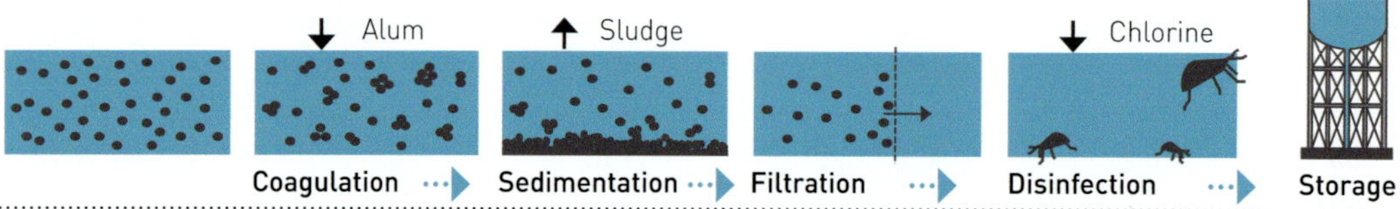

urban runoff from reaching the natural ecosystem. Urban runoff from buildings, impermeable open spaces, and especially motorways may contain several contaminants, and any outflow to the natural drainage system should be prevented. Integration of stormwater management with the sewage system should also be avoided as it unnecessarily increases the volume of water treated through energy-intensive sewage treatment plants and in many cases causes combined sewage overflows (CSOs) that highly compromise the quality of receiving waters. In general, a sustainable stormwater treatment system should make use of natural topography, increase infiltration, prevent surface flooding in a cost-effective way, remove pollutants, and create valuable public amenities. For runoff from buildings, the use of green roofs combined with a rain-harvesting system is the ideal solution. Water collected in a rainwater tank could be used for nonpotable purposes like toilet flushing or irrigation. For urban public space the use of permeable materials to increase on-site infiltration to the furthest possible extent and consequently aquifer recharge should be the first priority. The surface runoff that cannot be absorbed should ideally be directed to artificially constructed wetlands, retention ponds, and vegetated swales that can naturally infiltrate pollutants. Since the volume generated during storms can be tremendous, it is important to capture and treat at least the first flush of runoff, which is usually the most contaminated. As already mentioned, stormwater treatment and networks should be planned and resolved in synergy with Landscape infrastructure.

BENEFITS AND LIMITATIONS OF WATER TREATMENT PROCESSES

BENEFITS	LIMITATIONS
Water purification	
• Health improvement • Prevents waterborne diseases	• Energy consumption • Plants should be located between supply and consumption centers to reduce infrastructure network and energy needs for conveyance
Desalination	
• Endless water source for coastal water-stressed areas	• High energy consumption • High cost • Pollution by emissions and returning brine • Noise pollution • Buffer zone needed, especially in residential areas
Wastewater treatment	
• Potable water conservation through reuse for nonpotable uses • Prevents pollution of water bodies	• Energy consumption • Odors • Treatment of residual sludge • Sludge disposal • Buffer zone needed, especially in residential areas • Strong opposition by local communities
Stormwater treatment	
• Prevents flooding • Protects water bodies from urban runoff contaminants • Reduces erosion • Increases infiltration and groundwater recharge	• Low permeability in dense urban centers

SUSTAINABLE STORMWATER MANAGEMENT

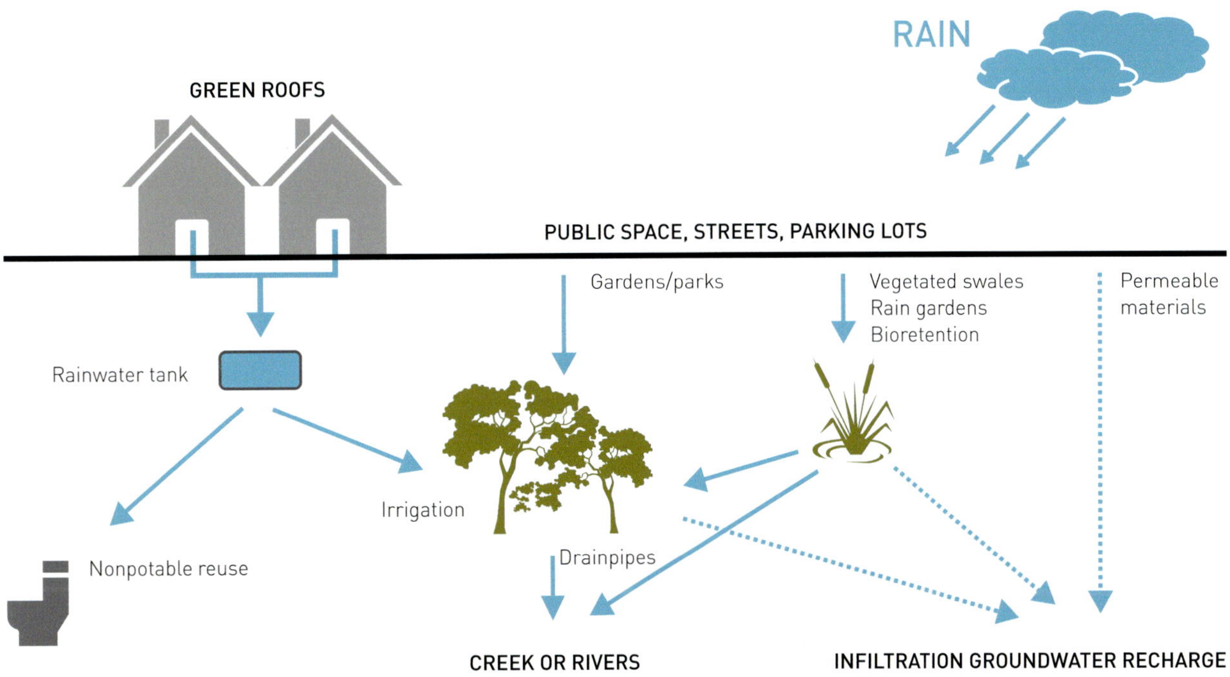

WATER NETWORKS

MAIN WATER DISTRIBUTION NETWORK SYNOPSIS

TYPE OF WATER	ORIGIN	DESTINATION
Potable water	Purification plant	Consumers
Recycled/ untreated freshwater	Wastewater plant/ supply sources	Consumers/ irrigation/discharge
Wastewater	Consumers	Wastewater plant/ discharge
Stormwater	Urban runoff	Contructed wetlands/ discharge

WATER DISTRIBUTION NETWORK CONFIGURATION

Loop topography

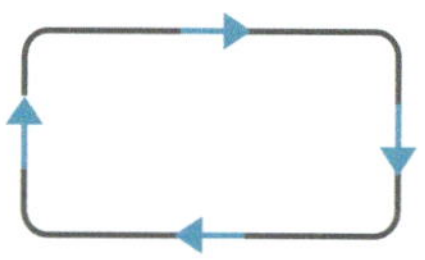

Branch topography

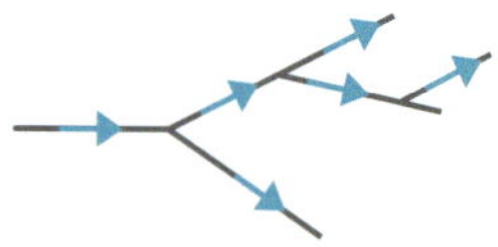

The Networks level examines the entire essential infrastructure used to convey and distribute different streams of water through a system of storage tanks, pipes, valves, and pumps. Networks should be able to provide uninterrupted service, ensuring that water delivered to consumers is always at appropriate quality and suitable for the intended use.

Four main water distribution/collection networks can be identified based on the quality and end use of the water they convey: potable water networks, recycled or untreated freshwater networks, wastewater networks, and stormwater networks.

Potable water networks convey water from purification plants to consumers; they are usually the most extended networks and surely the networks with the highest priority in the entire water infrastructure system.

Recycled water networks convey water from wastewater treatment plants to consumers for nonpotable uses, for public irrigation, or for discharge to the natural environment. Potable and recycled water supply networks should be able to protect water from contamination and preserve water quality at all stages, supply water at any intended point of the network at adequate pressure head, allow easy maintenance and repairs without disrupting service, minimize losses due to leakage, and finally be able to cover firefighting needs.

Wastewater networks convey sewage to wastewater treatment plants. Sewage networks should be able to quickly remove wastewater from source points, avoid any leakage that might contaminate soil, water bodies, or the water supply stream, and allow easy maintenance and repairs without disrupting service.

Stormwater networks manage urban runoff from the source points to constructed wetlands or discharge to the natural environment. Stormwater drainage networks should be able to manage stormwater flows and protect from flooding, avoid discharge of polluted urban runoff to rivers and lakes, and increase infiltration. Ideally, stormwater networks should strive for surface conveyance in green infrastructure rather than piped networks. Their design should aim as far as possible to mimic the natural runoff processes that would exist on site prior to development.

Water networks are generally underground, usually under pavement in urban areas combined with other utility service networks, and thus follow the road layout. They may have a loop or a branch network configuration, or a combination of the two. Loop configuration is the most favorable as it allows the isolation of one part of the system without disrupting service and avoids dead ends where stagnation, staleness, and possible damage to water quality can occur.

There are three methods of water distribution: gravity systems, pumping systems, and a combination of the two. Gravity systems take advantage of the topography; the supply source is located at the higher elevation of the city and water gets distributed by gravity. It is the most sustainable option since it requires a minimum amount of energy but can only be applied if there is suitable terrain.

In pumping systems water is pumped directly into the mains and distributed to households. It is a less sustainable solution since it requires large amounts of energy for pumping and water supply depends entirely on power supply.

The combined gravity and pumping system is the most popular and commonly used method. Water obtains the right pressure and gets distributed in the city through a combination of pumps and elevated reservoirs. It is the most cost-effective, efficient, and reliable system for water distribution when topography doesn't allow the use of the gravity method.

WATER DISTRIBUTION NETWORK HIERARCHY

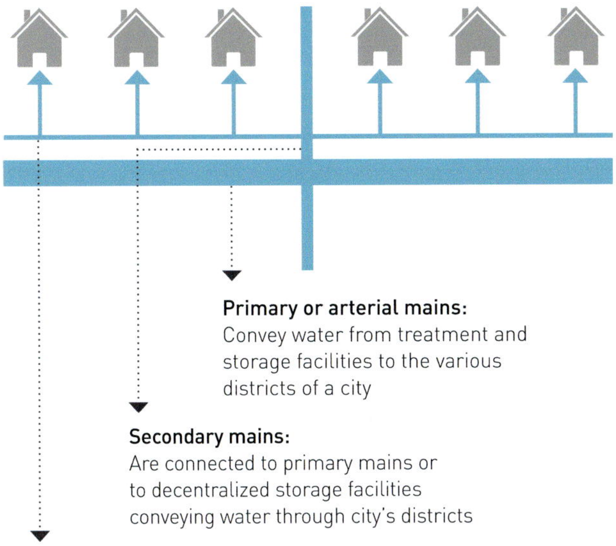

Primary or arterial mains:
Convey water from treatment and storage facilities to the various districts of a city

Secondary mains:
Are connected to primary mains or to decentralized storage facilities conveying water through city's districts

Distribution mains:
Convey water from secondary mains to individual consumers

WATER DISTRIBUTION METHODS

Gravity system

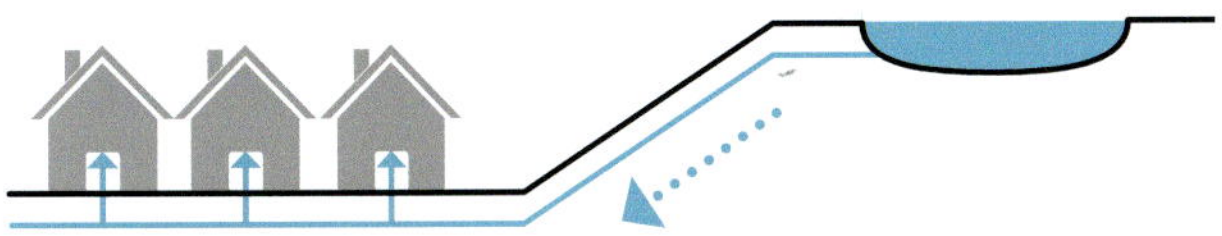

Pumping system

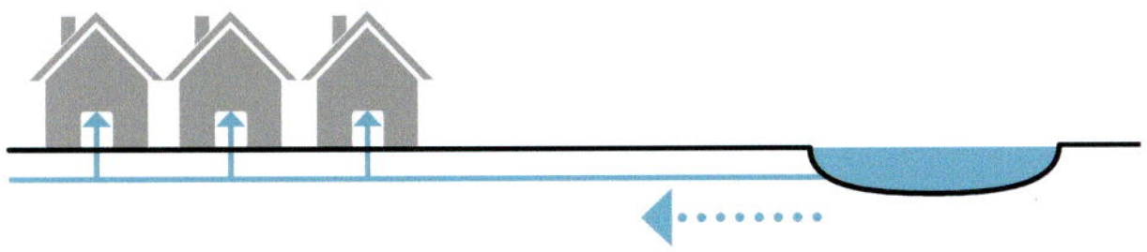

Combined gravity and pumping system

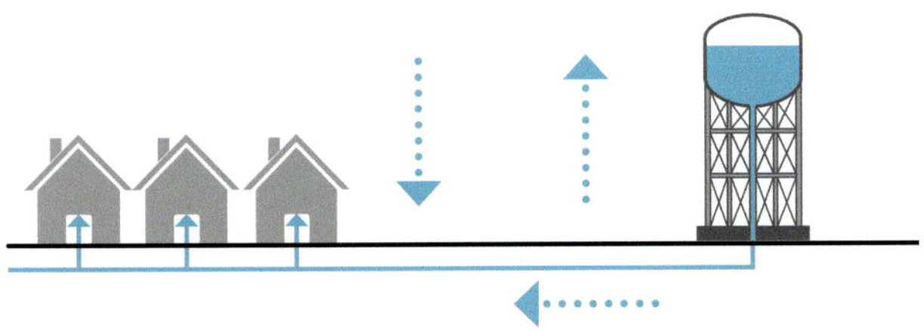

6.4 | OBJECTIVES AND GUIDELINES FOR A SUSTAINABLE WATER SYSTEM

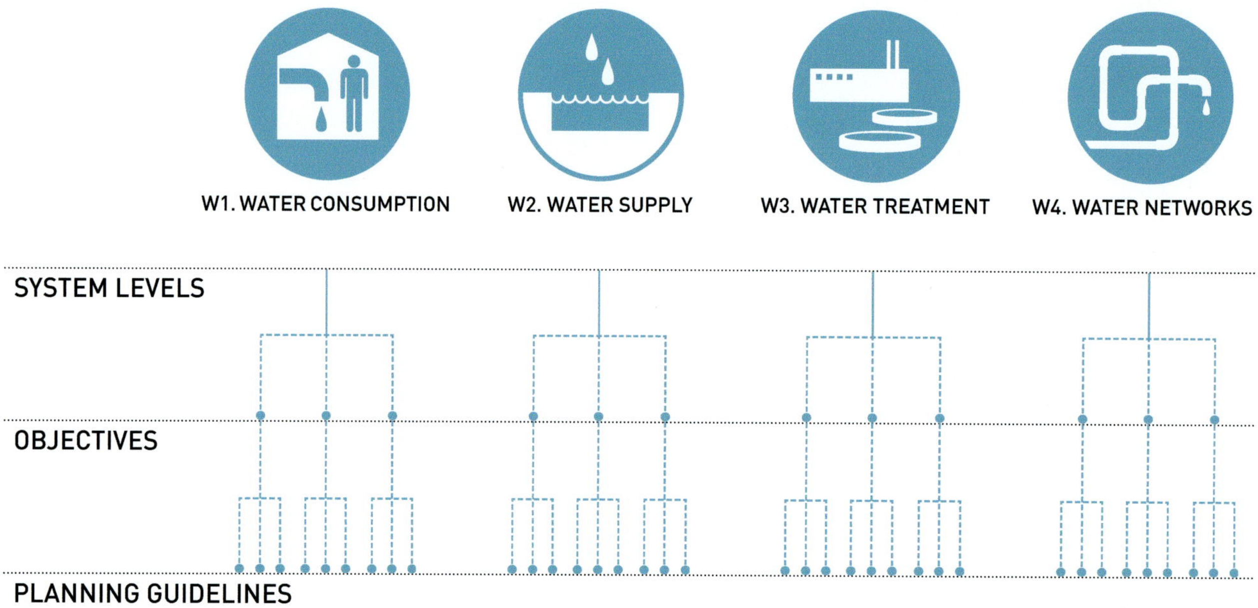

Following the decoding of the Water system on four system levels, planning objectives are defined for each level. They represent both challenges and strategies toward a sustainable system, expressing what should be the overall final outcome of water planning.

The objectives are associated with the five focus categories of the Zofnass Program, namely Leadership, Quality of Life, Natural World, Resource Allocation, and Climate and Risk. The Sustainable Planning Guidelines aim to lead toward these objectives, highlighting synergies and connections with the other six infrastructure categories of Energy, Information, Solid Waste, Landscape, Transportation, and Food.

The objectives and guidelines of the Water Consumption level are related to the Leadership category and primarily cover two factors: methods to reduce consumption and ways to monitor and regulate it. In the building sector, water reduction planning includes strategies such as the guiding and motivation of communities in rational use of water, the adoption of long-term goals and standards for water saving, and the promotion of on-site rainwater harvesting and reuse of gray water. In other water-intensive sectors like energy or food production and landscape maintenance, the main goal is the use of water-efficient technologies and techniques. It is obvious that strong synergies occur with the respective infrastructure systems of Energy, Landscape, and Food. Another important synergy at this level occurs with the Information infrastructure systems and has to do with monitoring water consumption trends and regulating water consumption rates and patterns. Guidelines for monitoring water consumption include the collection and organization of data on water consumption, the daily and seasonal monitoring of consumption fluctuations, as well as the identification of opportunities for use of different nonpotable water qualities. For better water consumption regulation, the reduction of peaks and the correlation of consumption rates according to climate pattern data are essential.

W1. WATER CONSUMPTION

- W1.1. Reduce Water Consumption in the building sector
- W1.2. Reduce Water Consumption in the Energy system
- W1.3. Reduce Water Consumption in the Landscape system
- W1.4. Reduce Water Consumption in the Food system
- W1.5. Reduce Water Consumption in other context-specific sectors with significant water demand
- W1.6. Monitor Water Consumption trends
- W1.7. Regulate Water Consumption rates and patterns

The Water Supply level primarily deals with the important strategic decisions that define a city's water supply sources and indicates how these sources can be exploited in the most efficient and sustainable way without degrading watersheds. Respectively, the objectives of this level refer to the categories of Leadership, Resource Allocation, and Natural World. The most important target of this level is to secure water supply, and in order to do this all available sources of water, preferably within or in proximity to the city's boundaries, should be identified, and their possible contribution to the city's supply mix should be evaluated in the long term. The promotion of alternative water supply options such as water recycling and recovery is equally important, while the financial viability of the system in the long term should not be overlooked. Clear goals to be achieved through water supply selection should be set, integrating long-term city planning and the city's sustainability performance goals with water supply planning. Another important objective of this level is the sustainable management of the city's watershed. Guidelines here call for a thorough understanding of the watershed's complexity and several different functions before developing an integrated management program. Basic sources of contaminants from any human activity within the watershed should be identified and controlled. The same applies to contaminants reaching the watershed through urban runoff water, where controlling methods like bioretention and bioinfiltration should be implemented. Of course these methods should be part of an integrated rainwater and stormwater management system that makes use of natural drainage, increases infiltration, and provides for aquifer recharge. Synergies here occur with the Transportation, Information, Solid Waste, Food, and especially Landscape infrastructure systems. Objectives connected to the Resource Allocation category provide guidelines for exploitation of available water resources within their regenerative capacities and ways to minimize energy consumed in providing water supply. Water supply, especially when remote sources are used, demands large amounts of energy, and obviously a synergy with Energy infrastructure occurs. To reduce energy consumption, efforts should be made to minimize the distance between water supply and consumption, mainly by exploiting local

W2. WATER SUPPLY

- W2.1. Secure city Water Supply in the long term
- W2.2. Define clear goals to be achieved through Water Supply planning
- W2.3. Manage the city's watershed
- W2.4. Exploit available Water resources within regenerative capacities
- W2.5. Minimize Energy consumed for Water Supply
- W2.6. Resolve issues of residues and by-products connected to the types of Water Supply used

sources, combined with the adoption of the most energy-efficient water supply methods.

The Water Treatment level has a dual significance, first to protect public health and second the natural environment from the adverse impacts of untreated supply water and wastewater streams. Objectives of the Treatment level refer to all five of the Zofnass Program focus categories. In the Leadership category the main objectives point to securing reliable and flexible water and wastewater treatment services in the long term, the importance of proper siting and distribution of treatment facilities for optimized performance of the water system, as well as the integration of water and wastewater treatment with other infrastructure systems. Integration with other infrastructure systems includes, for example, the management of treatment residues, like sludge, through the Solid Waste infrastructure system or any potential use of these by-products for energy generation. In the Quality of Life category, objectives focus on protecting community well-being from any health impacts, odors, or noise pollution caused by treatment facilities as well as protecting quality of space from fragmentation or any possible visual impact. It is a fact that poorly planned treatment facilities can cause significant nuisance to adjoining communities, leading to degradation and further reinforcing the not-in-my-backyard response. That is why it is important to ensure appropriate siting, the creation of adequate buffer zones, and the adoption of compensatory measures. The main objective in the Natural World category focuses on maintaining the integrity of natural elements and systems. Guidelines here call among other things for protection of natural habitats and wildlife corridors and compensation for their fragmentation or disruption by treatment facilities, and for minimizing landscape alteration during construction. In Climate and Risk, providing resilience against natural and man-made phenomena and especially periods of extreme climatic conditions is again a key for securing the continuous and smooth operation of the system. Equally important is controlling and compensating for greenhouse gas emissions, since many of the commonly used treatment technologies have a significant carbon footprint. In the Resource Allocation category, objectives mainly concern the management of residuals and energy conservation. Treatment residuals and by-products should ideally be processed on site; otherwise methods and routes for transfer to processing facilities should be carefully planned and resolved in synergy with Transportation infrastructure. Any material or energy recovery potential should be thoroughly examined in coordination with the Solid Waste and Energy infrastructure systems. Finally, optimization of energy consumption is highly important, since many of the commonly used treatment processes demand large amounts of energy to operate. The most energy-efficient technologies should be promoted.

W3. WATER TREATMENT

- W3.1. Secure the reliability and flexibility of Water Treatment services
- W3.2. Place and distribute Water Treatment facilities for optimized Water system performance
- W3.3. Integrate Water Treatment with other infrastructure systems
- W3.4. Promote community well-being
- W3.5. Enhance quality of space
- W3.6. Maintain the integrity of natural-world elements and systems
- W3.7. Prepare for short-term hazards and projected future threats
- W3.8. Control greenhouse gas emissions
- W3.9. Resolve issues of residues or by-products connected to each type of Water Treatment
- W3.10. Optimize Energy use in Water Treatment

On the Networks level, objectives relate again to all five of the Zofnass Program focus categories. In the Leadership category the importance of reliable and widespread water transfer and distribution networks is pinpointed. The guidelines focus on the provision of an adequate number of separate grids with the required coverage and length while always preserving the high quality of distributed water. Networks should be able to adapt and support decentralized water and wastewater systems and provide adequate capacity to cover long-term city needs. Storage and transfer facilities should also be provided adequate to cover peaks and any possible long-term expansion of the system. In a similar way rainwater and stormwater management needs should be covered through adequate networks, preferably by utilizing natural drainage. Guidelines also call for adequate administrative structures for network asset management and the long-term financial viability of the system. The stability of the network as well as the establishment of a maintenance plan combined with constant monitoring for leaks and overflow events are equally important. Synergy with Information infrastructure plays a key role for real-time monitoring and optimization of network operation. Another aspect of this category is integration with other infrastructure systems and coordination and compatibility with the city's service corridors, usually adapted to its transportation network. In the Quality of Life category, objectives focus again on protecting community well-being from health impacts or odors connected to wastewater collection and transfer. Protecting quality of space is also involved through reducing visual impact and minimizing land use and fragmentation of the urban fabric by water and wastewater networks. In the

Natural World category, the objectives point to minimizing disruptions of natural systems by avoiding landscape alterations and protecting habitats and wildlife corridors, maintaining the water cycle, and especially protecting soil, surface water, and groundwater from contamination due to wastewater network leakages or malfunctions. In Climate and Risk, providing resilience against natural and man-made phenomena is again key for securing the continuous and smooth operation of the system. Potential risks should be assessed, and the long-term adaptability of the water system to climate change trends and other hazards should be anticipated and addressed. Because of the extended physical infrastructure of the Networks level, Resource Allocation objectives are of special significance. Objectives call for optimization of energy and material use in water networks. Energy consumption for pumping and distribution can be notably reduced by taking advantage of landscape configuration. Besides energy saving, this strategy also increases water networks' resilience, since water supply becomes less dependent on power supply in case of an emergency. Materials required for water and wastewater networks should be sourced locally when possible, and a plan for reuse and recycling after decommission should be considered.

The Objectives of the respective Guidelines as well as the relevant Guidelines of other infrastructure systems are presented in the following tables. (TO) and (FROM) indicate that the guideline can assist or can be assisted by the other infrastructure system, where the arrow is placed. (WITH) indicates that the guideline might require planning for an Information or Landscape subsystem and should follow multiple Guidelines from these two systems.

W4. WATER NETWORKS

- W4.1. Secure the reliability and flexibility of Water transfer and distribution services
- W4.2. Prevent leaks and overflow events in Water Networks
- W4.3. Integrate Water Networks with other infrastructure systems
- W4.4. Promote community well-being
- W4.5. Enhance quality of space
- W4.6. Maintain the integrity of natural-world elements and systems
- W4.7. Provide for long-term adaptability of Water Networks to climate trends
- W4.8. Prepare for short-term hazards and projected future threats
- W4.9. Optimize Energy use in Water Networks
- W4.10. Optimize material use in Water Networks

GUIDELINES

W1 WATER CONSUMPTION

▲ TO
✚ WITH
▼ FROM

OBJECTIVES	GUIDELINES		SYNERGIES
W1.1. Reduce Water Consumption in the building sector	W1.1.1.	Guide, assist, and motivate communities for rational use of water	
	W1.1.2.	Set long-term goals and standards for reducing Water Consumption in the building sector	
	W1.1.3.	Promote on-site harvesting of rainwater and recycling of water	
W1.2. Reduce Water Consumption in the Energy system	W1.2.1.	Reduce water volume required for Energy Supply and Energy Generation	▼
W1.3. Reduce Water Consumption in the Landscape system	W1.3.1.	Reduce water required for Landscape Maintenance	▼
W1.4. Reduce Water Consumption in the Food system	W1.4.1.	Reduce water required for Food Production and Food Processing	▼
W1.5. Reduce Water Consumption in other context-specific sectors with significant water demand	W1.5.1.	Identify additional sectors with significant water demand and develop efficiency strategies	
W1.6. Monitor Water Consumption trends	W1.6.1.	Monitor daily and seasonal Water Consumption fluctuations	✚
	W1.6.2.	Regularly collect and organize quantitative and qualitative data on Water Consumption	✚
	W1.6.3.	Identify opportunities for use of different nonpotable water qualities	▼ ▼ ▼ ▼ ▼ ▼
W1.7. Regulate Water Consumption rates and patterns	W1.7.1.	Reduce peaks in water demand	
	W1.7.2.	Regulate Water Consumption rates based on climate patterns	

W2 WATER SUPPLY

▲ TO
✚ WITH
▼ FROM

OBJECTIVES	GUIDELINES		SYNERGIES
W2.1. Secure city Water Supply in the long term	W2.1.1.	Identify all available sources of Water Supply	
	W2.1.2.	Promote use of multiple and diverse water sources to cover city needs in the long term	
	W2.1.3.	Promote water recycling and recovery	
	W2.1.4.	Promote use of Water resources available within project boundaries	✚
	W2.1.5.	Identify opportunities to cover annual demand through seasonal water storage	
	W2.1.6.	Allocate available Water Supply according to varying quality requirements	

OBJECTIVES	GUIDELINES		SYNERGIES
	W2.1.7.	Trace correlation of alternative Water Supply sources with city competitiveness and economic sectors	
	W2.1.8.	Secure long-term financial viability of Water Supply plan	
W2.2. Define clear goals to be achieved through Water Supply planning	W2.2.1.	Set goals for Water Supply mix	
	W2.2.2.	Adopt quantitative indicators to benchmark and monitor the performance of the Water Supply plan	
	W2.2.3.	Integrate long-term city growth and development goals with Water Supply planning	
	W2.2.4.	Regularly update Water Supply planning according to community development and Water system performance	
W2.3. Manage the city's watershed	W2.3.1.	Map and monitor watershed and hydrological connections	+
	W2.3.2.	Develop a watershed management program	▲
	W2.3.3.	Identify and manage basic sources of contaminants within the watershed	+ ▼ ▼
	W2.3.4.	Control contaminants in runoff water through bioretention and bioinfiltration methods	+ ▼
	W2.3.5.	Make provisions for aquifer recharge	
	W2.3.6.	Provide for rainwater and stormwater management needs with natural drainage, infiltration, and floodplain management	+ ▼
W2.4. Exploit available Water resources within regenerative capacities	W2.4.1.	Adapt water resource usage to identified climate patterns and Water Consumption rates	
W2.5. Minimize Energy consumed for Water Supply	W2.5.1.	Minimize distance between Water Supply resources and Water Consumption locations	▲
	W2.5.2.	Promote energy-efficient Water Supply methods	▲
W2.6. Resolve issues of residues and by-products connected to the types of Water Supply used	W2.6.1.	Resolve Water Supply residues handling and disposal requirements	
	W2.6.2.	Define whether Water Supply and Water Treatment residues and by-products can be marketable	

W3 WATER TREATMENT

▲ TO
+ WITH
▼ FROM

OBJECTIVES	GUIDELINES		SYNERGIES
W3.1. Secure the reliability and flexibility of Water Treatment services	W3.1.1.	Develop decentralized Water Treatment systems	
	W3.1.2.	Adapt Water Treatment levels both to quality of supply and to varying demand requirements	
	W3.1.3.	Provide for adaptability to changes in quality of Water Treatment influent	
	W3.1.4.	Provide thorough environmental regulations for Water Treatment	
	W3.1.5.	Provide adequate Water Treatment capacity adaptable to long-term city needs	
	W3.1.6.	Prepare for Water Treatment failures or overflow events	▼

OBJECTIVES	GUIDELINES		SYNERGIES
	W3.1.7.	Develop smart, integrated management systems to optimize Water Treatment efficiency	+
	W3.1.8.	Secure long-term financial viability of Water Treatment systems	
	W3.1.9.	Optimize Water Treatment asset management through adequate administrative structures	
W3.2. Place and distribute Water Treatment facilities for optimized Water system performance	W3.2.1.	Optimize placement of Water Treatment facilities so as to minimize distances between treatment and Water Consumption	▲
	W3.2.2.	Optimize placement of Water Treatment facilities with regard to Network routes and flows	
	W3.2.3.	Reserve spaces for Water Treatment facilities taking into consideration long-term city development	
W3.3. Integrate Water Treatment with other infrastructure systems	W3.3.1.	Integrate Water Treatment facilities with Solid Waste infrastructure for leachate treatment	▲
	W3.3.2.	Integrate Water Treatment and Energy Generation facilities if applicable	▲
W3.4. Promote community well-being	W3.4.1.	Assess and resolve potential health impacts of Water Treatment facilities	
	W3.4.2.	Assess and resolve potential odor issues caused by Water Treatment facilities	+
	W3.4.3.	Assess and resolve potential noise pollution issues caused by Water Treatment facilities	+
W3.5. Enhance quality of space	W3.5.1.	Visually integrate Water Treatment facilities to the local context	+
	W3.5.2.	Minimize land use and fragmentation of urban fabric caused by Water Treatment facilities	
W3.6. Maintain the integrity of natural-world elements and systems	W3.6.1.	Ensure that quality of Water Treatment effluents is adequate for area of discharge	
	W3.6.2.	Minimize natural Landscape alteration caused by Water Treatment facilities	
	W3.6.3.	Protect natural habitats and wildlife corridors and compensate for their fragmentation or disruption by Water Treatment facilities	+
	W3.6.4.	Protect soil from contaminants found in Water Treatment facilities	+ ▲
W3.7. Prepare for short-term hazards and projected future threats	W3.7.1.	Assess and manage future threats from natural and man-made phenomena affecting Water Treatment facilities	
	W3.7.2.	Prepare action plans for periods of extreme water scarcity or high precipitation	+
	W3.7.3.	Plan response and recovery in case of emergency affecting Water Treatment facilities	
W3.8. Control greenhouse gas emissions	W3.8.1.	Use low-carbon technologies for Water Treatment processes	
	W3.8.2.	Compensate for emissions incurred in Water Treatment through carbon offset and sequestration practices	+

OBJECTIVES		GUIDELINES	SYNERGIES
W3.9. Resolve issues of residues or by-products connected to each type of Water Treatment	W3.9.1.	Promote on-site processing of Water Treatment residues	▲
	W3.9.2.	Investigate using Water Treatment residues for material recovery	▲
	W3.9.3.	Identify potential for Energy recovery and Generation through Water Treatment residues	▲
	W3.9.4.	Resolve routes and methods for the collection, treatment, and disposal of Water Treatment residues and by-products	▲▼ ▼
W3.10. Optimize Energy use in Water Treatment	W3.10.1.	Use energy-efficient Water Treatment technologies	▲

W4 WATER NETWORKS

▲ TO
+ WITH
▼ FROM

OBJECTIVES		GUIDELINES	SYNERGIES
W4.1. Secure the reliability and flexibility of Water transfer and distribution services	W4.1.1.	Provide required coverage and length for Water Networks	
	W4.1.2.	Provide adequate Water Networks capacity to cover long-term city needs	
	W4.1.3.	Provide adequate number of separate water grids based on Supply and Treatment needs	
	W4.1.4.	Preserve quality of distributed water	
	W4.1.5.	Integrate Water Networks with adequate storage facilities to cover demand peaks	
	W4.1.6.	Adapt Networks to support decentralized Water systems	
	W4.1.7.	Reserve spaces for Water transfer and storage facilities taking into consideration long-term city development	
	W4.1.8.	Secure long-term financial viability of Water Networks	
	W4.1.9.	Optimize Water Networks asset management through adequate administrative structures	
W4.2. Prevent leaks and overflow events in Water Networks	W4.2.1.	Provide for operational flexibility of Water Networks	+
	W4.2.2.	Optimize flow pressure in Water Networks	+
	W4.2.3.	Optimize Water Networks routes and distances covered	
	W4.2.4.	Provide operators with real-time monitoring of Water Networks	+
W4.3. Integrate Water Networks with other infrastructure systems	W4.3.1.	Integrate Water Networks with Transportation Networks	▼
	W4.3.2.	Plan Water Networks to support the operation of Energy Generation facilities	▲
	W4.3.3.	Plan Water Networks to support the operation of Information Nodes facilities	▲
	W4.3.4.	Integrate Water Networks with Landscape entities	▼
W4.4. Promote community well-being	W4.4.1.	Assess and minimize potential health impacts of Water Networks	
	W4.4.2.	Assess and minimize potential odor issues caused by Water collection and transfer facilities and processes	+

OBJECTIVES	GUIDELINES		SYNERGIES
W4.5. Enhance quality of space	W4.5.1.	Visually integrate Water Networks to the local context	+
	W4.5.2.	Minimize land use and fragmentation of urban fabric by Water Networks	
W4.6. Maintain the integrity of natural-world elements and systems	W4.6.1.	Protect habitats and wildlife corridors and compensate for their fragmentation or disruption by Water Networks	+
	W4.6.2.	Minimize natural Landscape alteration for the construction of Water Networks	
W4.7. Provide for long-term adaptability of Water Networks to climate trends	W4.7.1.	Design Water Networks capacity according to climate trends	
W4.8. Prepare for short-term hazards and projected future threats	W4.8.1.	Assess and manage threats from natural and man-made phenomena affecting Water Networks	+
W4.9. Optimize Energy use in Water Networks	W4.9.1.	Take advantage of natural landscape configuration to minimize needs for pumping in Water Networks	▲
W4.10. Optimize material use in Water Networks	W4.10.1.	Source materials required for Water Networks locally when possible	
	W4.10.2.	Promote use of reclaimed and recycled materials in Water Networks construction	
	W4.10.3.	Plan for the reuse of elements or recycling of materials after Water Networks decommission	

RELEVANT GUIDELINES

W1 WATER CONSUMPTION

FROM ▸ ▸ TO + WITH

GUIDELINES		RELEVANT GUIDELINES	
W1.2.1.	Reduce water volume required for Energy Supply and Energy Generation	E2.4.2.	Take into consideration available Water resources when planning Energy Supply mix
		E3.11.1.	Promote water-efficient technologies in Energy Generation facilities
W1.3.1.	Reduce water required for Landscape Maintenance	L4.8.1.	Set long-term goals and standards for reducing Water Consumption in Landscape Maintenance
		L4.8.2.	Use native plants of low water requirements
		L4.8.3.	Construct and operate efficient Landscape irrigation systems
		L4.8.4.	Harvest rainwater for Landscape Maintenance where possible
W1.4.1.	Reduce water required for Food Production and Food Processing	F2.8.2.	Promote efficient agriculture irrigation systems
		F2.8.3.	Promote crops suitable to local climate and Water resources
		F4.7.1.	Promote water-efficient technologies in Food Processing (Provide incentives for Water recycling in Food Processing)
W1.6.1.	Monitor daily and seasonal Water Consumption fluctuations	I1.3.1.	Identify infrastructure systems' processes to be monitored/documented through Information infrastructure
W1.6.2.	Regularly collect and organize quantitative and qualitative data on Water Consumption	I1.3.1.	Identify infrastructure systems' processes to be monitored/documented through Information infrastructure
W1.6.3.	Identify opportunities for use of different nonpotable water qualities	T4.12.1.	Use gray or reclaimed water for cleaning and irrigation in Transportation Networks
		SW3.13.2.	Use gray or reclaimed water where possible for Solid Waste Treatment and Disposal processes
		SW4.10.1.	Use gray or reclaimed water where possible for Solid Waste Collection and Transfer processes
		E3.11.2.	Use gray or reclaimed water where possible for Energy Generation processes
		L4.8.5.	Use gray or reclaimed water for cleaning and irrigation in Landscape entities
		F2.8.1.	Select alternative Water Supply sources of adequate quality for Food Production activities
		I3.9.2.	Use gray or reclaimed water where possible for the operation of Information Nodes facilities

GUIDELINES			RELEVANT GUIDELINES	
W2 WATER SUPPLY			FROM ▸ ▸ TO + WITH	
W2.1.4.	Promote use of Water resources available within project boundaries		L1.5.2.	Define strategies to enhance (benefit from) the productivity of landscape through Landscape infrastructure
W2.3.1.	Map and monitor watershed and hydrological connections		I1.2.1.	Identify opportunities to monitor and protect the natural context through Information infrastructure
W2.3.2.	Develop a watershed management program		F2.5.4.	Protect the foodshed's Water Supply
W2.3.3.	Identify and manage basic sources of contaminants within the watershed		SW3.8.3.	Protect soil from contaminants found in Solid Waste Treatment and Disposal facilities
			SW3.8.4.	Protect water elements and groundwater from contaminants found in Solid Waste Treatment and Disposal facilities
			SW4.6.2.	Protect soil from contaminants during Solid Waste Collection and Transfer
			SW4.6.3.	Protect water elements and groundwater from contaminants during Solid Waste Collection and Transfer
			F2.7.4.	Protect Water elements and groundwater from contaminants used in Food Supply and Production activities
			L1.2.4.	Define needs for restoration of contaminated areas through Landscape infrastructure
W2.3.4.	Control contaminants in runoff water through bioretention and bioinfiltration methods		L1.2.6.	Define needs for water infiltration/purification through Landscape infrastructure
			L4.4.6.	Control flows of fertilizer- and pesticide-laden runoff water from Landscape elements
			T3.7.4.	Develop permeable paving and vegetation configurations for Transportation Nodes to filter runoff
			T4.8.6.	Develop permeable paving and vegetation configurations for Transportation Networks to filter runoff
W2.3.6.	Provide for rainwater and stormwater management needs with natural drainage, infiltration, and floodplain management		T3.7.3.	Develop permeable paving and vegetation configurations for Transportation Nodes to provide natural drainage
			T4.8.5.	Develop permeable paving and vegetation configurations for Transportation Networks to provide natural drainage
			L1.3.1.	Define needs for rainwater/stormwater management through Landscape infrastructure
W2.5.1.	Minimize distance between Water Supply resources and Water Consumption locations		E1.3.2.	Minimize Energy Consumption for Water conveyance and distribution

GUIDELINES			RELEVANT GUIDELINES
W2.5.2. Promote energy-efficient Water Supply methods		E1.3.1.	Minimize Energy Consumption for Water Supply and Water Treatment

W3 WATER TREATMENT

FROM ▶ ▶ TO + WITH

W3.1.6.	Prepare for Water Treatment failures or overflow events	L1.7.1.	Define needs for mitigating and compensating for other infrastructure systems' impact on the natural world through Landscape infrastructure
W3.1.7.	Develop smart, integrated management systems to optimize Water Treatment efficiency	I1.3.2.	Identify other infrastructure systems' processes to be managed through Information infrastructure
W3.2.1.	Optimize placement of Water Treatment facilities so as to minimize distances between treatment and Water Consumption	E1.3.2.	Minimize Energy Consumption for Water conveyance and distribution
W3.3.1.	Integrate Water Treatment facilities with Solid Waste infrastructure for leachate treatment	SW3.4.2.	Provide all additional Treatment facilities required to support landfill operation
W3.3.2.	Integrate Water Treatment and Energy Generation facilities if applicable	E3.3.3.	Investigate potential of Energy Generation integration with wastewater Treatment facilities
W3.4.2.	Assess and resolve potential odor issues caused by Water Treatment facilities	L1.7.2.	Define needs for mitigating and compensating for other infrastructure systems' impact on community life through Landscape infrastructure
W3.4.3.	Assess and resolve potential noise pollution issues caused by Water Treatment facilities	L1.7.2.	Define needs for mitigating and compensating for other infrastructure systems' impact on community life through Landscape infrastructure
W3.5.1.	Visually integrate Water Treatment facilities to the local context	L1.7.2.	Define needs for mitigating and compensating for other infrastructure systems' impact on community life through Landscape infrastructure
W3.6.3.	Protect natural habitats and wildlife corridors and compensate for their fragmentation or disruption by Water Treatment facilities	L1.7.1.	Define needs for mitigating and compensating for other infrastructure systems' impact on the natural world through Landscape infrastructure
W3.6.4.	Protect soil from contaminants found in Water Treatment facilities	F2.5.5.	Identify and manage basic sources of contaminants within the foodshed
		L1.7.1.	Define needs for mitigating and compensating for other infrastructure systems' impact on the natural world through Landscape infrastructure
W3.7.2.	Prepare action plans for periods of extreme water scarcity or high precipitation	L1.7.3.	Define needs for protection of infrastructure systems against projected climate change and extreme events through Landscape infrastructure
W3.8.2.	Compensate for emissions incurred in Water Treatment through carbon offset and sequestration practices	L1.7.1.	Define needs for mitigating and compensating for other infrastructure systems' impact on the natural world through Landscape infrastructure

GUIDELINES		RELEVANT GUIDELINES	
W3.9.1.	Promote on-site processing of Water Treatment residues	SW1.1.3.	Promote on-site processing to reduce incoming Solid Waste
W3.9.2.	Investigate using Water Treatment residues for material recovery	L4.10.3.	Incorporate Solid Waste and wastewater Treatment residues and by-products in regular Landscape Maintenance if available
W3.9.3.	Identify potential for Energy recovery and Generation through Water Treatment residues	E2.1.1.	Identify all potential Energy sources
W3.9.4.	Resolve routes and methods for the collection, treatment, and disposal of Water Treatment residues and by-products	SW1.1.5.	Regulate use of hazardous substances bound to enter the Solid Waste management system
		SW3.5.4.	Solid Waste and Water systems to work in synergy for the Treatment and Disposal of Water Treatment residues
		SW4.3.1.	Solid Waste and Water systems to work in synergy for the Collection and Transfer of Water Treatment residues
		T3.3.2.	Investigate needs of Transportation Nodes integration with Water and wastewater Treatment facilities
		T4.3.2.	Investigate needs of Water and wastewater Treatment facilities for connection to Transportation Networks
W3.10.1.	Use energy-efficient Water Treatment technologies	E1.3.1.	Minimize Energy Consumption for Water Supply and Water Treatment

W4 WATER NETWORKS

FROM ▸ ▸ TO + WITH

W4.2.1.	Provide for operational flexibility of Water Networks	I1.3.2.	Identify other infrastructure systems' processes to be managed through Information infrastructure
W4.2.2.	Optimize flow pressure in Water Networks	I1.3.2.	Identify other infrastructure systems' processes to be managed through Information infrastructure
W4.2.4.	Provide operators with real-time monitoring of Water Networks	I1.3.2.	Identify other infrastructure systems' processes to be managed through Information infrastructure
W4.3.1.	Integrate Water Networks with Transportation Networks	T4.3.1.	Configure Transportation Networks to provide for needs of Water Networks and facilitate their maintenance
W4.3.2.	Plan Water Networks to support the operation of Energy Generation facilities	E3.3.2.	Integrate Energy Generation facilities with Water distribution facilities
W4.3.3.	Plan Water Networks to support the operation of Information Nodes facilities	I3.4.2.	Integrate Information Nodes with Water Networks
W4.3.4.	Integrate Water Networks with Landscape entities	L3.5.1.	Configure Landscape entities' vegetation and paving to provide for other infrastructure networks' needs and facilitate access and maintenance
W4.4.2.	Assess and minimize potential odor issues caused by Water collection and transfer facilities and processes	L1.7.2.	Define needs for mitigating and compensating for other infrastructure systems' impact on community life through Landscape infrastructure

GUIDELINES			RELEVANT GUIDELINES	
W4.5.1.	Visually integrate Water Networks to the local context		L1.7.2.	Define needs for mitigating and compensating for other infrastructure systems' impact on community life through Landscape infrastructure
W4.6.1.	Protect habitats and wildlife corridors and compensate for their fragmentation or disruption by Water Networks		L1.7.1.	Define needs for mitigating and compensating for other infrastructure systems' impact on the natural world through Landscape infrastructure
W4.8.1.	Assess and manage threats from natural and man-made phenomena affecting Water Networks		L1.7.3.	Define needs for protection of infrastructure systems against projected climate change and extreme events through Landscape infrastructure
W4.9.1.	Take advantage of natural landscape configuration to minimize needs for pumping in Water Networks		E1.3.2.	Minimize Energy Consumption for Water conveyance and distribution

NOTES

1. World Resource Institute (WRI), accessed on October 30, http://www.wri.org/our-work/topics/water.
2. World Water Week in Stockholm, 2008.
3. For further information, see United Nations website, accessed on April 10, 2014, http://www.un.org/News/Press/docs/2010/ga10967.doc.htm.
4. David Grey and C. W. Sadoff, "Sink or Swim? Water Security for Growth and Development," Water Policy 9 (2007), 545–571, accessed on April 10, 2014, http://cip.management.dal.ca/publications/Water%20security%20for%20growth%20and%20development.pdf.
5. Further reading: Peter H. Gleick, ed., The World's Water: The Biennial Report on Freshwater Resources, 8th ed. (Washington, DC: Island Press, 2014). Produced biennially by the Pacific Institute for Studies in Development, Environment, and Security.
6. Peter H. Gleick, ed.,The World's Water: The Biennial Report on Freshwater Resources, 7th ed. (Washington, DC: Island Press, 2011).
7. A. T. Wolf, J. Natharius, J. Danielson, B. Ward, and J. Pender, "International River Basins of the World," International Journal of Water Resources Development 15, no. 4 (1999), 387–427.
8. InterAction Water Council (IAC), August 2012 report, United Nations Environment Program (UNEP).
9. Wolf et al., "International River Basins of the World," 311.
10. Gleick, The World's Water, 8th ed., 4.
11. World Resource Institute (WRI), accessed on October 30, 2015, http://www.wri.org/our-work/topics/water.
12. "Polluted Runoff (Nonpoint Source Pollution)," US Environmental Protection Agency, accessed October 30, 2015, http://www.epa.gov/owow/nps/nps-conserve.html.
13. Robert A. Corbitt, *Standard Handbook of Environmental Engineering* (New York: McGraw-Hill, 1999), 347.

IMAGE CREDITS

P. 106: Photograph by Alberto Restifo, accessed on October 23, 2015, https://images.unsplash.com/photo-1419064642531-e5757283952?q=80&fm=jpg&s=d00c90764002592908e1cfbbe4e246dd. **P. 114:** International Water Management Institute. Source of scanned image: University of Texas Libraries, accessed on October 30, 2015, http://www.lib.utexas.edu/maps/world_maps/water_scarcity_2025.jpg. **P. 115:** Created with 'Map of World with Countries' by FreeVectorMaps.com based on data from World Resources 2000–2001 (Washington, DC: World Resources Institute, 2000), Table FW1, "People and Ecosystems: The Fraying Web of Life." **P. 116:** Adapted from "Trends in Global Water Use by Sector" in Vital Water Graphics: An Overview of the State of the World's Fresh and Marine Waters, 2nd ed. (United Nations Environment Programme, 2008), accessed on October 30, 2015, http://www.unep.org/dewa/vitalwater/article43.html. **P. 121:** (Global freshwater use) Data source: "Water Facts and Trends," World Business Council for Sustainable Development, 2005, p. 3, accessed on October 30, 2015, http://www.unwater.org/downloads/Water_facts_and_trends.pdf; (Percentage of increase in water withdrawals by 2025) "Water Scarcity," UN Water, accessed on November 24, 2015, http://www.unwater.org/fileadmin/user_upload/unwater_new/docs/Publications/water_scarcity.pdf. **P. 121:** (Domestic water consumption per capita) Data source: "Water Facts and Trends," World Business Council for Sustainable Development, 2005, p. 7, accessed on October 30, 2015, http://www.unwater.org/downloads/Water_facts_and_trends.pdf. **P. 121:** (Indicative indoor domestic water use in USA) Data source: Peter W. Mayer, William B. DeOreo, et al., Residential End Uses of Water (Denver: AWWA Research Foundation and American Water Works Association, 1999), p. xxvi, accessed on November 22, 2015, http://www.waterrf.org/publicreportlibrary/rfr90781_1999_241a.pdf. **P. 123:** (Center left) Data source: "Progress on Drinking Water and Sanitation," UNICEF and WHO, 2012 update, p. 8, accessed on October 30, 2015, http://www.unicef.org/media/files/JMPreport2012.pdf.

macys
SAFEWAY

Sustainable Infrastructure Planning Guidelines

Energy

Infrastructure

Chapter 7

Energy

converts and manages naturally available resources.

7.1 | INDUSTRY PERSPECTIVES

Planning Considerations for Energy Infrastructure

The act of planning involves defining a vision for the future, be it of an organization, a city, or a nation, and it is linked to a set of physical, ecological, economic, cultural, and political determinants and constraints that include people's expectations and aspirations. When it comes to energy, the expectation of the public in developed countries is to be able to continue to access cheap and reliable energy sources in order to enjoy the many commodities that modern technology has put at our disposal, and at the same time to achieve ambitious sustainability objectives in terms of pollution reduction and climate change mitigation. In developing countries, where access to energy sources is still often the privilege of a minority, the expectation is more to be able to access affordable and reliable energy, often moving away from traditional sources like firewood toward hydrocarbons and electricity.

If we try to express a general vision for the energy systems of the future, we think it should be built around the idea of providing a "sustainable energy supply for everyone,"[1] and consider mainly the following three components:

- Equity and affordability: since energy availability is such an important component of the efforts to lift people out of poverty, the issue of universal access to affordable energy sources should be considered a priority in developing countries and in economically marginal areas of developed countries.
- Renewable sources: while hydrocarbons and nuclear energy will still play a major role in the next decades, the overall impact on the environment and climate of the preponderant use of finite fossil fuels clearly points to the need for planners to prefer renewable energy sources.
- Technical and economic feasibility: since reliability and affordability of energy supply are so important, energy technologies need to be chosen with care in order to ensure that the energy systems they support can be built and maintained within certain economic, organizational, and institutional constraints.

In planning for energy systems, there are a number of dimensions to consider, such as:

- The geographic scale,
- The time scale,
- The interrelations between demand and supply,
- The interrelations with other systems.

Geographic scale

Energy sources and energy users naturally have different geographic patterns of distribution. While during the industrial revolution many cities were built close to local energy sources like hydropower, in modern times, with the dominant role of fossil fuels, the relation between sources and users has become more and more loose. In view of maximizing the use of renewable resources and reducing overall impacts, the first task in planning an energy system should be a close look at the available renewable energy sources in the area where the system needs to be planned and its surroundings. The scale for this analysis needs to be suitable for the various sources, including, for example, looking at the hydropower potential of the watercourses at a watershed scale; at forestry and agricultural practices, considering mainly logistical and transportation issues; or at the presence of geothermal resources and wind resources at a regional scale. The city itself needs to be considered an energy source: roofs and facades have potential to host solar power and micro wind turbines; in nearby mountainous areas water piping has potential for micro hydropower; organic waste can become a biomass source; coastal cities can use seawater as an energy buffer; low-enthalpy geothermal is practically everywhere, etc. In many situations, this analysis reveals the presence of neglected energy sources that can contribute to the energy mix and present opportunities for the development of local small and medium enterprises. Searching for local renewable energy sources is a useful exercise also to foster social and economic relationships between the city and the hinterland that in turn can provide opportunities for other infrastructure development, and to enhance social and economic capital. Increasingly, for example in Europe, citizens of rural areas are creating cooperative enterprises to develop small-scale solar and wind farms on their farmlands, and in some cases citizens of urban areas are making direct agreements with these enterprises to source their electricity instead of becoming small-scale producers themselves in more constrained urban areas.

Time scale

Time scale is important for different reasons. The most important is the long payback time of the investments that are needed to build the energy production and transport infrastructure system. Payback times for electricity-generating energy assets, for example, can range from 5–7 years for a small-scale photovoltaic plant to over 30 years for a nuclear power plant, with 15 years being the average for a medium-size gas-fired combined-cycle plant. Long payback times are also involved in developing large-scale hydrocarbon production, transport, and refining installations, or for the nuclear fuel industry from mining to refining. One important consequence of this is the so-called "asset inertia" that makes it extremely difficult to close down or stop operating plants as long as the return on the investment has been realized, or, often, as long as it is possible to make money. The longer the payback time for the energy system, the less its flexibility in adjusting to different demand patterns or environmental, economic, and societal pressures, or its ability to adjust to climate change. Another time factor is specific to biomass energy and the need to use biomass within sustainable boundaries. Forest biomass use means taking a long-term view. In temperate climates it is likely to take over a hundred years to ensure that the forest is used within its capacity to regenerate and maintain its ecological functions. In general, when planning an energy system the long-term view should consider the possibility that energy production and consumption patterns could change and fluctuate according to different drivers, and not always grow as they have in the past. Developing an energy production mix that includes elements with high flexibility, able to adjust to sudden changes in consumption patterns or even disruptions due to climatic events, will be key in the future.

Supply and demand

Energy demand has been increasing steadily for the past hundred years and is predicted to increase at a global level for at least the next 50–60 years, according to different estimates. However, it is clear that energy efficiency is also increasing and that the rate of improvement can be subject both to continuous incremental changes and to some disruptive technologies or radical policy decisions. Energy consumption patterns are heavily dependent on other systems, starting from land use planning and the geographic distribution of different functions within the city. Compact urban areas based on high-rise buildings and a mixed land use concept allow people to live, work, shop, and spend their free time within short commutes done by walking, biking, or public transport. These modes of transportation require much less energy than those associated with large urban sprawls. The quality and efficiency of buildings is another clear driver for reducing energy consumption; zero-energy buildings or even energy-positive buildings are already available in the market in Europe, North America, and part of Asia at competitive prices, and it is clear that, if widely adopted, these solutions could significantly change the energy demand pattern of future cities. Opportunities for the refurbishment of existing buildings can also drive energy consumption down by more than 30 percent, sometimes as much as 50 percent. Municipal and regional authorities can be very effective in defining standards for buildings' energy efficiency both by the use of bylaws and by creating a market for new technologies and solutions in their own buildings and services. Planning for energy systems should therefore carefully consider the interplay between demand and supply, and an integrated planning concept should be used, starting from scenario planning and simulation tools. Planning based on past consumption trends should always be avoided when the ultimate goal is to build a sustainable energy system.

Interrelation with other systems

Energy is used by all other systems within the city, and other systems can, in turn, become energy sources. The interrelation between the energy system and the other systems is therefore very strong and presents many aspects that can increase the overall sustainability of the city. At the city level, the transportation sector is generally the second largest user of energy after the residential sector, and it is clear that there are great opportunities to increase its energy efficiency. The most promising future relation between the two systems is probably the possibility of transforming the entire transportation network at the city level by use of electric vehicles integrated into a smart grid system. In this way electric vehicles could become both energy sources and energy storage devices, helping to buffer the system during peak periods. Waste-to-energy systems are already contributing to the energy mix in many cities, particularly in northern Europe, but it is very evident from the amount of waste still being sent to landfill at the global level that there is significant opportunity for expansion in this area. In the future, some further advances could be made through small-scale biodigesters for the production of biogas that could operate at the city or even at the neighborhood scale. Energy use in the water management system could also be optimized by designing the system in a way that maximizes the use of natural flow and minimizes pumping, and, of course, by introducing more water-efficient systems.

The planning of an energy system presents many opportunities to achieve sustainable development goals, particularly for new cities or cities that are under pressure to expand to accommodate most of the population growth of the next 50 years. Attention to locally available renewable sources and integration with the hinterland, introducing flexibility in the energy assets mix, smart land use planning, advanced building design, and integration with other systems at the city and regional levels will all be necessary elements of future energy planning for sustainability. Radical policy choices, both in existing and new cities, will need to drive the change, following a clear understanding of the financial and nonfinancial barriers to the implementation of innovative solutions.

Roberto Mezzalama
MSc, Global ESIA Technical Community Leader, Golder Associates

NOTES

1. "Ecofys": www.ecofys.com.

Developing a Sustainable Energy Plan

Designing energy infrastructure for a sustainable urban environment

Energy, including power generation, transmission and distribution, and renewable energy, is a critical component in all urban infrastructure planning. Today's popular call to be "green" and "sustainable" is commonly associated with renewable energy only. But the potential for sustainable design in urban and suburban settings goes well beyond just renewables. The focus on sustainability in the energy market has turned to the three main principles of environment, economics, and social impact.

The primary goals of a sustainable energy plan are:

- Reduce the demand for traditional energy through conservation and energy efficiency,
- Expand the supply of renewable energy and other forms of distributed (on-site) energy,
- Integrate energy infrastructure utilizing sound sustainable practices, and
- Leverage advanced technology.

Reduce the demand for traditional energy through conservation and energy efficiency

The first step in achieving a sustainable energy model is to reduce the amount of energy consumed, and to design and construct energy-efficient systems that minimize impact on the environment. From an urban development standpoint, there are many ways in which we can do this.

- Site selection: We can reduce urban sprawl by avoiding development of prime farmland or the habitat of endangered and critical species. In already developed areas, redevelopment sites such as parks or off-leash areas can be selected to blend in with the urban setting through co-development and/or providing benefit to the local community.
- Density development: We can focus new development on areas with existing infrastructure. This protects greenfield sites and areas with natural resources. We can also reuse existing infrastructure.
- Brownfield redevelopment: We can rehabilitate existing sites to reduce pressure on undeveloped land. We can also reuse existing infrastructure.
- Alternate transportation: We can locate development near existing public transportation to reduce pollution and land development from automobile use.
- Maximize open space: Vertical development provides more open space for biodiversity and concentrates infrastructure into a smaller area. This will reduce the overall impact on the environment by preserving green space.
- Site hardscape: We can select paving materials to reduce solar retention or shade hardscape areas to reduce the heat island effect. This reduces energy consumption.
- Green roofs: We can install highly reflective or green-vegetated roofs to reduce the heat island effect and lower the energy required to condition the building. Similarly, we can integrate solar PV and/or hot water.
- Light pollution: We can minimize light trespassing from a building or site in order to reduce the impact on nocturnal environments.
- Water use reduction: We can eliminate potable water use for landscape irrigation and employ high-efficiency water fixtures in buildings.

- Building materials: We can reuse/recycle building materials and use products that are extracted and manufactured within the region. This reduces energy consumption in the manufacturing and transportation of products.

Expand the supply of renewable energy and other forms of distributed (on-site) energy

During the planning process, sustainable energy infrastructure can be integrated in many ways. For example, large-scale renewable energy sources such as large wind farms and solar plants can be located in close proximity to urban centers or co-located with agriculture or similar land use types. Small-scale renewable energy can be integrated into urban planning by utilizing rooftops, transportation right-of-ways, stormwater areas, and other spaces where there will be minimal impact. Utilizing this type of distributed generation can minimize the overall impact on the urban environment, thus creating a sustainable infrastructure system.

Two important points need to be made here. It is critical to integrate these renewable energy sources in the urban planning stage. This will minimize the overall impact on the environment, since wind farms and solar fields can occupy a fair amount of green space. Because of the rapid changes in the renewable energy industry, it is important that design engineers keep informed on the latest developments in technology, finance, and social trends.

Growing energy demands indicate that renewable energy will not provide all of the energy supply that will be required. It will also be necessary to utilize highly efficient, environmentally friendly, nonrenewable generation sources as well.

Two examples are cogeneration and combined-cycle technologies. Cogeneration is the recovery and utilization of the normally wasted heat energy produced by a power plant or industrial process to generate electricity or drive some other industrial process. Combined-cycle technologies utilize a steam turbine to generate additional electricity from the wasted heat from a power-generating gas turbine's exhaust, thus increasing the overall thermodynamic efficiency of the power plant.

Once again, addressing these sustainability issues early with urban planners and developers will allow them to integrate solutions into their business model. This not only results in a better project financially, but also strengthens relationships with permitting and regulatory agencies.

Integrate energy infrastructure utilizing sound sustainable practices

From an urban planning perspective, there are three key strategies to integrating energy infrastructure in a sustainable manner: (1) co-location, (2) co-development, and (3) sustainable design.

Co-locating energy infrastructure facilities such as transmission and distribution lines with other utilities such as transportation right-of-ways, irrigation canals, etc., has obvious savings in green space. It is important to note, however, that packing all manner of facilities (transportation, water, sewer, electricity, food supply, housing, etc.) into a tight urban environment can cost a great deal more when compared to placing the same facilities in other environments. In these instances, the overall objective may not be to lower the cost of the facility or to reduce materials consumption. Rather, the goal may be multipurpose use of space and facilities.

The major challenge to developing multipurpose infrastructure and shared-purpose space is a political one. It requires a shared belief in the value of sustainability objectives and cooperation at the highest levels within government and private investment entities of every size. Such shared vision and cooperation aren't always achievable. Special interests and political differences are sometimes at the core of the inability to reach consensus.

Co-development may involve, for example, the merging of the local substation (see below) with another project. In dense urban areas, it is common practice to place substations within buildings, parking garages, or underground below city parks. This is a win for the public, since valuable space is not compromised with an unsightly substation. Underground or indoor installations provide buffers to sound, electromagnetic field, and light pollution.

In addition to co-location and co-development, the greatest contribution that a design engineer can make to address sustainable development is to offer a design that consumes less material (such as steel and concrete) and reduces resources associated with material transportation from site of procurement to the site of the project.

Denny Substation Rendering - Seattle City Light

In the specific case of transmission line design, there are significant opportunities for improvement in addressing sustainable development objectives within the urban environment. A set of guiding principles that builds upon co-location and sustainable design needs to be developed.

Substations play a vital role in distributing electric power from the point of generation to the point of use. Normally, electricity is generated at one voltage level, transmitted through long lines at higher levels, and transformed down to another level near the point of use at utility substations. The prime considerations in substation design are safety and reliability. However, there are significant opportunities to incorporate sustainable practices during the planning, designing, and construction phases. Some of the opportunities involve:

- Site selection: We can choose existing, non-greenfield sites that are close to existing infrastructure. We can avoid developing wetlands and restore plant and wildlife habitat in surrounding areas.
- Water runoff and management: We can control runoff from the site and minimize disruption to groundwater recharge. We can contain pollutants to prevent them from entering native streams.
- Construction materials: We can create construction/demolition waste management and recycling plans. We can use locally sourced materials when possible. We can use material with high recyclable content, including steel and other materials.
- Optimal layout and design: We can minimize the footprint of the site through the use of sustainable strategies in buildings, such as high-efficiency HVAC, improved insulation, and LED lighting.
- Public education: We can educate the community about sustainability integration and the role of the utility grid in everyday life. We can serve as an example of environmental awareness.

Leverage advanced technology

The planning and development of our future energy systems should include careful evaluation of advanced and emerging technologies. A city is clearly intended to last for centuries. Many major cities have already been in existence for a few centuries, if not millennia. Yet the technologies they use change significantly by the decade, if not sooner. Even the financial models that are employed with a current technology change at a rapid rate.

Since we can't make use of tomorrow's technology today, we should aim to incorporate the ability to easily transition from current, available technology to a technology not yet envisioned.

The designer's role in creating awareness

Not all sustainability solutions come at the cheapest price or with the shortest time commitment. While improved sustainability should always be top-of-mind, the engineering community needs to recognize that cost and/or time may be a client's first priority. The challenge is to create awareness of a project's effects on the environment and provide clients with options to mitigate those effects. Some benefits, like saving a life with a cleaner-operating power plant, are intangible but priceless. Designers have to be flexible, innovative, and solution-oriented when it comes to presenting sustainability solutions to clients.

Creating awareness can be as simple as asking how aspects of design will impact the environment, businesses, transportation, and the cityscape. Early in the process, the designer needs to answer one question: can sustainable design elements be incorporated into the design features of the power plant, substation, or transmission and distribution system, and not take away from existing infrastructure but rather enhance it? Can we utilize a rating system such as ISI/Envision® or LEED to investigate, evaluate, and incorporate sustainability into our energy infrastructure projects? Addressing sustainability issues with urban planners, business and transportation contacts, community residents, and developers early will allow them to integrate sustainable solutions into their model, which will benefit the community and help lead to global solutions to sustainability.

Richard Corolewski
PE, Electrical Engineer/Federal Division Director
POWER Engineers, Inc

7.2 | SUSTAINABLE ENERGY PLANNING

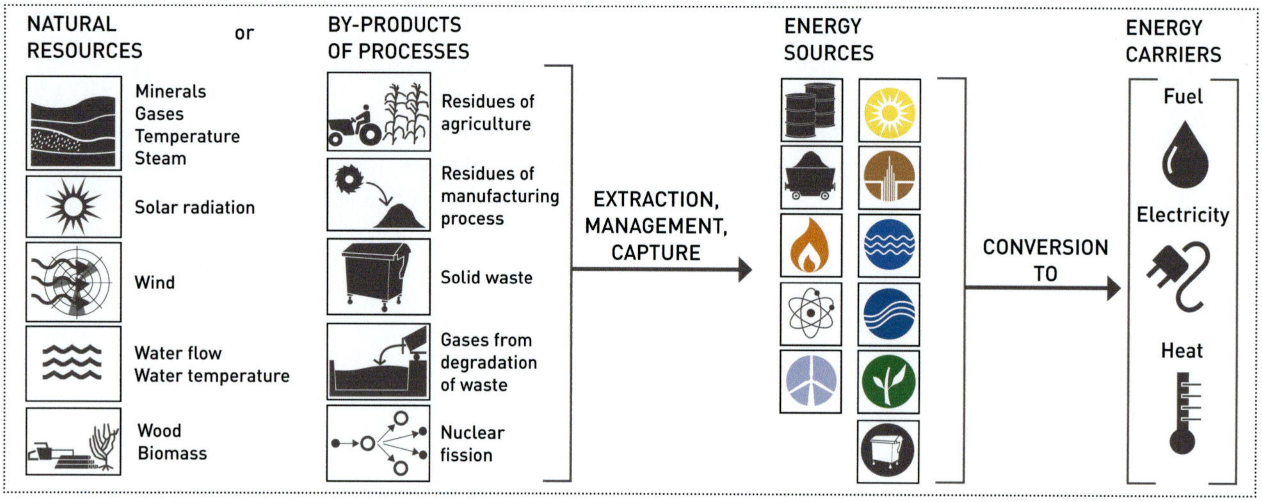

"Energy is what keeps us—and our way of life—going. It is something we tend to take for granted."[1]

Energy is the driving force of innumerable practices of everyday life. It is so tightly connected with everyday operations that its importance is taken for granted and becomes apparent only in case of a shortage. It is required for practically every process, for individuals, businesses, manufacturing processes, and for the operation of other infrastructure systems. The extent to which energy infrastructure affects a contemporary community's everyday life is unquestionable, making energy security fundamental for economic and social prosperity.

Energy infrastructure provides the service of harvesting and converting naturally available resources into energy carriers (electricity, heat, and fuel) and securely delivering them to end users. However, the extraction, capture, and conversion of these resources to useful energy could be responsible for dramatically adverse impacts on the natural world. The availability of natural resources is finite, and their overexploitation to cover continuously growing needs is not sustainable at its core.

Sustainable Energy infrastructure procures the use of natural resources within their capacity for regeneration, providing renewable and low-emission alternatives to fossil fuels. It secures constant and equal access to energy carriers for all individuals while promoting rational use and efficient practices. It aims to regulate and compensate for the adverse impact of conversion processes on the natural world and community life through clean technologies and control of emissions and side effects.

The sustainable energy system to be examined here does not include resource extraction, such as the mining or drilling process along with oil, coal, or gas refining processes. It focuses instead on those aspects of the system more related to the urban context and aims to map and make visible the linkages between the two. It offers an overall visualization of the system to highlight how a decision on energy source provision, for example, affects technologies, sizing, and distribution of generation plants and how the technologies implemented and processes carried out affect overall performance and include short-term, long-term, and cumulative impacts.

The central contribution of renewables to achieving a sustainable energy system is unquestionable, but there are numerous challenges within a complex system of many actors: "the question is no longer whether renewables have a role to play in the provision of energy services, but rather how we can best increase the current pace of deployment to achieve a 100% renewables future with full energy access for all. For this to become reality, current thinking needs to change: continuing with the status quo of a patchwork of sometimes contradictory policies and actions is no longer sufficient. Instead, technology and market developments, finance models, as well as stable and predictable renewable energy policies need to be systematically linked across the public and private sectors in order to support and drive the transition process."[2]

ENVIRONMENTAL IMPACTS OF ENERGY SYSTEM

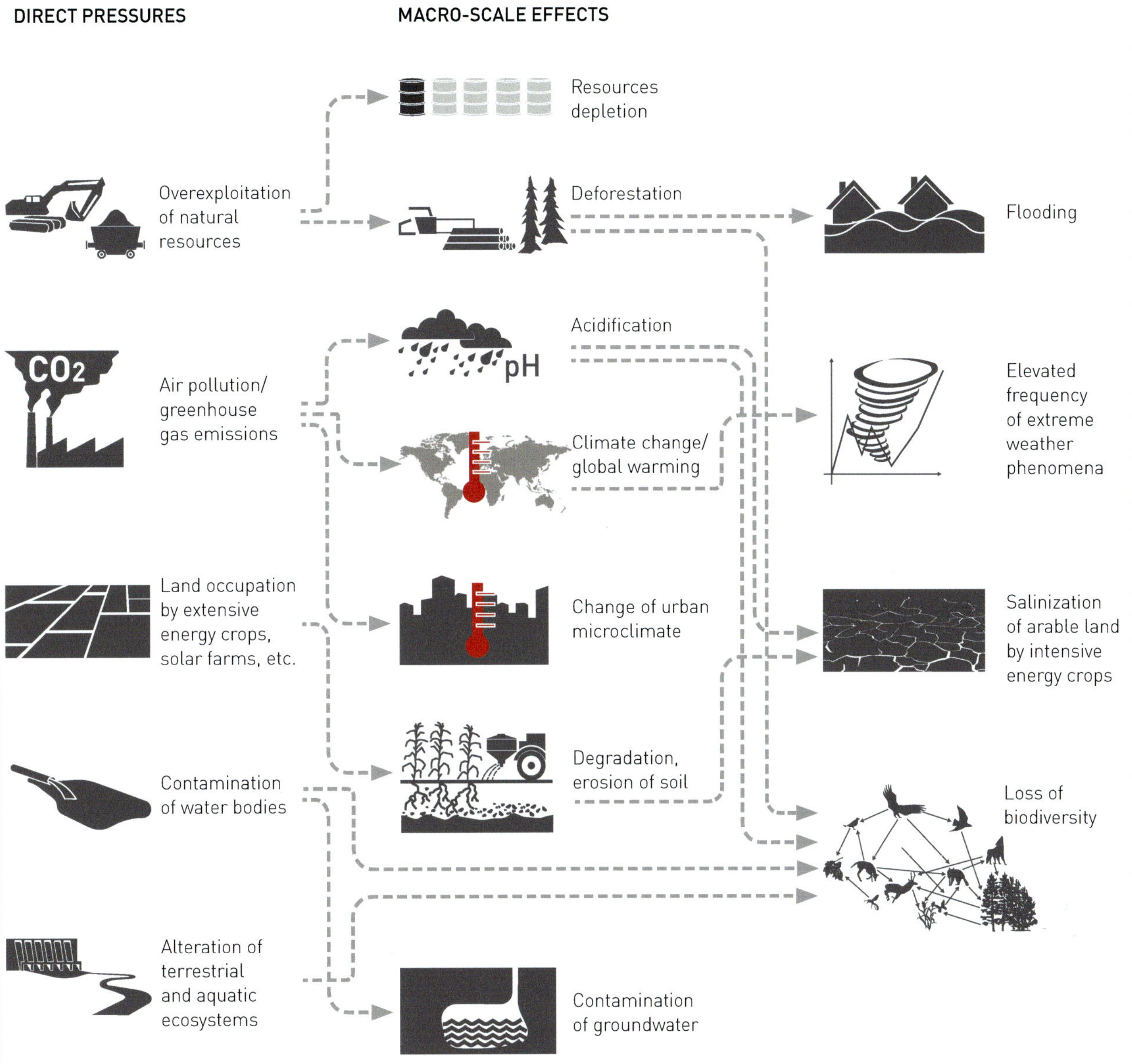

CHALLENGES TOWARD A SUSTAINABLE ENERGY SYSTEM

The issues that need to be addressed to achieve a sustainable energy system can be classified as environmental, social, and economic.

Challenges reside in aligning the system's social and economic goals (providing services and opportunities for a community) with its environmental objectives that affect the community in the long term. The challenge is to identify, regulate, and prioritize in order to counterbalance conflictive adopted strategies so that one does not invalidate the other. Strategies need to be consistent to meet multiple targets; energy infrastructure sustainability has to be approached in a systematic, overall way and within the framework proposed in relation to other infrastructure systems.

A shift toward renewables is on the rise worldwide, fueled by technological advances, policy changes, and the urgent need to reduce both our dependence on carbon-intensive sources of energy and the emission of greenhouse gases to the atmosphere. However, the use, for example, of renewable energy sources does not guarantee by itself a sustainable energy system. Though their sustainability potential is great, they still have to match certain framework conditions that need to be critically weighed, reviewed, and monitored continuously.

Traditional energy systems overlooked many issues to deal with the challenge of providing cheap, reliable, and secure service to meet demand. This initiated a chain of environmental and financial consequences leading to the current energy crisis and climate change vulnerability, all global-scale phenomena.

Energy production is the main factor responsible for air pollution, greenhouse gas emissions, and consequent human-induced climate change, with immeasurable impacts on the natural world and quality of life. The exact type and intensity of impacts vary depending on the specific technology used, the scale of production, the geographic location of facilities, and a number of other factors. Impact can range from contamination of water bodies, acidification, alteration of terrestrial and aquatic ecosystems, degradation of habitats, and the consequent loss of biodiversity to the space requirements of large-scale facilities with their extended land occupation and land cover change.

The challenge for energy systems is to avoid or minimize environmental impact, or else to provide mitigation and compensation measures to counterbalance it. Various reports show that decarbonizing energy and enhancing end use efficiency can help meet climate change mitigation targets.

Additionally, the constantly growing demand for energy has led to intensified resource extraction, resulting in the gradual production decline of oilfields, natural gas wells, and coalfields. Fossil fuels production has already reached its peak and is entering a decline phase with high costs that inevitably affect future projections for energy security, giving further urgency to the need to shift to renewable sources of energy.

Decarbonization is a major challenge still to be met. As climate concerns become an essential component of energy policymaking, the generation and use of electricity will be subject to increasingly strong policy actions by governments to reduce their associated CO_2 emissions. Despite these actions, and despite very rapid growth in renewable energy generation, significant technology and policy challenges remain if this essential transition is to be achieved.[3]

Increased consumption has to be moderated and managed for energy saving. Energy production must result from a balancing of supply and demand to regulate adverse impact. Environmental challenges can be approached with optimized technologies that focus, for example, on losses during conversion and transmission, so that the primary energy input that feeds the system is not unnecessarily wasted. A system's performance is evaluated both through its internal function and its externalized impact, and reconsidering the former can significantly reduce the latter.

The transition from centralized to distributed energy systems is an example of a shift aiming to optimize the system's internal function; however, it entails another challenge for the system itself, that of instability. A stable system that responds continuously to the demand for service is another key objective for a sustainable energy infrastructure.

A shift in energy practices can be promoted on a global scale through policies, incentives, and finance mechanisms, but for it to be effective implies in many cases the upgrading of existing structures and aging infrastructure to support it. Advances in the field of energy research provide alternative practices and technologies to be incorporated to make the shift feasible, such as evolved methods to increase the use of renewable sources of energy, energy storage systems, etc.

Sustainability is central to the overall planning of an energy system, whether in terms of practices, commitments, or goal setting. In order to avoid conflicts between adopted strategies and other local priorities, a balance has to be planned between enablers, opportunities, barriers, and defined capacities for a system's sustainable strategy. Policy commitment to reduce local carbon emissions cannot be disconnected from energy generation practices as well as energy sources selected. A city's overarching aim to be recognized as a leader in the climate change and/or sustainability fields has to be supported by the entire energy system's configuration. If it is to be met, this target needs to be supported by energy consumers' demand for low-carbon and/or decentralized energy. Negative public perceptions due to absence of constructive alliances with the community can undermine the application of a plan.

The potential linkages with programs related to other systems and activities with shared goals can also prove beneficial, highlighting the importance of synergies. Programs that seek to address related agendas, such as climate change mitigation or fuel poverty, have to be seen as helpful partners in a program for more local, urban energy that can supply the needed skills for implementation, install the adequate technologies, and overcome barriers to the application of a strategy.

"Adequate and affordable energy services have been critical to economic development and the transition from subsistence agricultural economies to modern industrial and service-oriented societies."[4]

The use and production patterns of energy, as well as

the quality of energy services, affect progress in economic development, and the status of the energy sector and its trends in a country might improve the chances for economic development to be sustainable in the long term. Energy is vital for eradicating poverty, improving human welfare, and raising living standards.[5]

All sectors of an economy—residential, industrial, or commercial, as well as transportation, services, and agriculture—require energy. These energy services in turn foster economic and social development at the local level by raising productivity and facilitating local income generation. Availability of energy affects jobs, productivity, and development. Electricity is the dominant form of modern energy for telecommunications, information technology, manufacturing, and services. The stability of electricity networks thus has a direct economic dimension[6] and entails another challenge, especially within the transition from centralized to distributed energy systems that cannot fully support the system's ability to respond continuously to demand.

All the above apply on local, national, and international scales at the same time, and there are cross-linkages between the different dimensions of a sustainable system. It is a challenge for the system to define the energy-related priority areas for assessment.

As mentioned in the chapter on Landscape infrastructure, the causal links between the environment and human health and well-being are multiple and often indirect. Impact on the natural world inevitably affects the ecosystems that provide certain services that support human well-being, in the form of clean air, freshwater, regulation of local and regional climates, and control of the range and transmission of certain diseases. The five main aspects that approximate an assessment of human well-being—materials, health, social relations, security, and freedom of choice[7]—are vulnerable both to interruptions in the provision of energy as well as to negative impacts of energy provision. The challenge resides in providing equal access to energy services for all, affordability, and availability of alternatives while at the same time securing safety, quality of urban environment, and quality of the natural world.

ENVIRONMENTAL

Challenges for environmental sustainability include:

- Preserving habitats and wildlife corridors
- Protecting water cycle processes and water routes
- Protecting soil and water quality
- Minimizing greenhouse gas and air pollutant emissions
- Resolving noise pollution and visual issues
- Using natural resources within their regenerative capacity
- Safely managing hazardous residues

SOCIAL

Challenges for social sustainability include:

- Securing access to energy services for all social groups
- Ensuring system stability and reliability
- Protecting public health and safety
- Promoting public awareness for energy saving
- Maintaining affordability while providing financial incentives for energy saving

ECONOMIC

Challenges for economic sustainability include:

- Adapting to economic sectors' needs
- Securing investments for continuous upgrades to grids and networks
- Securing long-term economic viability
- Integrating renewable energy sources while maintaining affordability
- Ensuring system stability and reliability

STRATEGIES TOWARD A SUSTAINABLE ENERGY SYSTEM

SYSTEM PURPOSE: RELIABLE ACCESS TO ENERGY	
	ENERGY SAVING: Avoid unnecessary energy consumption
	SYSTEM EFFICIENCY: Increase efficiency in generation, conversion, distribution, etc.
	USE OF RENEWABLE ENERGY SOURCES: Increase renewable energy use
	MINIMIZATION OF FOSSIL FUELS IMPACT: Minimize impact of conventional-fuel energy generation

The primary purpose of the Energy infrastructure system is to provide reliable access to its services. The structures and operations put in place create several sustainability challenges, as discussed in the previous section. The strategies listed here suggest basic steps that could be included in the planning of an energy system. They do not aim to outline all the requirements for a sustainable system, but instead highlight approaches that if implemented could resolve multiple issues simultaneously.

Energy saving is the first strategy listed. The main target of each energy system should be to avoid and reduce unnecessary energy consumption, so that supply and demand are balanced to a perceived optimum. Customer-side data are a necessary input to managing and reducing energy use, as the energy users' behavior is the first factor to be reconsidered in efforts to reduce demand. Lifestyle is responsible for extensive use of energy; what is at stake is finding ways to achieve same quality of life using less energy.

Energy saving must be taken into consideration in the process of planning, both at a building scale and in urban-scale planning. The design process entails both globally recommendable and site-specific practices to be incorporated in order to achieve energy savings. Due to their wide scale of application, understanding the importance of these strategies is essential for a more sustainable urban model.

The configuration of the urban fabric affects overall energy consumption. Adaptation to the local climate determines the dependence (or not) on artificial ventilation, cooling or heating, and lighting methods. Natural shading and consideration of prevailing wind directions can improve the urban microclimate with heat sinks. Urban compactness, or the optimum location of land use attractors within the city, can reduce transportation needs and thus fuel consumption, proving that energy consumption and urban fabric configuration are linked. The extended scale of urban public transportation networks calls for an overall approach in terms of energy consumption, making the selection and operation of vehicles and the modal split crucial.

To improve energy performance it is also essential to aim for the retrofitting and refurbishment of the existing building stock, in a way consistent with the energy system. In this way the energy needed within the buildings can be substantially reduced, aligning energy use with the system's goals.

Energy system efficiency is the second strategy and encompasses all phases of an energy system, from generation and conversion to transmission, distribution, and end use.

Optimizing the technologies utilized through research and innovation can improve the performance of traditional energy generation methods, exploiting their potential to the maximum. Combined-cycle generation of electricity and cogeneration of power and heat (CHP) are examples of such practices.

Transmission and distribution losses can be resolved through a reconsideration of the spatial relation between energy production and end use, but also through monitoring and efficient management. Generating power locally avoids energy loss that occurs when delivering power over long distances. Centralized generation systems waste more than two-thirds of their original energy input, primarily as heat. Distributed energy generation systems are a relevant practice, as will be further analyzed in the following section. Finally, technological advances in the manufacturing of electric devices allow for the optimized

ENERGY SAVING PROVISION

BUILDING SECTOR

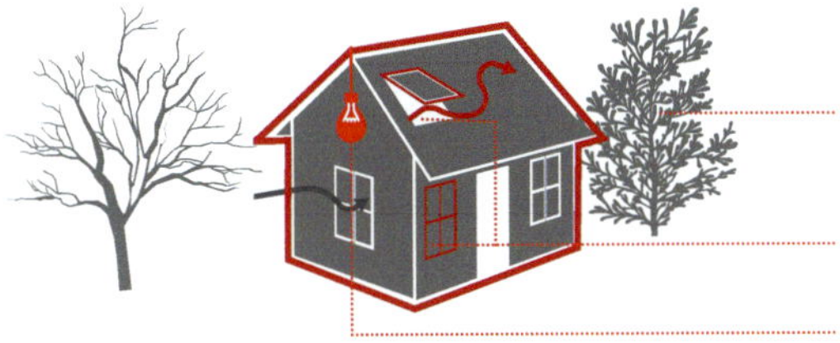

- Energy upgrade of existing buildings
- Improved microclimate
- Optimized building envelope (thermal properties/solar control)
- Optimized openings (sizing/thermal properties/natural ventilation/daylighting, solar control)
- Efficient systems (for lighting/appliances/HVAC/smart technologies)

URBAN PLANNING

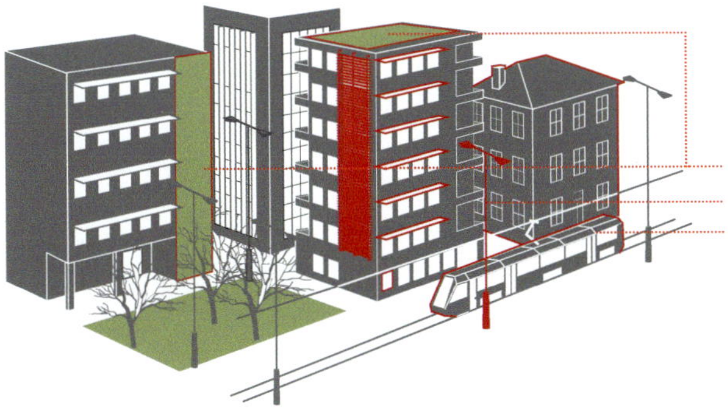

- Compactness of urban fabric
- Land use distribution and density
- Large-scale application of green roofs/green walls
- Efficient urban equipment (e.g., street lighting)
- Mass public transport
- Improved microclimate (vegetated areas/permeable surfaces)
- Planning considering of prevailing wind direction

performance of their assigned service, with less energy wastage at the time of final use of energy.

The third strategy listed is that of increasing renewable sources in the supply of the system. An elevated share of the energy supply must come from renewable energy sources, which is necessary to fulfill the goal of reduced greenhouse gas emissions. However, increased energy demand cannot be met from renewable sources alone. "Securing 'sustainable energy for all' involves the development of systems that support the optimal use of energy resources in an equitable and socially supportive manner while minimizing environmental impacts."[8]

Renewable energy can be promoted through a framework of regulatory policies, fiscal incentives, public finance mechanisms, and climate-led policies that seek to encourage the shift to cleaner technologies. "Regulatory policies include feed-in tariffs, quotas or portfolio standards, priority grid access, building mandates, and biofuel blending requirements. Fiscal incentives refer to tax policies and direct government payments such as rebates and grants. Public finance includes mechanisms such as loans and guarantees. Climate-led efforts include carbon pricing mechanisms, cap and trade, emission targets, and others."[9]

Fossil fuels should be a choice of last resort as supply for an energy system. Coal, oil, and natural gas do substantially more harm than renewable energy sources in most respects, including air and water pollution, damage to public health, wildlife and habitat degradation and loss, water use, land occupation, and carbon emissions. Still, their use has not yet been abandoned. The decarbonization of the energy system is a target yet to be reached, and a sustainable system that is in the process of approximating its goal for clean energy sources must compensate and mitigate the adverse impact of conventional practices, e.g. by use of carbon sequestration and storage.

ENERGY LOSSES DURING CONVERSION AND TRANSMISSION

Conventional centralized generation systems waste more than two-thirds of their original energy input

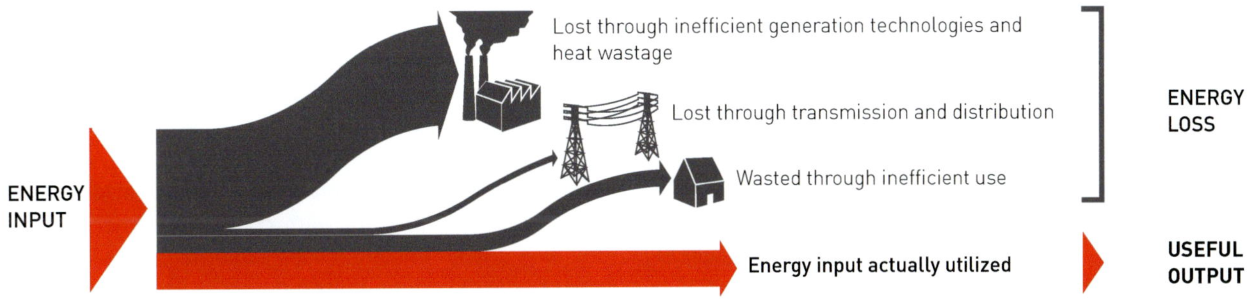

7.3 | ENERGY SYSTEM DECODING

In line with the general framework of the Guidelines, the Energy infrastructure system is decoded at four system levels: Consumption, Supply, Generation, and Distribution and Transmission.

The first level examines the demand which the infrastructure system is to serve. This level of decoding aims to promote understanding and optimization of the needs for energy. It addresses the amounts of energy carriers, electricity, heat, and fuel that need to be available for customers.

The second level, Supply, represents the basic strategic decisions concerning the mix of energy sources (and therefore resources) and technologies that will be used to meet the demand for energy. The supply mix will determine to a large extent the organization and planning of the entire energy system and its impact on community, economy, and natural systems. The benefits, limitations, availability, feasibility, and environmental impacts in the specific context have to be examined and resolved prior to planning the individual components of the system. The breakdown of sources available comprises the city's energy mix, an important indicator of sustainability performance in urban areas.

The third and fourth levels, Generation and Distribution and Transmission networks respectively, represent the main infrastructure entities that form the system. Energy generation includes the nodes of the system where different energy sources are converted and managed to produce energy in the form of fuel, electricity, and heat. The efficient function of energy generation entities determines whether or not the entire system will operate properly, and most importantly how accessible it will be for users. The fourth level includes the network for long-distance transmission and distribution to end users of electricity and heat, as well as the transfer of fuel through a hierarchized system of pipes, overhead or underground cables, and vehicles. Both Generation and Transmission and Distribution levels include storage, an important component of the energy system, as will be further analyzed.

The interconnection of all system levels is also obvious in the case of end users, with consumers who produce their own energy in some cases achieving total autonomy from the system or/and contributing to the system's needs.

ENERGY CONSUMPTION

The Consumption level examines the demand for energy by user, land use type, and urban sector. The consumption of energy in the form of fuel, heat, and electricity depends on a community's demographic and economic characteristics, but also on each user's preferences. It reflects requirements for general standards of living but also weather conditions, and varies accordingly. The processes that use electricity, for example, could make a long list depending on the average equipment of a home or business.

The accelerating increase of demand for energy, particularly in developing countries, results from the rapid economic growth in emerging economies and projected increases in the world's population.

The system's planning is based on data extracted through monitoring of consumption, analyzed to provide daily, monthly, and annual consumption patterns in order to sufficiently meet the needs of a community. Energy consumption varies during the day, creating extremes of high use (peak hours) and low use (off-peak hours).

These sharp changes are illustrated as steep ramps on the electricity load charts that call for flexible operational capabilities to ensure the system's reliability. The system has to react quickly to match this variable demand and meet the expected operating levels both in terms of supply and expected generation capacity. This is more apparent in emerging operating conditions, due to the growing penetration of renewable on the grid.[10]

This level includes customer-side data to manage and reduce energy use. A local power company collects these data and informs the customers of the system's function, guiding and motivating them to rational energy use. Awareness of one's own consumption patterns, apart from optimizing one's consuming behavior economically, benefits the overall system through energy conservation. The efficient management of energy consumption requires dealing with the varying electrical loads, managing the level and timing of demand for electricity, and decreasing the difference between generation output and demand. Time-dependent pricing negotiated ahead of time, or even at minute-by-minute spot prices, may allow those users with monitoring equipment to detect demand peaks as they occur (dynamic demand) and shift loads[11] to off-peak hours sold at lower prices. Involving the customers encourages them to be more energy-conversant and aware of their energy options.

The process of energy consumption regulation requires the systematic collection and organization of data on energy use. This involves monitoring of daily and seasonal energy consumption fluctuations and specific metrics to benchmark a building's energy consumption, such as energy use intensity. As already mentioned, the building envelope design and the urban context's configuration predetermine to a degree the needs for electricity for lighting, heating, or cooling purposes or fuel for transportation needs. To achieve a sustainable energy system the planning process has to identify all actors in the overall demand, the initial input of the system, and its final useful output at the same time.

FORMS OF ENERGY NEEDED

Fuel   Electricity   Heat

ENERGY CONSUMERS AND PRIMARY NEEDS

RESIDENTIAL
- Heating/cooling
- Interior/exterior lighting
- Interior equipment/appliances
- Other

COMMERCIAL
- Heating/cooling
- Interior lighting/exterior lighting
- Interior equipment/exterior equipment
- Other (pumps, refrigeration...)

PUBLIC SPACE
- Street lighting
- Public space lighting
- Traffic lights
- Other

AGRICULTURE INDUSTRIAL
- Motor-driven physical processes (cutting, forming, assembly)
- Heat treating, drying and bonding
- Other chemical processes

TRANSPORTATION
- Fuel for public and private vehicles
- Electricity for electric/plug-in vehicles

ELECTRIC POWER
- Processes supporting energy generation

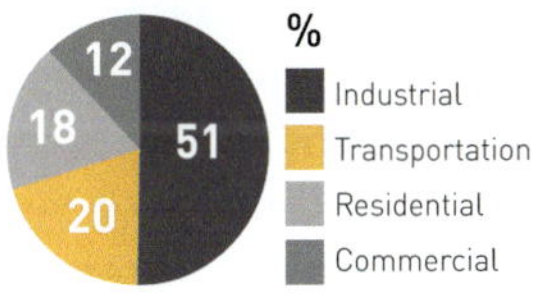

World energy consumption by sector (EIA)

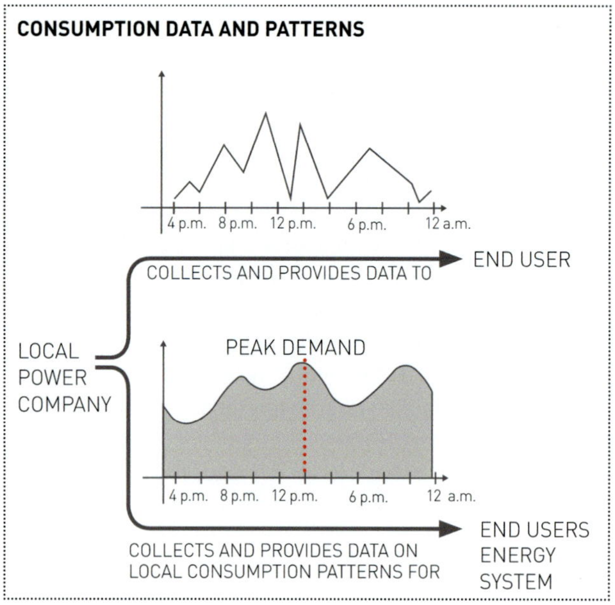

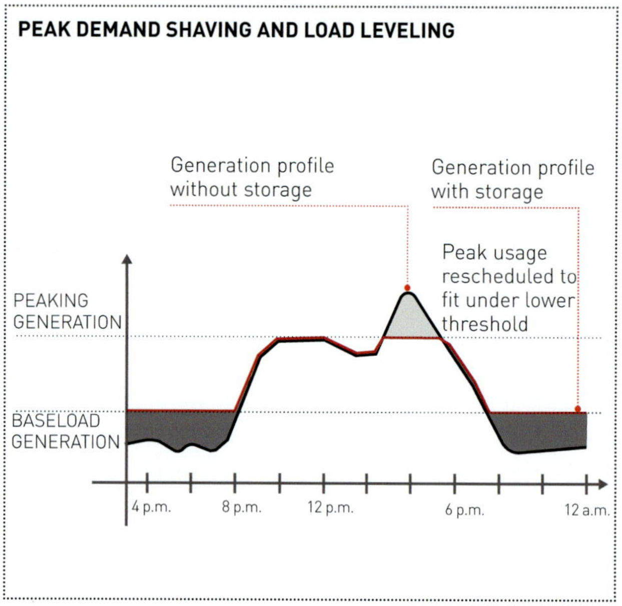

ENERGY SUPPLY

The Supply level examines the sources that will be the input for the generation of energy, and the maintaining of sufficient supply to meet baseline energy needs.

Energy sources are the direct or indirect outcome of natural resources or processes, as well as by-products of man-made processes (in the case of nuclear), and have an intrinsic relation with availability of supply and with short-term and long-term provision. Energy security is a central topic for this level as well as for the energy system in general.

Energy sources can be classified in various ways: conventional and nonconventional, renewable and nonrenewable, alternative sources and fossil fuels, etc. Conventional sources are those that have already been used widely for a long period of time (such as fossil fuels, but also hydropower), while renewables are the ones that are constantly being replenished due to their origin in natural processes (such as wind, solar power, etc.). An alternative source is any source that constitutes an alternative to fossil fuel and more specifically that can be produced or recovered without the undesirable carbon emissions of coal, oil, or gas. These classifications create strong opposites in relation to their capacity for regeneration or not, or their contribution to carbon emissions or not.

The current approach opts for a classification in terms of fossil fuels, nuclear fuel, and renewable sources in order to group together similar attributes focusing on sustainability. Within these categories, further groupings still coexist that reflect certain dissimilarities.

Though every community is adapted to its own local context, the energy sources that are asked to serve its needs belong to a rather global context. Resources that come from fossil fuel reserves are obviously not equally distributed, as they depend on geological processes that cannot be artificially replicated. This results in significant distance between the points of source extraction and use, with high additional costs accompanied by monopoly of goods on one side and dependence on the other. In the face of their decline, a growing awareness of the consequences of globalization of fossil fuels is leading to the use of localized, site-specific alternatives. Energy security is and has to be a profoundly local issue for a given community. Natural conditions may be conducive to the use of renewables, and possible opportunities have to be explored for a community's supply, liberating it from high dependence on imports.

The selection of sources to be used as feedstock to the energy system, and the definition of their contribution to the energy mix, imply the selection and implementation of adequate technologies. The use of varied sources provides the security of availability of alternate sources in the case of shortage in one source. For renewable sources to become major contributors to the energy mix, the problem of these sources' intermittence has to be dealt with (wind and sun, for example, are dependent on natural phenomena, thus hard to predict or schedule in the short term). The importance of selecting various sources to supplement each other as backup is again highlighted. A system "needs a source mix that can react quickly to adjust electricity production to meet the sharp changes in electricity net demand"[12] to ensure reliability.

Terms such as urban energy, local energy, low or zero carbon, carbon-intelligent, clean energy, flexible resources, etc. are constantly proliferating in the global energy-related vocabulary to describe the spreading awareness on issues mentioned above.

On the margin of the main challenges for energy provision, stands municipal solid waste as an alternative energy source. Constant and growing waste generation, as well as by-products of waste processing that need to be managed, may potentially be recovered in the form of energy.

WORLD ELECTRICITY PRODUCTION BY SOURCE, 2014 (EIA)

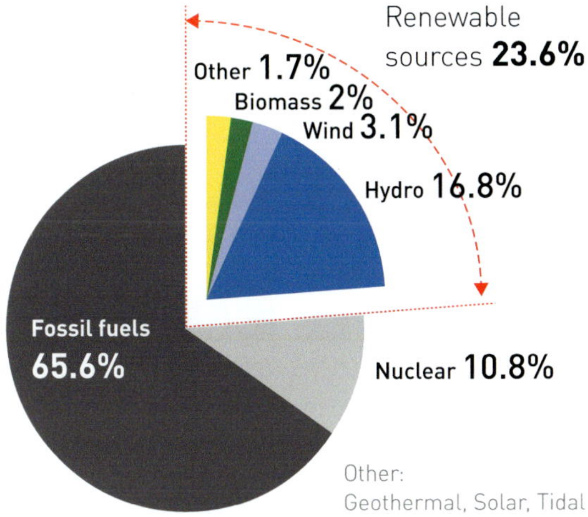

	ENERGY SOURCES	RELATED RESOURCES OR PROCESSES
FOSSIL FUELS	Crude oil and petroleum products	Indigenous reserves Imports Stocks
	Coal	Indigenous reserves Imports Stocks
	Natural gas	Indigenous reserves Imports Stocks
NUCLEAR	Nuclear	Nuclear fission
RENEWABLE SOURCES	Wind	Natural phenomena: wind
	Solar	Natural phenomena: solar radiation
	Geothermal	Geological phenomena: geothermal resources Ground thermal properties Groundwater resources
	Hydro	Natural phenomena: rainfall Water bodies: falling water/flowing water
	Tidal energy	Phenomena: ocean circulation/tide/waves Ocean thermal properties
	Biomass	Agricultural crops and residues Forestry crops and residues Energy crops Stock farming Aquatic species (e.g. algae) Industrial residues
	Waste	Municipal solid waste Organic and food waste Landfill gas Sewage sludge

ELECTRICITY GENERATION EMISSIONS (tonnes CO_2eq /GWh_{el}) PER SOURCE (WEC 2004)

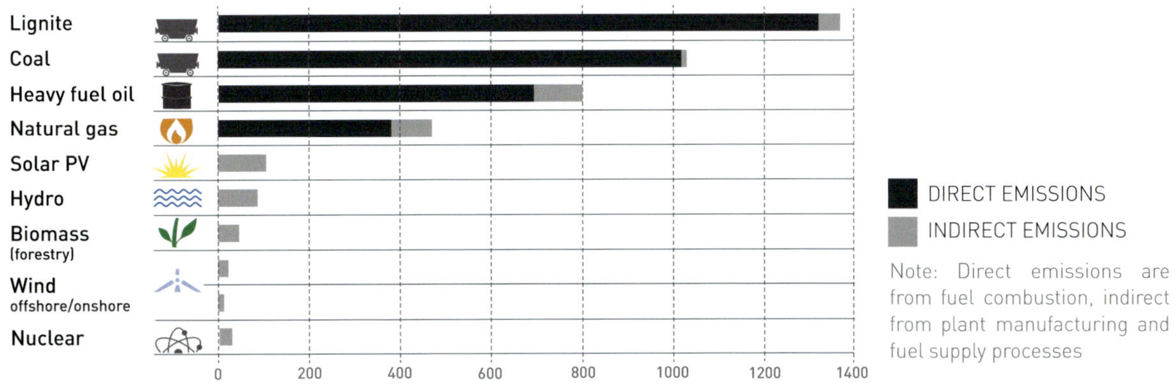

ENERGY GENERATION

SCALE OF GENERATION

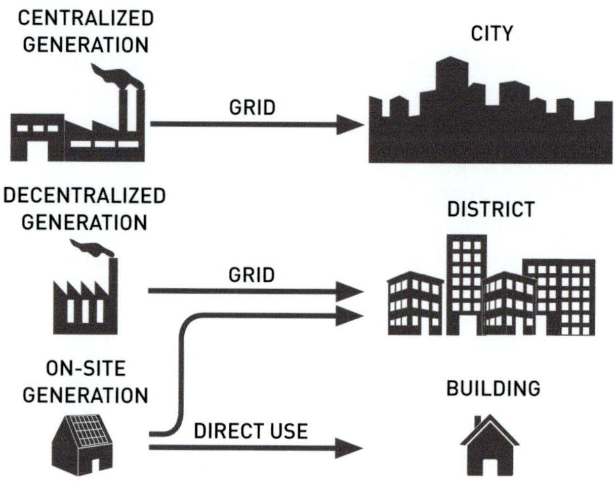

BASIC PROCESSES OF GENERATION OF FUEL, ELECTRICITY, HEAT

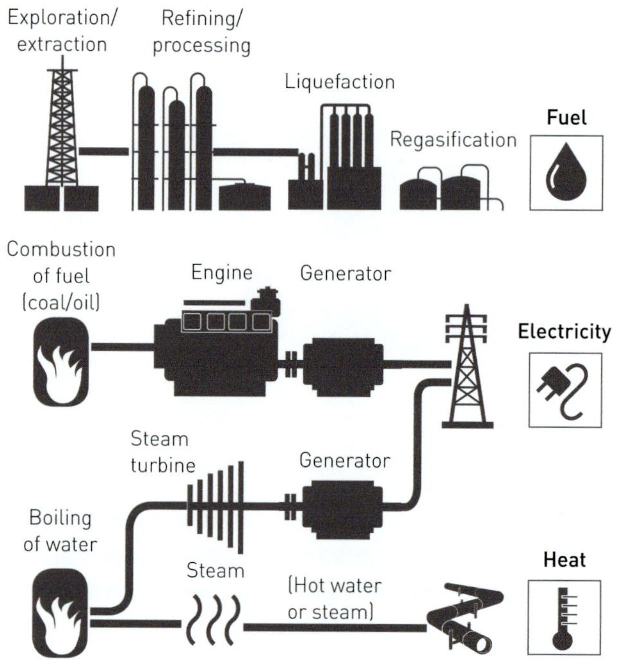

The Generation level examines the process of conversion from energy sources to marketable and fit-for-use forms of electricity, heat, and fuel. Energy generation units are principally defined by their source of supply and the technologies they utilize. This level has a strong spatial dimension, given that a main characteristic of generation entities is the connection of their generation capacity with space requirements and land cover. The siting of these entities is also an important attribute to be examined, given that, to allow proximity to the natural energy resource, a facility is sometimes sited within completely undisturbed, pristine natural habitats and causes significant land cover change.

For nonrenewable sources, the basic processes of electricity or heat generation are combustion and nuclear fission. The former process has major environmental impacts, given that the burning of energy input releases carbon to the atmosphere, with direct and indirect consequences as already mentioned. This is the process of conversion applied to fossil fuels, as well as to biomass and solid waste if incineration is the recovery process used.

One of the topics on this level involves the scale of generation, ranging from large-scale generation plants to decentralized, district-scale generation to cover a local community's energy needs, or even on-site generation in buildings to cover their individual needs. The scale of generation to be adopted is a strategic decision during the process of planning an energy system.

Large-scale or centralized generation is the traditional and familiar way of generating energy. Decentralized energy is "usually taken to mean the technology that enables the generation and distribution of energy within the boundaries of, or located near to and directly connected to, a building, a group of buildings, or a community. It can also include local networks, such as district heating."[13] The scale of generation is connected with the capacity to generate certain amounts of energy, but also with conversion efficiency.

Another topic crucial to and principally connected to this level concerns the processes and thus technologies utilized in energy generation and their output and by-products. "Some generation methods such as coal-fired power plants release the majority of GHGs during operation. Others, such as wind power and nuclear power, release the majority of emissions during construction and decommissioning."[14]

Increasing investments in clean technologies are producing a shift in energy generation toward environment-friendly applications and enabling the optimization of inefficient processes. The examination of processes within energy generation and their side effects, for example, has led to models such as combined-cycle generation and

SOURCE	GENERATION ENTITIES PER SOURCE	MICRO-SCALE GENERATION
	• Conventional power plants • Conventional thermal plants • Combined heat and power plants	• Micro combined power and heat plant (micro CHP) • Natural gas generators • Micro turbines
	• Nuclear power station	
	• Offshore wind farms • Onshore wind farms	• Micro wind turbines
	• Photovoltaic farms (PV) • Concentrating solar power farms (CSP) • Solar thermal technologies	• Building-adapted PV systems (BAPV) • Building-integrated PV systems (BIPV) • Solar heaters
	• Geothermal power plant	• Ground source heat pumps
	• Hydroelectric power plant	• Micro hydro • Mini hydro
	• Tidal energy parks • Tidal power station • Wave farm	
	• Biomass power stations • Combined heat and power plants (CHP) • Anaerobic digestion plant	• Micro digester
	• Waste-to-energy plants (anaerobic digestion plant/incineration plant) • Landfills with energy production	• Small-scale incinerator

cogeneration that are based on the exploitation to the maximum of the original energy input. Cogeneration, using heat wasted during energy conversion to electricity, proves the high potential for further optimization of conventional practices.

Both the scale and type of a generating process determine its impact on the natural world and the community, and require a corresponding set of control measures. Additionally, both scale and type are highly connected to the efficiency of generation, determining the factor that will be applied to the primary energy input to define the net energy output.

Hybrid generation introduces the possibility of a mix of fuels as input for combustion and the combination of different types of energy within the system.

Distributed energy is related both to the scale and the type of generating process and is tightly connected with renewable energy systems. The inclusion of distributed energy in national energy policies has a direct relation to the promotion of renewable energy systems and the transition from carbon-based fuels and fossil-fuel grid-based electricity to clean and local resources.[15] The transition from centralized to distributed systems has yet to deal with issues of system stability to match demand on a constant basis. It requires "the ability to start and stop multiple times per day"[16] to support demand variability.

The combination of multiple different generators, including backup mechanisms as well as the possibility of energy storage, reduces the need for centralized fossil-fuel-based units for constant and secure energy generation.

GENERATION PROCESSES PER ENERGY CARRIER

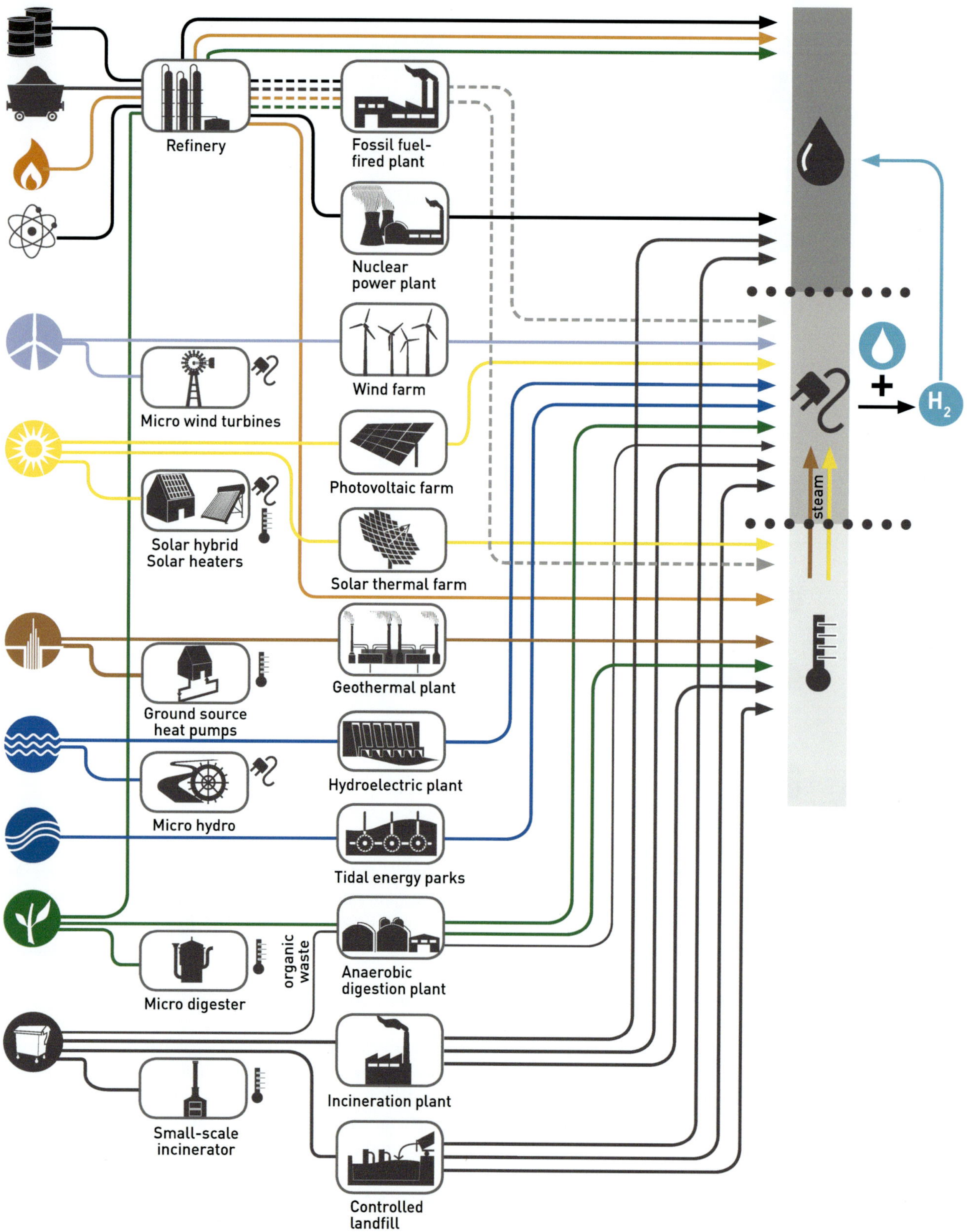

ENERGY DISTRIBUTION AND TRANSMISSION

The Distribution and Transmission level examines the transfer of energy generated to the end users.

The network of energy transmission and distribution varies by energy carrier. All networks, though, share a similar differentiation according to the scale they are applied at: international, national, regional, local or district, and site scale. Within the structure of energy transfer, transmission and distribution are further divided in subcategories based on the system's scale.

Fuel distribution varies from home heating delivery by service companies to large wholesale fuel distribution, with the corresponding variation in means of transfer. Electricity is initially transmitted at higher voltage to cover long distances and then distributed at lower voltage to end users, through overhead or underground cables. Thermal energy is transmitted and distributed in the form of steam or heated or cooled water, through a network of pipelines.

The transfer of energy is facilitated and secured by certain transfer stations, nodes within these networks. They act as points of control and regulation that provide the appropriate conditions for constant flow, the connectivity and stability of the system's function.

Storage is an essential part of the process of energy transfer, introducing the possibility of time offset for the delivery of energy. It is traditionally practiced as the storage of fossil fuels in barrels or tanks, offering the possibility of overseas export-import as well as a means of storing a stock of supply. In the case of renewable energy sources, storage offers the additional opportunity of dealing with the inherent intermittence of these sources. Storage makes possible peak demand shaving strategies, through charging and discharging in accordance with daily patterns of energy consumption.

In the transition from centralized to distributed systems, storage provides stability, reducing steep ramping issues in demand.

Storage ranges from long-term to seasonal and short-

ENERGY CARRIERS' DISTRIBUTION

- Fuel distribution varies from home heating delivery by service companies to large wholesale fuel distribution (oil, gas, propane, and diesel)
- Electricity is initially transmitted in higher voltage to cover long distances and then distributed in a lower voltage to end users
- Thermal energy is distributed in the form of steam or heated or cooled water through a network of pipelines.

ENERGY FORMS GENERATED		DISTRIBUTION MEDIUMS	END USE
Fuel	Centralized generation	Network of dispensing stations	Transportation fuels
		Home delivery of fuel	Household heating etc.
		Underground pipeline network	District heating and other household needs
Electricity	Centralized generation	Overhead and underground wire network	District electricity
	On-site generation	Export of autogenerated electricity to grid/import from grid	Household electricity/ district electricity
Heat	Centralized generation	Underground pipeline network	District heating/cooling
	On-site generation	NO DISTRIBUTION	On-site heating/cooling
Off-grid systems		NO DISTRIBUTION	Direct use and storage

ELECTRICITY DISTRIBUTION

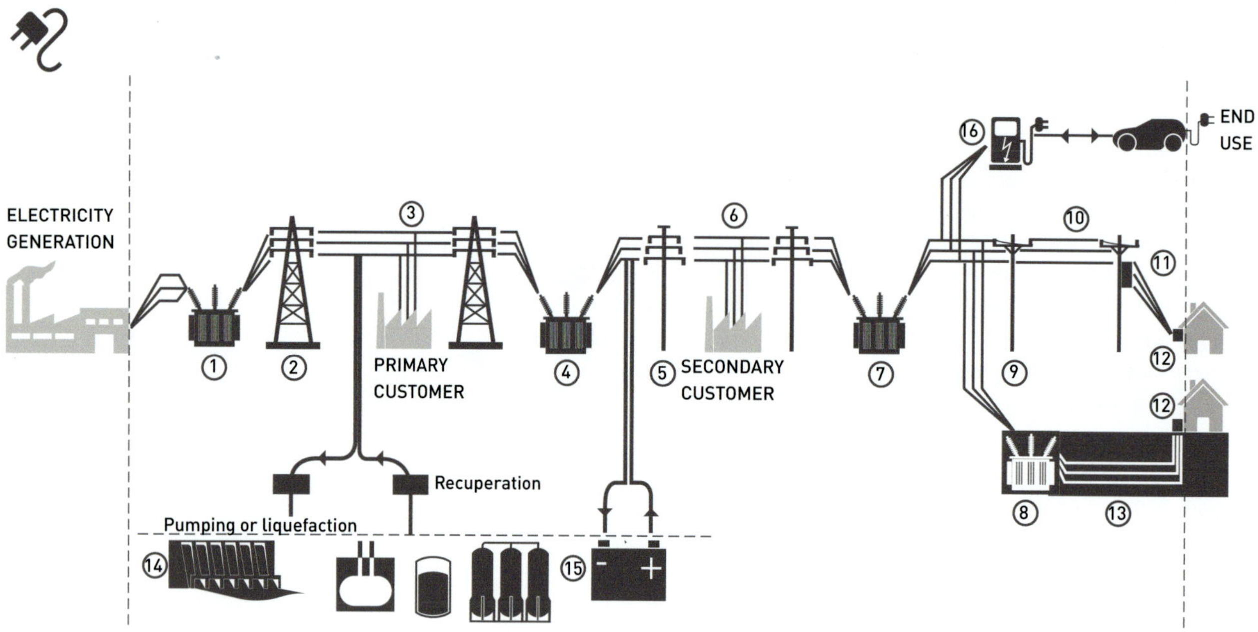

1 Generator's step-up transformer/switchyard
2 Transmission towers
3 Long-distance/high-voltage transmission lines
4 Transmission substations/step-down transformers
5 Sub-transmission poles
6 Sub-transmission lines
7 Distribution ground substations/transformers
8 Distribution underground substations/transformers
9 Distribution poles
10 13 Distribution lines, overhead/underground
11 Pole transformers
12 Meters
14 15 Long-term/short-term storage
16 Charging stations for electric vehicles

HEAT / COLD DISTRIBUTION

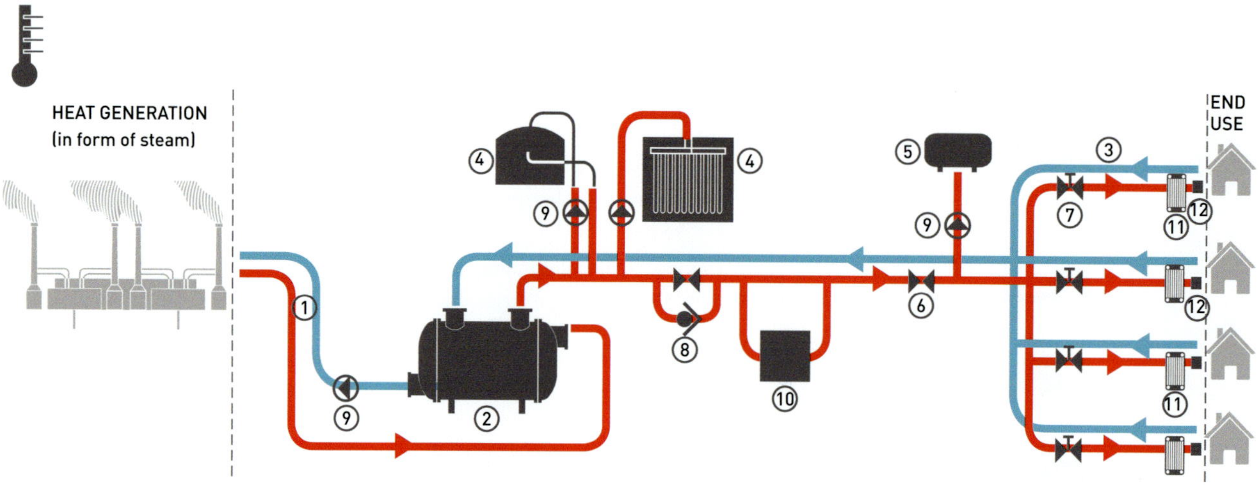

1 Primary pipeline network
2 Substation/heat exchanger
3 Secondary pipeline network (hot feed and cold return piping)
4 5 Long-term/short-term storage tanks
6 Pressure and flow controllers/control valves
7 Block valves
8 Bypass valves
9 Pumps
10 Back-up boiler/stand-by/peak load boiler
11 Building level substations
12 Heat meters

ENERGY STORAGE (LONG-TERM AND SHORT-TERM)

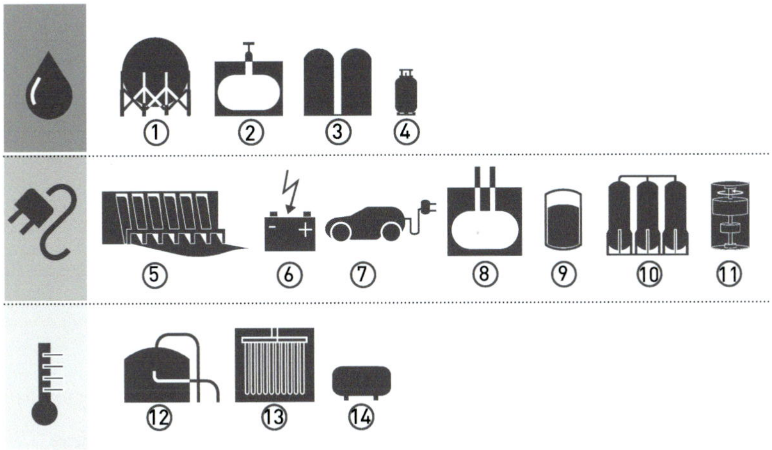

① Ground storage tanks
② Underground storage fields
③ Liquified gas tanks
④ Bottled gas
⑤ Pumped-storage plant/hydroelectricity
⑥ Battery banks
⑦ Electric vehicles
⑧ Compressed air energy storage
⑨ Liquid air and cryogenic energy storage
⑩ Hydrogen storage
⑪ Flywheel energy storage (rotational energy)
⑫ Hot water storage tank
⑬ Bore-hole long-term/seasonal storage
⑭ Short-term storage

term. It is essential to the use of renewable energy sources to exploit the natural capacity and frequency of phenomena whose excess otherwise would end as wastage.

The impossibility of synchronizing energy generation and demand, along with the losses in the process of energy transfer, call for a more flexible system of transmission and distribution.

The smart grid is an integrated system of digital meters, batteries, and communication that distributes power more efficiently and conserves energy by directly connecting demand, supply, and demand response. It offers the ability to monitor in real time and remotely control and adjust components, balancing interactions between consumers, generators, and the networks themselves. These systems driven by information are more adaptive and can reconfigure themselves in response to use and supply or in case of an emergency. Over time, information allows for more predictive capability and optimization of performance, since models of the systems can be created and used.

Smart technology is essential for the upgrading of energy networks to support and more readily integrate the renewable energy and autogenerated energy at homes and businesses, as well as new energy storage methods and hybrid vehicles, into a distribution grid. At the same time it allows for an immediate response to incidents and disturbances, protecting and stabilizing the grid.

SMART GRID

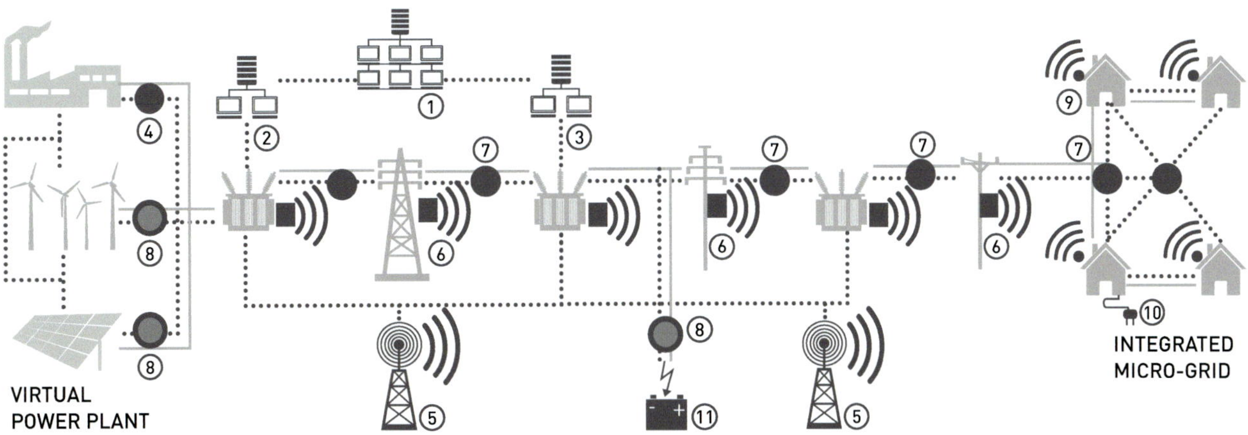

·········· Communication network
——— Power grid

① Network operation center
② Transmission substation control
③ Distribution substation control
④ Processors
⑤ Wireless base stations
⑥ Pole-mounted nodes
⑦⑧ Wireless sensors activated or on stand-by
⑨ Smart meters
⑩ Smart appliances
⑪ Intelligent energy storage

7.4 | OBJECTIVES AND GUIDELINES FOR A SUSTAINABLE ENERGY SYSTEM

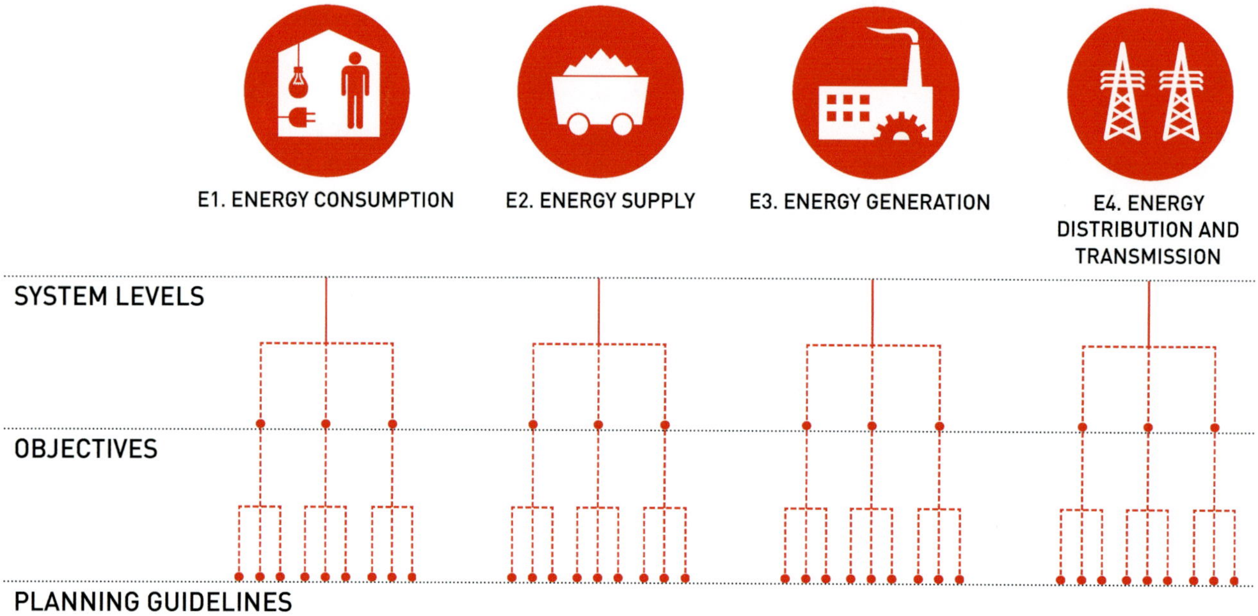

Following the decoding of the Energy system at four system levels, planning objectives are defined for each level. They represent both challenges and strategies toward a sustainable energy system, expressing what should be the overall final outcome of energy planning.

The objectives are associated with the five focus categories of the Zofnass Program, namely Leadership, Quality of Life, Natural World, Resource Allocation, and Climate and Risk. The Sustainable Planning Guidelines aim to lead toward these objectives, highlighting synergies and connections with the other six infrastructure categories of Transportation, Water, Solid Waste, Landscape, Information, and Food.

The objectives and guidelines of the Energy Consumption level are primarily associated with the Leadership category. They do not refer directly to sustainable infrastructure planning, but instead aim to create a solid base for the system through user engagement and monitoring mechanisms. Energy saving is a central strategy at this level. Objectives call for the reduction of energy consumption in the identified principal energy-consuming sectors (the building, transportation, and water management sectors), not by discouraging use but by detecting behaviors and conditions that lead to unnecessary overuse. Urban planning strategies such as the promotion of mixed-use development and urban compactness that reduce transportation needs, as well as building design practices adapted to natural conditions, minimize a city's overall energy needs. Other objectives require the provision of monitoring for energy consumption trends and regulation of their rates and patterns, reducing peaks in energy use. A thorough understanding of the demand for energy services allows the coordinating of demand and generation, but also provides a strong base of collected data upon which to develop future action plans. The Energy Consumption level features strong synergies with Transportation and Water infrastructure, identified as among the most energy-consuming sectors. Information infrastructure as a subsystem of Energy can be used to optimize, monitor, and regulate energy consumption through smart technologies and databases of collected data.

E1. ENERGY CONSUMPTION

- E1.1. Reduce Energy Consumption in the building sector
- E1.2. Reduce Energy Consumption in the Transportation system
- E1.3. Reduce Energy Consumption in the Water system
- E1.4. Reduce Energy Consumption in other context-specific sectors with significant energy demand
- E1.5. Monitor Energy Consumption trends
- E1.6. Regulate Energy Consumption rates and patterns

The Energy Supply level aims to aid the strategic planning of the city's energy supply mix. Objectives are related to the Leadership, Climate and Risk, and Resource Allocation categories. The importance of the appropriate selection of energy sources has already been highlighted in the Energy Supply level, directly affecting the availability of natural resources both on a local and a global level as well as determining the emissions of the energy sector. The main objective of this level is to ensure security of supply in the long term and promote multiple and diverse energy sources. This involves exploiting all potential energy resources primarily on a local level to minimize dependence on imported sources. Supply is the initial input to the system, and thus the supply mix selection determines technologies and practices that affect the whole system's performance. The performance of a system can only be evaluated within a framework of specific goals, and this level calls for strategic decisions on specific, measurable, attainable, and relevant goals, both in reducing greenhouse gas and air pollutant emissions as well as in promoting renewable energy sources.

Guidelines here call for a strong planning connection to the city's economic sectors and to projections for long-term development. Specific benchmarks should be adopted in order to monitor the system's performance and the implementation of planning. Strong synergies are created with Solid Waste, Landscape, and Food infrastructure, as by-products of these systems can act as potential alternative energy sources. Integration of these systems and alignment of their goals, especially with the solid waste system, is essential.

E2. ENERGY SUPPLY

- E2.1. Secure city Energy Supply in the long term
- E2.2 Define clear goals to be achieved through Energy Supply planning
- E2.3. Minimize greenhouse gas and air pollutant emissions
- E2.4. Resolve conflicts of Energy Supply with resources used by other infrastructure systems
- E2.5. Resolve issues of residues connected to the types of Energy Supply used

On the Energy Generation level, objectives cover all five of the Zofnass Program focus categories. Energy generation plays a significant role within the energy system as the conversion of sources to useful energy, but simultaneously it has the most adverse impact. In terms of the Leadership category, the main objectives concern high-level decisions for optimized application

E3. ENERGY GENERATION

- E3.1. Secure the reliability and flexibility of Energy Generation services
- E3.2. Place and distribute Energy Generation facilities for optimized Energy system performance
- E3.3. Integrate Energy Generation with other infrastructure systems
- E3.4. Promote community well-being
- E3.5. Enhance quality of place
- E3.6. Maintain the integrity of natural-world elements and systems
- E3.7. Control greenhouse gas emissions
- E3.8. Control air pollutant emissions
- E3.9. Prepare for short-term hazards and projected future threats
- E3.10. Resolve issues of residues or by-products connected to each type of Energy Generation
- E3.11. Optimize Water use in Energy Generation

of technologies and spatial distribution of energy plants and their correlation to load centers, as well as ensuring adequate coverage of demand and stability. As already mentioned, there has been a shift from conventional methods depending on centralized systems to more flexible ones. Flexibility is mainly related to the promotion of hybrid systems that efficiently integrate intermittent renewable energy sources that depend on natural phenomena such as wind and solar power, which are unpredictable in the short term, exploiting them to the maximum at the time they occur. Generation facilities should be planned with capacity compatible with long-term city needs, supported by adequate administrative structures and operated based on real-time information to achieve the optimum balance with demand. Energy generation entities are defined by their input source and their capacity, which is connected with space requirements and land occupation. They have a very strong spatial presence that can create a significant barrier to a community's everyday life, and planning has to consider the potential land fragmentation of the urban fabric caused as well as the visual impact of generation facilities. The Quality of Life category is directly connected to quality of place and public spaces and is affected by the location of generation. In the Climate and Risk and Natural World categories, objectives include the resilience of the system in the face of climate change risks, and the protection of natural processes. The guidelines focus on assessing and managing potential risks from natural and man-made phenomena, addressing response and recovery times in case of emergency, and mitigating the heat island effect. The objectives for Resource Allocation address the need for optimized material use and by-products of processes management. They are expressed through guidelines promoting the use of reclaimed and recycled water or water-efficient technologies in the generation process and the management of waste according to Solid Waste infrastructure requirements. At this system level, synergies are noted with all other infrastructure systems. Information infrastructure as a subsystem can support the real-time monitoring and management of hybrid systems. Some of the most important synergies, however, link with Landscape infrastructure, which represents both a mitigating and a compensatory measure for the negative impacts of energy generation facilities, such as air pollution or contaminant runoff, but also can help protect facilities themselves from extreme weather phenomena through the creation of barriers, buffers, etc.

The Energy Distribution and Transmission system level is guided by objectives that relate to all five focus categories of the Zofnass Program. The Leadership category includes most of the objectives for this level, mainly concerning the reliability, flexibility, and stability of networks that enable electricity, heat, and fuel distribution along long or short distances and integrating them with other infrastructure networks. Necessary considerations include providing adequate coverage and length of networks for all energy carriers and combining and integrating with other networks. Another aspect of Leadership is to establish adequate administrative structures and the financial viability of networks throughout construction and operation. In the Quality of Life category, objectives focus again on improving public space and minimizing impacts such as urban fragmentation, while ensuring safety for the community. In Climate and Risk, providing resilience against natural and man-made phenomena is again critical for securing the continuous and smooth operation of the system. Potential risks should be assessed so that fast response and recovery can be addressed. In the Natural World category, the objectives point to minimizing disruptions of natural systems by avoiding landscape alterations, maintaining the water cycle, and protecting habitats and wildlife corridors. The Resource Allocation category has considerable importance, including objectives for consumption of materials. It is important to source materials locally and promote the use of reclaimed and recycled materials at a high level. Achieving integration with other infrastructure networks is central for a successful systemic approach that minimizes the visual impact of energy system components. Synergies with Information and Landscape can assist Energy infrastructure to achieve sustainability goals. Information applications can transform transmission and distribution networks into smart, integrated management systems for optimized balance between energy generation and demand with the use of real-time use data.

The planning of an integrated energy system, like

that of any infrastructure system, is objective-dependent, and it is the setting of objectives that determines the requirements the system has to meet. The four levels of the energy system have their specific objectives with specific guidelines to follow, but mainly they supplement each other to build an overall sustainable system. The objectives with the respective guidelines as well as the relevant guidelines of other infrastructure systems are presented in the following tables.

E4. ENERGY DISTRIBUTION AND TRANSMISSION	
● E4.1.	Secure the reliability and flexibility of Energy Distribution and Transmission services
● E4.2.	Secure electricity network stability
● E4.3.	Minimize energy losses during Energy Distribution and Transmission
● E4.4.	Integrate Energy Distribution and Transmission entities and facilities with other infrastructure systems
● E4.5.	Promote community well-being
● E4.6.	Enhance quality of place
● E4.7.	Maintain the integrity of natural-world elements and systems
● E4.8.	Prepare for short-term hazards and projected future threats
● E4.9.	Optimize material use in Energy Distribution and Transmission networks

GUIDELINES

E1 ENERGY CONSUMPTION

▲ TO
+ WITH
▼ FROM

OBJECTIVES		GUIDELINES	SYNERGIES
E1.1. Reduce Energy Consumption in the building sector	E1.1.1.	Guide, assist, and motivate communities in rational use of energy	
	E1.1.2.	Set long-term goals and standards for reducing Energy Consumption in the building sector	
	E1.1.3.	Adapt the urban fabric configuration to the local climate	
	E1.1.4.	Improve the urban microclimate	▼ +
E1.2. Reduce Energy Consumption in the Transportation system	E1.2.1.	Optimize the Transportation modal split for reduced Energy Consumption	▼
	E1.2.2.	Optimize the selection and operation of Transportation vehicles for reduced Energy Consumption	▼
	E1.2.3.	Plan the urban fabric to provide for short trips and public transportation options	▼
E1.3. Reduce Energy Consumption in the Water system	E1.3.1.	Minimize Energy Consumption for Water Supply and Water Treatment	▼
	E1.3.2.	Minimize Energy Consumption for Water conveyance and distribution	▼
E1.4. Reduce Energy Consumption in other context-specific sectors with significant energy demand	E1.4.1.	Identify additional sectors with significant Energy Consumption and develop efficiency strategies	
E1.5. Monitor Energy Consumption trends	E1.5.1.	Monitor daily and seasonal Energy Consumption fluctuations	+
	E1.5.2.	Regularly collect and organize data on Energy Consumption	+
E1.6. Regulate Energy Consumption rates and patterns	E1.6.1.	Reduce peaks in energy demand	

E2 ENERGY SUPPLY

▲ TO
+ WITH
▼ FROM

OBJECTIVES		GUIDELINES	SYNERGIES
E2.1. Secure city Energy Supply in the long term	E2.1.1.	Identify all potential Energy sources	+ ▼ ▼ ▼
	E2.1.2.	Promote use of multiple and diverse Energy sources to cover city needs in the long term	
	E2.1.3.	Minimize dependence on imported Energy sources	
	E2.1.4.	Minimize use of nonrenewable Energy sources	
	E2.1.5.	If waste is selected as an alternative Energy source, integrate with overall Solid Waste management planning	▼
	E2.1.6.	Trace correlation of alternative Energy sources with city competitiveness and economic sectors	

OBJECTIVES		GUIDELINES	SYNERGIES
	E2.1.7.	Secure long-term financial viability of Energy Supply plan	
E2.2. Define clear goals to be achieved through Energy Supply planning	E2.2.1.	Set goals for Energy Supply mix	
	E2.2.2.	Adopt quantitative indicators to benchmark and monitor the performance of the Energy Supply plan	
	E2.2.3.	Integrate long-term city growth and development goals with Energy Supply planning	
	E2.2.4.	Regularly update Energy Supply planning according to community development and Energy system performance	
E2.3. Minimize greenhouse gas and air pollutant emissions	E2.3.1.	Prioritize and promote clean Energy sources	
	E2.3.2.	Set long-term goals and standards for reducing emissions in the Energy sector	
E2.4. Resolve conflicts of Energy Supply with resources used by other infrastructure systems	E2.4.1.	Avoid conflict between crops used both for Food Supply and Energy Supply	▲
	E2.4.2.	Take into consideration available Water resources when planning Energy Supply mix	▲
E2.5. Resolve issues of residues connected to the types of Energy Supply used	E2.5.1.	Resolve Energy Supply residues handling and disposal requirements	▲
	E2.5.2.	Define which Energy Supply residues can be marketable	

E3 ENERGY GENERATION

▲ TO
+ WITH
▼ FROM

OBJECTIVES		GUIDELINES	SYNERGIES
E3.1. Secure the reliability and flexibility of Energy Generation services	E3.1.1.	Develop distributed Energy Generation systems	
	E3.1.2.	Promote hybrid Energy Generation technologies and systems	
	E3.1.3.	Plan backup systems for non-firm-distributed Energy Generation	
	E3.1.4.	Integrate thermal and electricity systems for increased efficiency in Energy Generation	
	E3.1.5.	Develop smart, integrated management systems to optimize Energy Generation efficiency	+
	E3.1.6.	Secure long-term financial viability of Energy Generation system	
	E3.1.7.	Optimize Energy Generation asset management through adequate administrative structures	
E3.2. Place and distribute Energy Generation facilities for optimized Energy system performance	E3.2.1.	Place Energy Generation facilities close to energy load centers	
	E3.2.2.	Promote district heating/cooling systems where efficient	
	E3.2.3.	Optimize the placement of Energy Generation facilities with regard to network routes	
	E3.2.4.	Reserve spaces for Energy Generation facilities taking into consideration long-term city development	

OBJECTIVES	GUIDELINES	SYNERGIES
E3.3. Integrate Energy Generation with other infrastructure systems	E3.3.1. Investigate potential of Energy Generation integration with Solid Waste Treatment and Disposal facilities	
	E3.3.2. Integrate Energy Generation facilities with Water distribution facilities	
	E3.3.3. Investigate potential of Energy Generation integration with wastewater Treatment facilities	
E3.4. Promote community well-being	E3.4.1. Assess and resolve potential health impacts caused by Energy Generation facilities	
	E3.4.2. Assess and resolve potential visual discomfort and glare caused by Energy Generation installations	
	E3.4.3. Assess and resolve noise pollution issues caused by Energy Generation facilities	
E3.5. Enhance quality of place	E3.5.1. Visually integrate Energy Generation facilities to the local context	
	E3.5.2. Minimize land use and fragmentation of urban fabric caused by Energy Generation facilities	
E3.6. Maintain the integrity of natural-world elements and systems	E3.6.1. Minimize natural Landscape alteration caused by Energy Generation facilities	
	E3.6.2. Protect natural habitats and wildlife corridors and compensate for their fragmentation or disruption by Energy Generation facilities	
	E3.6.3. Protect soil from contaminants found in Energy Generation facilities	
	E3.6.4. Protect Water elements and groundwater from contaminants found in Energy Generation facilities	
E3.7. Control greenhouse gas emissions	E3.7.1. Implement carbon capture and storage solutions in Energy Generation facilities	
	E3.7.2. Compensate for Energy Generation emissions through carbon offset and sequestration practices	
E3.8. Control air pollutant emissions	E3.8.1. Promote clean technologies in Energy Generation facilities	
E3.9. Prepare for short-term hazards and projected future threats	E3.9.1. Assess and manage threats from natural and man-made phenomena affecting Energy Generation facilities	
	E3.9.2. Plan response and recovery in case of emergency affecting Energy Generation facilities	
E3.10. Resolve issues of residues or by-products connected to each type of Energy Generation	E3.10.1. Resolve routes and methods for the collection, treatment, and disposal of Energy Generation residues and by-products	
E3.11. Optimize Water use in Energy Generation	E3.11.1. Promote water-efficient technologies in Energy Generation facilities	
	E3.11.2. Use gray or reclaimed water where possible for Energy Generation processes	

OBJECTIVES	GUIDELINES		SYNERGIES
E4 ENERGY DISTRIBUTION AND TRANSMISSION			▲ TO + WITH ▼ FROM
E4.1. Secure the reliability and flexibility of Energy Distribution and Transmission services	E4.1.1.	Provide required coverage for the Distribution and Transmission of all Energy sources and carriers selected	
	E4.1.2.	Provide adequate Energy Distribution and Transmission capacity to cover long-term city needs	
	E4.1.3.	Reserve spaces for Energy Distribution and Transmission facilities taking into consideration long-term city development	
	E4.1.4.	Maintain compatibility of Energy Distribution and Transmission with Energy Supply sources	
	E4.1.5.	Adapt Energy Distribution and Transmission to support absorption of electricity generated by renewable sources	
	E4.1.6.	Secure long-term financial viability of Energy Distribution and Transmission networks	
	E4.1.7.	Optimize Energy Distribution and Transmission asset management through adequate administrative structures	
E4.2. Secure electricity network stability	E4.2.1.	Develop smart, integrated management systems to optimize balance of electricity generation and demand	+
	E4.2.2.	Promote connectivity within electricity grids	
	E4.2.3.	Provide adequate energy storage facilities	
	E4.2.4.	Adapt Distribution and Transmission to support decentralized Energy Generation systems	
	E4.2.5.	Provide real-time monitoring of electricity grids	+
E4.3. Minimize energy losses during Energy Distribution and Transmission	E4.3.1.	Optimize Distribution and Transmission routes and distances	
	E4.3.2.	Develop smart, integrated management systems to optimize Energy Distribution and Transmission performance	+
E4.4. Integrate Energy Distribution and Transmission entities and facilities with other infrastructure systems	E4.4.1.	Integrate Energy Distribution and Transmission networks with Transportation Networks	▼
	E4.4.2.	Plan Energy Distribution and Transmission networks to support the operation of Information Nodes	▲
	E4.4.3.	Integrate Energy Distribution and Transmission networks with Landscape entities	▼
E4.5. Promote community well-being	E4.5.1.	Assess and resolve health impacts of Energy Distribution and Transmission networks	
E4.6. Enhance quality of place	E4.6.1.	Visually integrate Energy Distribution and Transmission facilities to the local context	+
	E4.6.2.	Minimize land use and fragmentation of urban fabric by Energy Distribution and Transmission facilities	

OBJECTIVES	GUIDELINES		SYNERGIES
E4.7. Maintain the integrity of natural-world elements and systems	E4.7.1.	Minimize natural Landscape alteration caused by Energy Distribution and Transmission facilities	
	E4.7.2.	Protect natural habitats and wildlife corridors and compensate for their fragmentation or disruption by Energy Distribution and Transmission	+
E4.8. Prepare for short-term hazards and projected future threats	E4.8.1.	Assess and manage future threats from natural and man-made phenomena affecting Energy Distribution and Transmission	+
	E4.8.2.	Plan Energy Distribution/Transmission services response and recovery in case of emergency	
E4.9. Optimize material use in Energy Distribution and Transmission networks	E4.9.1.	Source materials required for Energy Distribution and Transmission locally when possible	
	E4.9.2.	Promote use of reclaimed and recycled materials in Energy Distribution and Transmission networks construction	
	E4.9.3.	Plan for the reuse of components or recycling of materials after Energy Distribution and Transmission networks decommission	

RELEVANT GUIDELINES

GUIDELINES		RELEVANT GUIDELINES	
E1 ENERGY CONSUMPTION		FROM ▸ ▸ TO + WITH	
E1.1.4.	Improve the urban microclimate	L1.4.1.	Define needs for climate adaptation and microclimate improvement through Landscape infrastructure
		T4.9.3.	Configure Transportation Networks so as to mitigate the urban heat island effect
E1.2.1.	Optimize the Transportation modal split for reduced Energy Consumption	T2.3.1.	Provide and promote reliable public and mass transit
E1.2.2.	Optimize the selection and operation of Transportation vehicles for reduced Energy Consumption	T2.4.2.	Promote nonmotorized Transportation Modes
		T2.3.2.	Provide car sharing and bike sharing options
		T2.4.3.	Use efficient and/or alternative-fuel vehicles
		T2.4.4.	Put in place mechanisms to ensure regular maintenance of vehicles
		T2.4.5.	Educate users on the efficient operation of vehicles
		T4.2.3.	Organize road networks to promote use of high-occupancy vehicles
E1.2.3.	Plan the urban fabric to provide for short trips and public transportation options	T1.2.3	Promote compactness in urban fabric
E1.3.1.	Minimize Energy Consumption for Water Supply and Water Treatment	W2.5.2.	Promote energy-efficient Water Supply methods
		W3.10.1.	Use energy-efficient Water Treatment technologies
E1.3.2.	Minimize Energy Consumption for Water conveyance and distribution	W2.5.1.	Minimize distance between Water Supply resources and Water Consumption locations
		W3.2.1.	Optimize placement of Water Treatment facilities so as to minimize distances between treatment and Water Consumption
		W4.9.1.	Take advantage of natural landscape configuration to minimize needs for pumping in Water Networks
E1.5.1.	Monitor daily and seasonal Energy Consumption fluctuations	I1.3.1.	Identify infrastructure systems' processes to be monitored/documented through Information infrastructure
E1.5.2.	Regularly collect and organize data on Energy Consumption	I1.3.1.	Identify infrastructure systems' processes to be monitored/documented through Information infrastructure
E2 ENERGY SUPPLY		FROM ▸ ▸ TO + WITH	
E2.1.1.	Identify all potential Energy sources	SW2.2.2.	Identify potential for Energy recovery and Generation through Solid Waste management
		W3.9.3.	Identify potential for Energy recovery and Generation through Water Treatment residues

GUIDELINES			RELEVANT GUIDELINES	
			F2.10.1.	Identify potential for using Food Supply and Production by-products as feedstock for other infrastructure systems
			F4.9.2.	Identify potential for using Food Processing and Distribution residues and by-products as feedstock for other infrastructure systems
			L1.5.1.	Identify opportunities for the provisioning of food, fiber, and feedstock provided by the natural context
			L4.10.2.	Use residues of Landscape maintenance practices as supply resources for other infrastructure systems
E2.1.5.	If waste is selected as an alternative Energy source, integrate with overall Solid Waste management planning		SW2.3.3.	Match Solid Waste Diversion Model residues and by-products with resource and feedstock needs of other infrastructure systems
E2.4.1.	Avoid conflict between crops used both for Food Supply and Energy Supply		F3.4.2.	Resolve potential conflicts with food resources used as supply for other systems
E2.4.2.	Take into consideration available Water resources when planning Energy Supply mix		W1.2.1.	Reduce water volume required for Energy Supply and Energy Generation
E2.5.1.	Resolve Energy Supply residues handling and disposal requirements		SW1.1.5.	Regulate use of hazardous substances bound to enter the Solid Waste management system

E3 ENERGY GENERATION

FROM ▶ ▶ TO
+
WITH

E3.1.5.	Develop smart, integrated management systems to optimize Energy Generation		I1.3.2.	Identify other infrastructure systems' processes to be managed through Information infrastructure
E3.3.1.	Investigate potential of Energy Generation integration with Solid Waste Treatment and Disposal facilities		SW3.5.2.	Integrate Solid Waste Treatment and Disposal with Energy Generation facilities if applicable
E3.3.2.	Integrate Energy Generation facilities with Water distribution facilities		W4.3.2.	Plan Water Networks to support the operation of Energy Generation facilities
E3.3.3.	Investigate potential of Energy Generation integration with wastewater Treatment facilities		W3.3.2.	Integrate Water Treatment and Energy Generation facilities if applicable
E3.4.3.	Assess and resolve noise pollution issues caused by Energy Generation facilities		L1.7.2.	Define needs for mitigating and compensating for other infrastructure systems' impact on community life through Landscape infrastructure
E3.5.1.	Visually integrate Energy Generation facilities to the local context		L1.7.2.	Define needs for mitigating and compensating for other infrastructure systems' impact on community life through Landscape infrastructure
E3.6.2.	Protect natural habitats and wildlife corridors and compensate for their fragmentation or disruption by Energy Generation facilities		L1.7.1.	Define needs for mitigating and compensating for other infrastructure systems' impact on the natural world through Landscape infrastructure

GUIDELINES		RELEVANT GUIDELINES	
E3.6.3.	Protect soil from contaminants found in Energy Generation facilities	L1.7.1.	Define needs for mitigating and compensating for other infrastructure systems' impact on the natural world through Landscape infrastructure
		F2.5.5.	Identify and manage basic sources of contaminants within the foodshed
E3.6.4.	Protect Water elements and groundwater from contaminants found in Energy Generation facilities	L1.7.1.	Define needs for mitigating and compensating for other infrastructure systems' impact on the natural world through Landscape infrastructure
E3.9.1.	Assess and manage threats from natural and man-made phenomena affecting Energy Generation facilities	L1.7.3.	Define needs for protection of infrastructure systems against projected climate change and extreme events through Landscape infrastructure
E3.10.1.	Resolve routes and methods for the collection, treatment, and disposal of Energy Generation residues and by-products	SW3.5.3.	Solid Waste and Energy systems to work in synergy for the Treatment and Disposal of Energy Supply and Energy Generation residues
		SW4.3.3.	Solid Waste and Energy systems to work in synergy for the Collection and Transfer of Energy Generation residues
E3.11.1.	Promote water-efficient technologies in Energy Generation facilities	W1.2.1.	Reduce water volume required for Energy Supply and Energy Generation
E3.11.2.	Use gray or reclaimed water where possible for Energy Generation processes	W1.6.3.	Identify opportunities for use of different nonpotable water qualities

E4
ENERGY DISTRIBUTION AND TRANSMISSION

FROM ▸ ▸ TO
+
WITH

E4.2.1.	Develop smart, integrated management systems to optimize balance of electricity generation and demand	I1.3.2.	Identify other infrastructure systems' processes to be managed through Information infrastructure
E4.2.5.	Provide real-time monitoring of electricity grids	I1.3.1.	Identify infrastructure systems' processes to be monitored/documented through Information infrastructure
E4.3.2.	Develop smart, integrated management systems to optimize Energy Distribution and Transmission performance	I1.3.2.	Identify other infrastructure systems' processes to be managed through Information infrastructure
E4.4.1.	Integrate Energy Distribution and Transmission networks with Transportation Networks	T4.3.3.	Configure Transportation Networks to provide for needs of Energy Distribution and Transmission networks and facilitate their maintenance
E4.4.2.	Plan Energy Distribution and Transmission networks to support the operation of Information Nodes	I3.4.1.	Integrate Information Nodes with Energy Distribution and Transmission facilities
E4.4.3.	Integrate Energy Distribution and Transmission networks with Landscape entities	L3.5.1.	Configure Landscape entities' vegetation and paving to provide for other infrastructure networks' needs and facilitate access and maintenance

GUIDELINES			RELEVANT GUIDELINES	
E4.6.1.	Visually integrate Energy Distribution and Transmission facilities to the local context	+	L1.7.2.	Define needs for mitigating and compensating for other infrastructure systems' impact on community life through Landscape infrastructure
E4.7.2.	Protect natural habitats and wildlife corridors and compensate for their fragmentation or disruption by Energy Distribution and Transmission	+	L1.7.1.	Define needs for mitigating and compensating for other infrastructure systems' impact on the natural world through Landscape infrastructure
E4.8.1.	Assess and manage future threats from natural and man-made phenomena affecting Energy Distribution and Transmission	+	L1.7.3.	Define needs for protection of infrastructure systems against projected climate change and extreme events through Landscape infrastructure

NOTES

1. David Buchan, *The Rough Guide to the Energy Crisis* (London: Rough Guides, 2010).
2. "Renewables 2014 Global Status Report," WNA, accessed on December 10, 2014, http://www.ren21.net/Portals/0/documents/Resources/GSR/2014/GSR2014_full%20report_low%20res.pdf.
3. "Climate and Electricity Annual 2011," International Energy Agency, accessed on December 10, 2014, http://www.iea.org/publications/freepublications/publication/climate-and-electricity-annual-2011.html.
4. "Indicators for Sustainable Energy Development," UNDP, et al., 2000, accessed on December 10, 2014, http://www.un.org/esa/sustdev/publications/energy_indicators/chapter2.pdf.
5. "Indicators for Sustainable Energy Development," UNDP, et al., 2000.
6. "Indicators for Sustainable Energy Development," UNDP, et al., 2000.
7. Millennium Ecosystem Assessment, *Ecosystems and Human Well-Being: Health Synthesis* (Washington, DC: World Resources Institute, 2005).
8. "Promotion of New and Renewable Sources of Energy," United Nations General Assembly, 2011, accessed on December 10, 2014, http://www.un.org/esa/dsd/resources/res_pdfs/ga-66/SG%20report_Promotion_new_renewable_energy.pdf.
9. "Promotion of New and Renewable Sources of Energy," United Nations General Assembly, 2011.
10. "What the Duck Curve Tells Us about Managing a Green Grid," California ISO, accessed on June 15, 2015, https://www.caiso.com/Documents/FlexibleResourcesHelpRenewables_FastFacts.pdf.
11. "Grid Matters," accessed on August 1, 2015, http://windturbineins.com/MemberLiterature.
12. "What the Duck Curve Tells Us about Managing a Green Grid," California ISO, 3.
13. "Energy: Looking to the Future: A Tool for Strategic Planning," The CLUES project, UCL, accessed on December 10, 2014, http://www.ucl.ac.uk/clues/files/CLUES_Tool_2013.
14. "Comparison of Lifecycle Greenhouse Gas Emissions of Various Electricity Generation Sources," World Nuclear Association, accessed on December 10, 2014, http://www.world-nuclear.org/WNA/Publications/WNA-Reports/Lifecycle-GHG-Emissions-of-Electricity-Generation/.
15. "Renewables 2014 Global Status Report."
16. "What the Duck Curve Tells Us about Managing a Green Grid," California ISO, 3.

IMAGE CREDITS

P. 142: Photograph by Edewaa Foster, accessed on October 01, 2015, https://unsplash.com/photos/VZJMcJwUMP8. **P. 148:** Photograph courtesy of Seattle City Light. **P. 155:** (Bottom) Adapted from "Energy [R]evolution: A Sustainable World Energy Outlook," 4th ed., 2012, p. 28, accessed on February 28, 2013, http://www.greenpeace.org/international/en/campaigns/climate-change/energyrevolution. **P. 157:** (World energy consumption by sector) Data source: U.S. Energy Information Agency, 2011, accessed on September 14, 2015, http://www.eia.gov/tools/faqs/faq.cfm?id=447&t=1. **P. 158:** (Top right) Adapted from "system demand chart," Prudent Energy, accessed on September 10, 2014, http://www.pdenergy.com/grid-services.php. **P. 159:** (World electricity production by source, 2014) Data source: U.S. Energy Information Agency, accessed on September 1, 2015, http://www.tsp-data-portal.org/Breakdown-of-Electricity-Generation-by-Energy-Source#tspQvChart. **P. 159:** (Electricity generation emissions) Adapted from "Comparison of Energy Systems Using Life Cycle Assessment: A Special Report of the World Energy Council," July 2004, p. 4, accessed on November 12, 2015, https://www.worldenergy.org/wp-content/uploads/2012/10/PUB_Comparison_of_Energy_Systens_using_lifecycle_2004_WEC.pdf. **P. 160:** (Basic processes of generation) Adapted from "Combined Heat and Power CHP, "WSE Cogeneration, accessed on February 18, 2013, http://www.solar-window.ca/cogen.php. **P. 164:** (Top) Adapted from "Basic Structure of the Electric System," United States Department of Energy, accessed on February 22, 2012, http://www.ferc.gov/industries/electric/indus-act/reliability/blackout/ch1-3.pdf. **P. 165:** (Top) Data sources: "Energy Storage Technologies," Energy Storage Association, accessed on June 20, 2015, http://energystorage.org/energy-storage/energy-storage-technologies; Lavinia Gabriela Socaciu, "Seasonal Sensible Thermal Energy Storage Solutions," in Leonardo Electronic Journal of Practices and Technologies, no. 19 (July-December 2011): 49–68. **P. 165:** (Bottom) Adapted from "EPRI Smart Grid R&D," Overview Electric Power Research Institute (EPRI), 2008, page 8, accessed on January 2, 2015, http://web.stanford.edu/group/peec/cgi-bin/docs/behavior/workshop/2008/presentations/05-01_EPRI_Smart_Grid_R_and_D_Overview.pdf.

Solid Waste
Infrastructure

Chapter 8

Solid Waste

manages material resources, should restart their life cycle.

8.1 | **INDUSTRY PERSPECTIVES**

From Wasteful to Waste-less: Transforming Perspectives on Urban Waste Management

There are two key aspects of waste management that strike us in the matter of community sustainability and infrastructure development: the need for a shift in public attitude and understanding of what waste really is, and the need for communities to plan for waste infrastructure needs in the context of broader community planning initiatives.

The *Oxford English Dictionary* defines waste as "an act or instance of using or expending something carelessly, extravagantly, or to no purpose," or "material that is not wanted." In either case these definitions point to an underlying attitude of either carelessness or abandonment, and in the case of waste both terms can apply. If we believe resources are limitless, easily acquired, carelessly used, and disposed of without much consequence, then the large-scale accumulation of waste is inevitable. Within this world view "waste" can always be discarded and replaced because, after all, there really is an abundance of ground, even in dense societies, upon which to discard our detritus, assuming we manage to accommodate the aftereffects of emissions to air, ground, and waters and the negative impacts on quality of life in surrounding communities. In the manner of "reality" television, we can seemingly afford to be careless, extravagant, and without purpose if we live on limitless trust funds. But the global environmental reality is that we never had such a resource trust fund, and having lived as if we did, we are now facing impending bankruptcy. Therefore, the imperative of sustainability requires that we radically shift our perspective from managing waste to shepherding and protecting resources as nutrients in a continual cradle-to-cradle cycle.

This transformation may be most challenging in so-called "advanced" economies where resources are abundant. Where resources are less abundant, as in most parts of the world, waste is not seen by those closest to it as the unfortunate by-product of human existence but as a resource to be "mined" for its commodity value and/or its beneficial end use. In developing countries, landfill picking and/or scavenging at collection points is a common way to glean the economic value of some recyclable commodities (e.g., metals); in some cases it is a cottage industry. In developed countries it is becoming more common for governments to view waste as a resource, and we see more and more jurisdictions setting targets with the concept of zero waste in mind. Regulatory and policy tools are evolving to encourage or force product industries to minimize waste through mechanisms that enforce extended producer responsibility for the upstream and downstream costs of their full product life cycle. These approaches are slowly advancing production and material efficiencies, packaging reduction, and product cycles that result in deconstruction of used materials as the precursor to their fresh reuse as industrial inputs. Technology has also rapidly evolved over the last few decades to support the processing of waste for "resource extraction," be it in the form of optical sorting in recycling facilities, evolutions in organic waste processing through aerobic composting and anaerobic digestion, mixed construction and demolition waste recycling facilities, electronics recovery, or biofuels from waste; these technologies are enabling the concept of zero waste to evolve to plausibility.

Despite these advances, we still have considerable waste to attend to. While policymakers and "pickers" see the value of waste, this is lost on the vast majority of society. We still don't act in recognition of the value or costs associated with materials considered waste. Even in very well-developed solid waste management systems, incentives are still needed for waste to be segregated in order to be processed for beneficial end use. In fact, more advanced waste systems that cite higher waste diversion rates rely on programs like user-pay, bag-limits, and clear garbage bag programs (mandatory recycling) as key components of their success. Incentives for industrial, commercial, institutional, construction, and demolition sectors to segregate waste streams are typically driven by economics, regulation, industry standards or accreditation, or some combination of the above. These incentives are necessary to drive the desired behavior in the management of waste, in the absence of inherent understanding of its value.

The saying goes that "one man's trash is another man's treasure," and this is a truism in the waste management industry. In cradle-to-cradle management of waste lies tremendous value both socially and economically in the form of employment (construction, collection, processing, marketing, shipping, administration, regulatory oversight, and the like) and environmentally, as described in this chapter. An apple is plucked from the tree, eaten to the core, the core separated and collected, processed and turned into compost as the perfect substitute for chemical fertilizers. The compost helps return nutrients and organic matter to the soil, a valuable material for agriculture, horticulture, and landscaping, enabling a new apple to grow. The same analogy can apply to inorganic products, as design-for-environment models are evolving to the point where industries benefit by reclaiming their own products to use their constituents as raw material for new ones.

Self-defined "natural capitalists" believe that these changes can become central to a new industrial revolution that turns the making of things into a positive force for society, economy, and the planet. Such thinking leads us to the view that we each need to look at every object we intend to discard with a sense of value and say "I constructed you ... now how can I find a way to deconstruct you, reuse or repurpose you, or send you 'back to the earth' from whence you came?"

The right infrastructure in the right place needs to exist to support this thinking.

Historically waste has been managed and planned for in isolation from other planning activities undertaken in most jurisdictions. In many cases it is not planned for at all in the context of anticipated future population growth, associated increases in waste generation, and necessary infrastructure. Community sustainability plans are now more commonly being developed that speak on high levels to sustainable waste management, but this planning does not specifically address the matter of the infrastructure needed to support associated goals and objectives. Government land use planning agencies, in forming official plans and zoning bylaws, do not typically consider future community waste management infrastructure needs. Conversely, jurisdictions that now develop comprehensive long-term solid waste management strategies do not typically engage planning departments in the results of those plans which might identify longer-term siting needs for new facilities, for example, facilities that could be planned for sooner rather than later and not on an ad hoc basis.

There are many areas that have seen tremendous population and industrial growth but have not previously planned for the necessary waste sites and infrastructure to support it. While governments may develop long-term plans for the construction of new road infrastructure, more wastewater treatment capacity, and more stormwater management capacity, future waste infrastructure needs are not typically considered. So while a road is planned, what's at the end of it isn't always. History has seen these approaches to planning result in the shoehorning of new facilities into existing and permitted waste management sites, which is not always optimum from a cost, processing capacity, or operating perspective. In some cases siting does not occur at all. This can greatly increase the cost of doing business for any level of government responsible for the management of waste.

The other matter that arises from this reactive rather than proactive waste approach is the "not in my backyard" or NIMBY phenomenon, which can add considerable time and cost to new waste infrastructure projects and in many cases bring the project to an end. Urban design, official planning processes, and appropriate zoning should appreciate and incorporate long-term waste infrastructure needs as an integrated part of the overall urban fabric. If this is done, subsequent processes can be undertaken in a more sensible and defensible context and can be linked to preliminary site selection and screening to potentially avoid or minimize land use and community conflict or incompatible use. It could enable consideration of adjacent land uses, suitable site size, hydrogeology, buffer zones, transportation networks, and other factors affecting the suitable siting of this type of infrastructure. In many jurisdictions it's already too late.

An integrated planning approach should enable a better balance and better adherence to the sustainable approach that considers social, environmental, and financial aspects of both waste systems and their relationship to the totality of infrastructure and community systems. This attitude will also prompt us to identify synergies such as taking advantage of the embodied energy within waste streams that can be captured, linked, and contribute to other infrastructure such as district heating schemes, energy, or steam production. If waste is seen through such a lens of sustainability, it becomes clear that it is a piece of a much bigger system, and that early planning for it and its infrastructure promotes sustainable management. Proper attitudes, vision, and planning can drive the understanding of the inherent value of waste and the thinking that these materials, now seen as resources, are simply too valuable to carelessly discard.

Cathy Smith
HBA, MA, National Waste Sector Leader, Stantec Corporation
Marty Janowitz
MES, ENV SP, President & Practice Leader Sustainable Development, Stantec Corporation

Integrated Solid Waste Management Systems

Solid waste management is the one service that almost every municipal or local government provides for its residents. Managing these wastes in an environmentally sound manner is a critical part of a local government's ability to ensure public health and livable cities.

Differing levels of solid waste management services are provided around the world. In developed countries we have seen an evolution toward integrated waste management systems. These integrated systems have developed for a variety of reasons, which may include legislative requirements, environmental performance, or public demand for more sustainable solutions. The components typically include some or all of the following: recycling, organic treatment, thermal treatment, and landfill.

A number of jurisdictions in the developed world have created a waste hierarchy to guide decisions about waste management. The United States Environmental Protection Agency (EPA) states that the purpose of the hierarchy is to place the emphasis on reducing, reusing, and recycling the majority of the wastes generated. The EPA's Sustainable Materials Management Program (SMM) is "an effort to protect the environment and conserve resources for future generations through a systems approach that seeks to reduce materials use and the associated environmental impacts over their entire life cycles, starting with extraction of natural resources and product design and ending with decisions on recycling or final disposal."[1]

With the evolution of integrated waste management systems, a core component of the system is the engineered landfill site to contain waste and prevent the release of pollutants. What has changed over time is the amount of waste requiring landfilling. In many developed countries landfills are now looked on as a means of last resort in handling waste.

The result has been the development of systems to divert, recycle, and compost certain products from the waste stream. In a number of countries, energy-from-waste (EfR) technologies have been constructed to burn waste and produce energy as an alternative to landfilling waste. We are seeing the development of integrated waste systems that incorporate all the above elements to balance environmental protection, public health, and financial costs. Over time it has become evident that integrated waste systems will not be the same for all jurisdictions. Factors like local climate, social acceptability, environmental drivers, and financial ability vary from jurisdiction to jurisdiction.

Added to this is the growing interest in extended producer responsibility (EPR) programs around the world to help reduce the financial burden faced by local governments. Producers of the products that eventually end up in the waste stream are being asked to help finance the operating costs of diversion programs run by local governments that collect these materials for reuse or recycling.

The philosophy of zero waste is also being discussed in a number of countries. The Zero Waste Alliance defines the idea as follows: "Zero Waste is a goal that is ethical, economical, efficient and visionary, to guide people in changing their lifestyles and practices to emulate sustainable natural cycles, where all discarded materials are designed to become resources for others to use. Zero

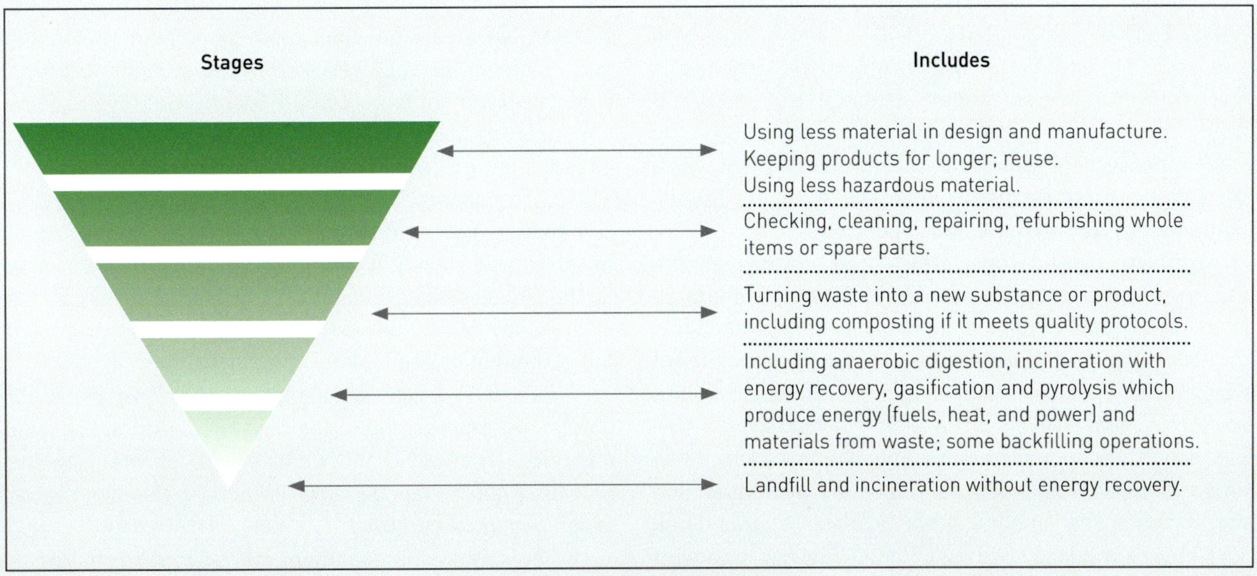

The Waste Hierarchy by the United States Environmental Protection Agency (US EPA)

Waste means designing and managing products and processes to systematically avoid and eliminate the volume and toxicity of waste materials, conserve and recover all resources, and not burn or bury them. Implementing Zero Waste will eliminate all discharges to land, water or air that are a threat to planetary, human, animal or plant health."[2]

Many local governments look on zero waste as an aspirational goal and have set long-term targets to divert 50, 60, 70, and as high as 80 percent of waste from landfill.

A World Bank study points out that, globally, cities currently generate 1.3 billion tonnes of solid waste per year and that this is expected to increase to 2.2 billion tonnes in 2025. A significant amount of this increase will occur in developing countries. As the world population migrates to cities, the pressure to provide reliable waste management systems will increase. As these cities develop, consideration needs to be given to how solid waste will be handled in the buildings and homes citizens frequent on a daily basis.

A significant amount of this growth is going to occur in developing countries, which in most cases have not yet developed integrated waste management systems and rely heavily on landfill. With the growth of larger cities, the need to develop integrated waste management systems will be key for local governments.

Michael Cant
Principal and Global Waste Practice Leader, Canada, Golder Associates

NOTES

1. US Environmental Protection Agency, "Municipal Solid Waste (MSW) in the United States: Facts and Figures," February 2014, http://www.epa.gov/solidwaste/nonhaz/municipal/msw99.htm (accessed April 6, 2015).
2. Zero Waste Alliance, "ZW Definition," www.zwia.org.

Sustainable Cities and Solid Waste

Human activity increasingly burdens our natural systems and built environment. As worldwide population grows, so does the quantity of wastes that requires management. William Rathje, the renowned garbologist and founder of the Garbage Project, once commented that garbage is "in sight, out of mind." To those concerned with planning and sustainability, of course, garbage is never "out of mind." Regardless of one's view, there is a clear nexus between urban planning and sustainable solid waste management. However, success of integrated waste management systems often hinges on changing human behaviors through effective public policies.

There are myriad considerations for urban planners and solid waste professionals developing an integrated solid waste management system. While well established for planning and solid waste professionals, the principles of avoiding waste creation, extending producer responsibility, developing innovative processing technologies, and fostering continuous public engagement add new dimensions to traditional urban and solid waste planning and minimize the amount of waste requiring traditional "end-of-pipe" options such as landfills. Public involvement is a critical success factor; it is best leveraged at key opportunities throughout the planning process and can serve an ongoing role following policy decisions.

Public Involvement—transparent, inclusive, frequent

A program that engages stakeholders is an essential element for integrated solid waste management planning and execution. At each key step in the planning process, public involvement features prominently. But for this to be optimally effective, there are a few key aspects of the public engagement process that must be maintained.

1. Transparency: Information sharing must be complete, honest, and straightforward. Interactions should be structured such that each participant contributes value to the process, builds trust and cooperation across the stakeholder community, and facilitates results that meet the community's needs.

2. Inclusivity: It is essential that all impacted stakeholder groups are represented in the development process. This minimizes friction in the later stages of the planning process when challenges and directional changes can be expensive and disruptive.

3. Frequency: Engagement on a routine basis, at a minimum at each key decision stage, ensures that

stakeholders are heard and minimizes corrective actions and rework.

Below are suggestions on how to engage the community at many levels of the planning process.

Form a stakeholder steering committee

Before beginning the technical evaluation, baseline planning, or other components, a representative group of stakeholders—a steering committee—must be formed. This steering committee should be large enough to include the full range of stakeholders, yet small enough to function as an effective partner in the planning process. Members should be willing and able to engage repeatedly, and to put in the sometimes long hours needed to achieve consensus.

The steering committee's first task is to hold a chartering to create a common vision, agree on operating guidelines, and establish short- and long-term goals for this new team. Outcomes are frequently captured in a team chartering document, which each member signs to signify agreement and commitment. This process sets the stage for all the engagements to come, building a sense of teamwork and common purpose.

Use the public to inform your baseline planning

Establishing a planning baseline is a good starting point for developing an integrated solid waste management system and serves as the condition against which all future alternatives are compared. A comprehensive baseline is key to defining current waste system needs as well as how those needs are likely to change in the future, and helps participants understand how the waste system will evolve over time. It is also essential that the baseline include an articulation of the community's and elected officials' goals in order to maintain credibility and viability over the long term. Public input can inform the baseline assessment and set simple goals such as a reduction in per capita waste generation or an increase in the recycling rate, or more complex and/or quantitative goals such as social impact, natural capital, and systemic sustainability. Goals can be solicited through public engagement, going beyond the steering committee to include other residents, commercial property owners, business owners, professional staff, and elected officials. The goals should be defined in a way that facilitates measurement and reporting as the waste system operates, and includes provisions for adaptive management over the system's life cycle to adjust performance and/or the goals as more operational data become available, so that these objective measures can be reported back to the public.

Plan integrated waste system components

To effectively analyze the possible components of an integrated solid waste management system, a consistent methodology to assess materials management options with a broad focus should be utilized. Public input is critical here to cultivate an understanding of the system complexities among nontechnical stakeholders.

The solid waste collection options present some unique considerations in terms of materials collected, types of vehicles used, frequency of collection, and ultimate disposition of the collected materials; each of these factors can be used to guide customer behaviors in alignment with the system goals. However, planners should also separately consider waste generation from residential, multifamily, commercial, and institutional sources, as each sector will have its own unique challenges and opportunities. For example, recent innovative changes in collection methods include the use of alternative fuels to power traditional collection vehicles.

Solid waste collection is integrally linked to processing, so available processing options should inform evaluation of collection options. For example, if a materials recovery facility is being considered to process recyclables, the type of facility matters in terms of how the materials are collected (i.e., single vs. multiple streams). Processing technologies such as optical sorting are continuously being improved and employed to allow collection of more materials in a single stream, resulting in higher recovery rates that increase recycling and minimize disposal needs.

Landfills will continue to be an important part of an integrated solid waste management system over the short and medium term. However, with the implantation of progressive solid waste management principles, the quantity of wastes requiring disposal will be greatly reduced. This can be especially true if traditional commodities (e.g., glass, plastic, metals, etc.) and organic waste are recycled or otherwise processed. This leaves only essentially inert materials as wastes being put in landfills, which means that landfill odors are minimized, helping address tensions which tend to accompany decisions surrounding processing and disposal options. Issues of this nature are controversial and can often become wedge issues in a community, as few people want to live near such a facility and health, safety, and environmental concerns are raised as the basis for opposition. In addition, solid waste facilities have a long service life, so flexibility is an important consideration to accommodate innovation and technology evolution.

Plan for integrated solid waste system alternatives

Once the solid waste system elements are analyzed, it is important to identify four to six alternatives. Using the steering committee and engaging community stakeholders when evaluating solid waste system alternatives can be especially helpful in building consensus, validating the evaluation criteria, and ultimately endorsing the evaluation process and recommendation. In evaluating collection, the alternatives vary with the types of materials collected, the types of containers used, the mode of collection, the collection frequency, and even the basis for payment. For example, a growing trend in residential collection is an approach known as "pay-as-you-throw," in which the pricing scheme serves as an incentive to recycle more and waste less. Processing alternatives may include different types of recycling and conversion facilities. Disposal options should address whether the landfill site is developed within the municipality or beyond. Each alternative should be evaluated in terms of compatibility with the system planning goals. At this point, preliminary capital and operational cost estimates should be developed for each alternative. The alternatives are measured based on the evaluation criteria and ranked to form the basis of a recommendation. There are still many considerations embedded within the traditional planning process that must be developed to align the goals of the solid waste management system with the day-to-day operations. These considerations include, but are not limited to, how education and training will be communicated, how the collection containers will be appropriately screened, how extended producer responsibility will be implemented, and how compliance will be monitored and enforced. Again, engaging stakeholders will provide important feedback for the planning process.

Develop the integrated solid waste system implementation plan

Once the technical work is completed and alternatives identified, an implementation plan can be prepared. The implementation plan should detail the capital investment schedule, regulatory approval requirements and process, and the other steps necessary to implement the recommended alternative(s). As with the planning process, public engagement is a key ingredient in building community consensus and support for the implementation plan.

Throughout the planning process, consideration should be given to local regulations that support the goals and needs of the integrated solid waste management system. These regulations can address operational considerations such as litter and illegal dumping ordinances and commercial waste container screening requirements.

Implement the integrated solid waste system

Following the development of the implementation plan, the next step is to seek policy direction from the governing body. Building upon a credible and participatory planning process, the focus shifts to crafting the "story," including the technical, financial, and social considerations. It is essential that the decision makers recognize the process that resulted in the presented recommendations. The steering committee can also play an ongoing role as advocates as the policies are implemented. At the same time policies can be developed to guide system users in terms of acceptable materials, set-out procedures, and management of special wastes such as household hazardous waste, electronic waste, and medical waste. These details help round out the integrated solid waste management system, because they are the most visible to residents on a day-to-day basis, as opposed to the processing and disposal facilities that are often sited away from population centers. It should be noted that policy and regulatory frameworks can also be useful for driving behaviors around waste minimization and recycling, commensurate with community goals and values.

Integrated solid waste management planning has changed considerably over the past few decades, and continues to evolve today. Along with this transition, many communities around the world have established aspirational goals to manage wastes as resources and reduce the amount of wastes requiring landfilling. When developing a sustainable and integrated solid waste management system, following a defined and participatory process is a critical success factor, and, as with most plans, if one starts with the end in mind, the likelihood of a positive result is greatly increased.

Daniel Dietch
MBA, MPS, Client Service Manager, CH2M
John Wood
PE, Global Practice Director for Integrated Waste Solutions, CH2M
Robert M. Beinstein
PE, ENV SP, EFCG

8.2 | SUSTAINABLE SOLID WASTE PLANNING

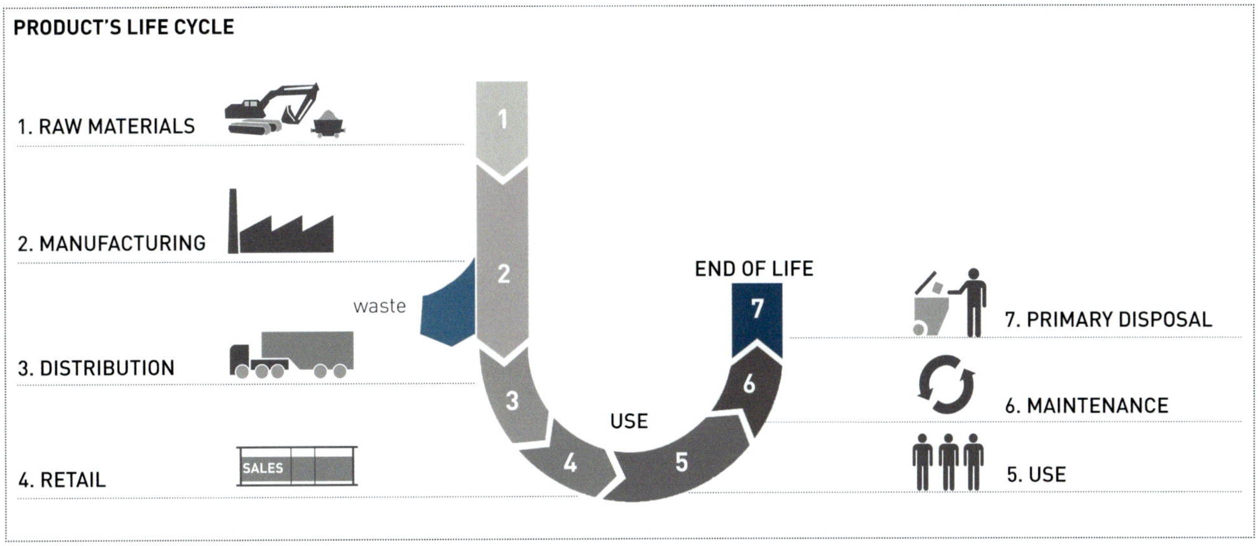

One of the most important services provided by a local government is solid waste management. A conventional approach to this service is to carefully and safely remove unwanted or unusable material, substances, or by-products, namely waste, from the urban environment.

Waste is generated during manufacturing processes or with the end of the useful life of products. Right after its generation, waste should be properly managed, taking into consideration a complex range of environmental, social, and financial parameters.

The environmental and health impacts of poor waste management can be grave. Waste decomposition in landfills generates methane, a very potent greenhouse gas. Littering and leaching can lead to extended water and soil contamination, affecting habitats and food chains. The absence of waste collection on the other hand, i.e., the accumulation of uncollected waste in residential areas, may cause public health problems such as respiratory ailments, diarrhea, dengue fever, chemical poisoning through chemical inhalation, cancer, low birth weight, congenital malformations, etc. Uncollected waste may also interrupt stormwater flows and potentially cause flooding by clogging natural and artificial drainage systems.

The importance of waste management is evident and the role of authorities crucial. Waste management practices are very indicative of the living standards and development of a community.

As is well documented, the amount of solid waste generated globally increases each year. This rapid increase is a result of population growth, modernization, urbanization, and the increase of industrial manufacturing. "Solid waste (MSW), one of the most important by-products of an urban lifestyle, is growing even faster than the rate of urbanization."[1]

Considering the projections for future increase of world population and particularly of urban population, the amount of solid waste generated might become much harder to manage. It is critical that local governments adopt effective strategies toward waste prevention and change the current common practices of waste management. The latter commonly involve the use of landfills, but, for social and environmental reasons, the continuous designation of new sites for this purpose is particularly difficult, and it becomes further complicated by the continuous expansion of cities.

The total prevention of waste generation is obviously impossible. As long as people are consumers of goods, waste will always be generated. What actually is possible is the reduction of the amounts of waste that end up in landfills; in other words, the increase of waste diversion.

Several strategies can contribute to increased waste diversion. The first is reducing the overall consumption of goods or choosing to consume goods that use less materials, such as products with minimal packaging.

PROCESSES AND TECHNOLOGIES IN RELATION TO BY-PRODUCTS

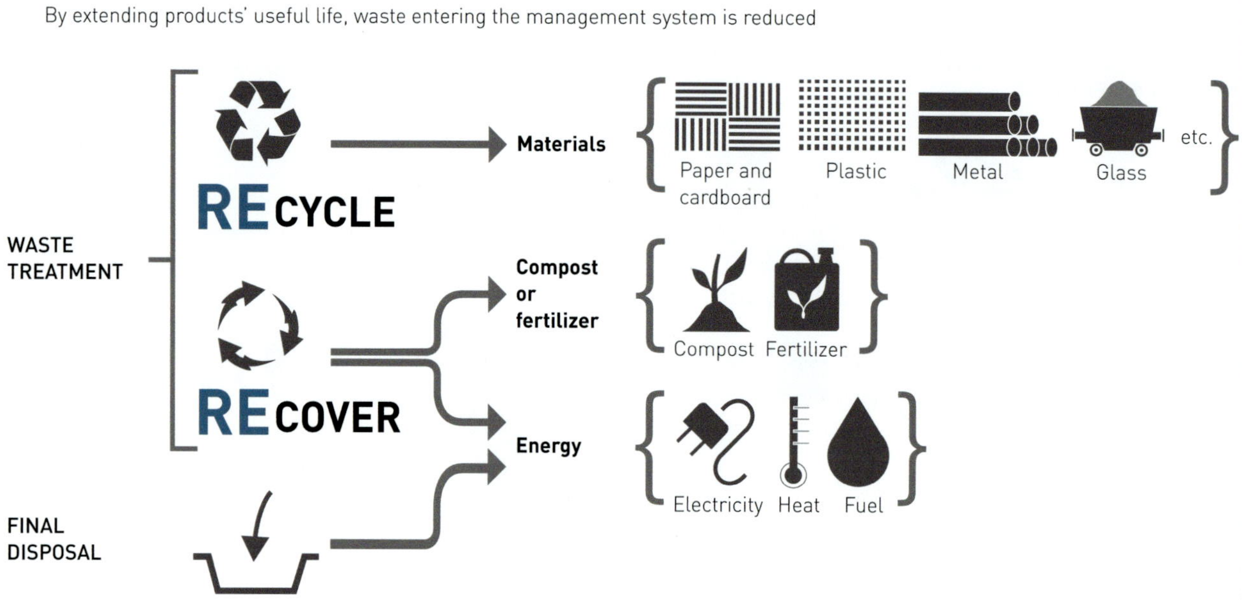

Recycling and reusing products extends materials' and products' life cycles, respectively, and therefore delays the final disposal and the manufacture of new ones to replace them. It should be noted that the degree of success of these practices is strongly related to people's consumption behaviors and waste sorting practices. Even if a municipality's solid waste infrastructure includes a detailed operation plan and all the necessary technical equipment, the whole effort will eventually falter without the participation of the population. Recycling programs are a typical example. The challenge here is to engage people in actively participating in the waste management system.

CHALLENGES TOWARD A SUSTAINABLE SOLID WASTE SYSTEM

Sustainable waste management should provide services that are environmentally efficient, economically affordable, and socially acceptable.[2] Nowadays, waste management systems should be aiming to become as "green" as possible, that is, to function with the least impact on the environment and human health. However, the numerous social and environmental benefits of a sustainable waste management system often come in conflict with economic interests. For example, although it has been proven that waste separation at source contributes to higher recycling rates than those of mixed waste collection (single stream), the collection of multiple streams also increases the cost of the waste collection service. Similarly, the installation of high-technology solutions in waste treatment facilities with the aim of reducing the environmental impact is also associated with economic cost.

Sometimes solid waste management may be one of the most costly items in a local government's budget. While economic cost was the main factor affecting waste management decisions in the past, nowadays these decisions should be taken after environmental and social consequences have also been considered. Although a cost-benefit analysis could address all social, economic, and environmental parameters, contributing to making the right choice regarding, for example, a waste treatment option, any decision with social cost could raise objections and reactions. Politics are seriously involved in waste management decisions, and in some cases local politicians adopt the NIMTOO ("not in my term of office") attitude to avoid jeopardizing their political position, thus putting on hold a potential upgrading of their cities' solid waste management systems.

A sustainable waste infrastructure system should be safe and sanitary, but it should also be reliable, financially viable, adaptable, and able to manage all types and amounts of waste generated by the specific context. Another challenge for waste management relates to the system's social acceptance. Due to their odors, visual impacts, and accumulation of contaminants, waste facilities are commonly met with NIMBY ("not in my backyard") reactions by local communities.

It takes considerable effort by local governments to eliminate negative impacts and more importantly to engage and inform residents from the early stages of a waste management plan's development. Public involvement in the government's waste management plan is absolutely essential. For instance, people's consumer behavior affects the waste mix, and their involvement in waste separation at source affects recycling rates.

Communication strategies developed by local governments can contribute to improving residents' participation in recycling. Also, objections regarding location of waste facilities can be minimized by a considerate facility design and public consultation. Another factor that contributes to public support of waste management is the stability of the system and minimization of changes in waste sorting habits or waste collection methods. People prefer stable systems with minimum alterations, as well as systems with high integrity which leave no doubt that all materials sourced separately are actually sent for recycling. Also, residents are more likely to accept the location of waste facilities if they are aware that pollution control technologies with high standards are used in them.[3]

Besides the challenges that should be addressed when aiming at a sustainable waste system, there are also opportunities that emerge when applying green strategies. Environmental benefits include natural resource and energy savings, reductions in GHG emissions, and energy generation, but it has also been proven that investing in sustainable waste management creates new businesses and jobs. According to the United Nations Environment Programme, it is a fact that "sorting and processing recyclables alone sustain ten times more jobs than land

PRODUCT'S LIFE CYCLE

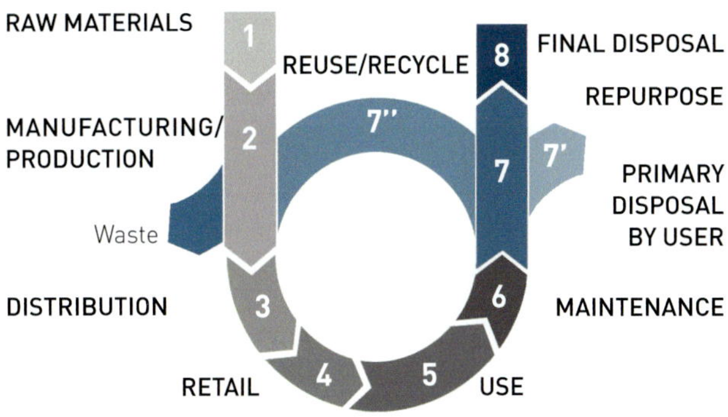

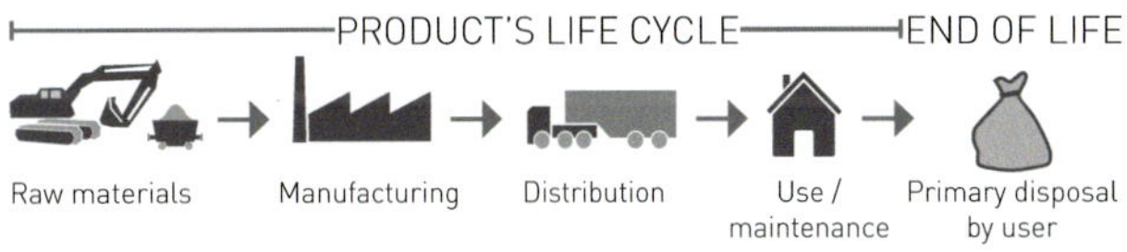

filling or incineration on a per tonne basis." It is also interesting that "the overall net employment appears to be positive." Organic agriculture also profits from the generation of compost.[4]

The main target for solid waste should be to turn the waste into a resource. The waste infrastructure system should not only offer regulating services, through disposing and containing contaminants, but should also focus on developing provisioning services. The continuous increase in consumption of goods and the subsequent disposal of their waste products lead to an inversely proportional effect: reduced availability of raw materials and resources. Sustainable solid waste management should therefore aim to adopt a twofold role: regulating waste's impact on the natural world and communities through required processing and disposal, while also exploiting the possibilities of waste to recover resources and feed them back into supply chains. This closed-loop system is feasible due to technologies of waste treatment and disposal that exploit the opportunity for waste management by-products to actually become valuable resources, either as materials, as energy in the form of fuel (biogas from anaerobic digestion, heat from combustion in incinerators, landfill gas), or as compost and fertilizer from biodegradable waste. In other words, the negative side effects of waste decomposition, such as the production of gases, are simultaneously managed (captured) and converted to useful materials or energy sources.

Overall, sustainable waste management goes beyond the target of removing unwanted and useless products from urban areas. It seeks to protect the natural environment, resources, and human health. In this context, the Zofnass Guidelines framework refers to Solid Waste as the infrastructure system whose aim is to remove waste with the least possible impact on the environment (natural world), public health and society (quality of life), and resources (resource allocation). In addition, waste infrastructure aims to maximize the benefits of waste by extending products' life cycles through appropriate management and treatment.

ENVIRONMENTAL

Challenges for environmental sustainability include:

- Managing methane generated in landfills
- Resolving water and soil contamination from littering and leaching in landfills
- Minimizing natural landscape alterations caused by treatment and disposal facilities
- Safely managing all hazardous residues

SOCIAL

Challenges for social sustainability include:

- Protecting public health and safety
- Maintaining license to operate by the public
- Inducing the public to participate in reuse and recycling programs
- Increasing public acceptability of waste management
- Assisting in users' adaptation to changes in waste management practices
- Engaging the public in decisions on waste management
- Resolving noise pollution, odor, and visual issues with regard to waste facilities
- Fairly distributing the cost of waste management

ECONOMIC

Challenges for economic sustainability include:

- Constantly updating the waste management system with improved technologies
- Managing high costs of social and environmental protection
- Applying cost-benefit analysis for important decisions
- Planning with regard to future projections in order to secure long-term economic viability
- Ensuring absorption of products and by-products by the market

SYSTEM PURPOSE: DIVERSION FROM LANDFILLS	
	SOURCE REDUCTION: Minimization of waste generation
	REUSE: Further use of products in their original form or repurposing to extend their useful life
	RECYCLING: Breaking down used products into materials
	FINAL DISPOSAL / RESOURCE RECOVERY: Bulk reduction, nonrecyclable waste disposal through sanitary landfill and waste-to-energy conversion

The primary service of sustainable solid waste management is to collect and consolidate waste to minimize human health and environmental impacts (for countries where these systems are currently inadequate) and then to safely implement resource recovery strategies for that consolidated waste (in some countries those strategies are not safe, e.g., landfill picking)

The strategies listed above suggest basic steps that could be included in the planning of a solid waste system. They do not aim to outline all the requirements for a sustainable system, but instead highlight approaches that if implemented could resolve multiple issues simultaneously.

When aiming to reduce waste, the emphasis should be placed on source reduction. Reducing the sources of waste will lead to an overall reduction of the volumes and contaminants to be processed by the infrastructure system. It includes actions that require manufacturers and consumers to extend the useful life of products and minimize the generation of waste volume and toxic content. Additionally, attention should be given to manufacturing products that can be easily repurposed or disassembled. The role of manufacturers and users is very important at this stage. The public should be engaged with the concept of source reduction through awareness strategies promoting responsible consumer behaviors.

The second strategy in the hierarchy of sustainable waste management practices is reusing products instead of discarding them. Reuse of products can be achieved either by passing them on to a new user or by giving them a new, different use from the one they were initially designed to have. Sometimes the new use and purpose of a product can even be more valuable than the previous one. This is the concept of "upcycling" products, creating a new product of higher value than the one that was intended for discarding.

The reuse of products allows the extension of their life cycles through repeated use with minimum reprocessing (reuse of intact items). The value of items is thus maintained. Reuse of products can be promoted either by individuals who are aware of the benefits or through organized programs with deposit or virtual exchange centers for donation or sale. Closed-loop programs can encourage cyclical use of resources by providing incentives toward zero waste.

Recycling is one of the most well-established strategies aiming to prevent increased amounts of waste from ending up in landfills. Recycling comes after reuse in the hierarchy, as it takes place when an item is no longer suitable for use and its useful life finally ends. The process of recycling includes breaking down products into raw materials to be used for the production of new items. It also involves mechanical or manual sorting and cleaning. For recycling programs to be successful, a market must be secured for the distribution and sale of recycled materials. It is evident that recycling is a highly participatory process as it is closely connected to consumers' behavior and involvement (e.g., separation at source). However, the role of the manufacturer is also important. A complex life cycle analysis is often required during a product's design phase to determine the efficiency of reuse, recycling, or neither and to design accordingly. In this context, the introduction of extended producer responsibility provides incentives for manufacturers to produce environmentally friendly products, as they become responsible for the costs of the products' management (recycling and disposal).

Landfilling should be a process of last resort due to its multiple adverse impacts on the environment and community. However, it is currently the most common practice in waste management. The ideal aim is for zero

WASTE MANAGEMENT STRATEGIES

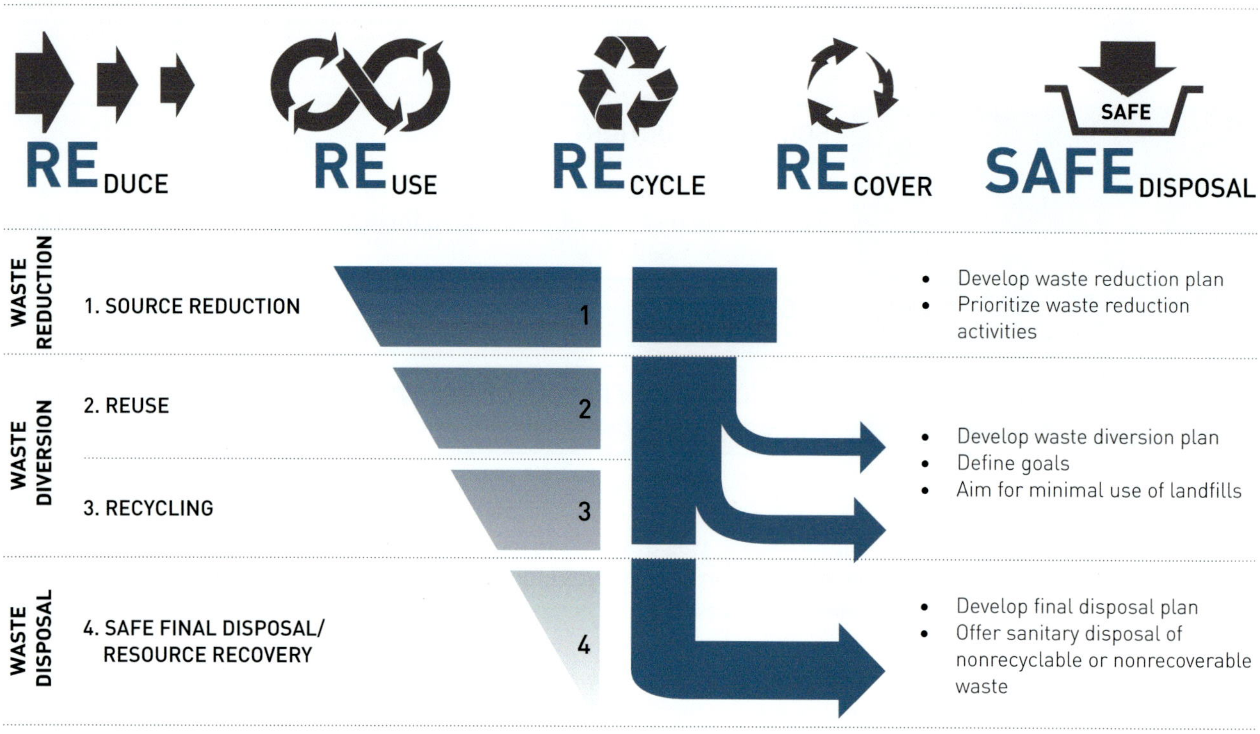

waste to go to landfills. Nevertheless, even if a high degree of waste diversion is achieved, eventually a certain amount of waste will reach a landfill. For this reason compensatory measures must be implemented. Landfill sites should be selected after careful consideration. They should be strictly controlled and configured to include sanitary measures, leachate treatment facilities, and gas capturing, as significant amounts of energy can be generated through the collection of emitted gases. Additionally, provisions should be made for the restoration of decommissioned landfill sites.

Waste recovery is the processing of waste for the extraction of materials or energy. It refers to the conversion of nonrecyclable materials into a different state. The transformed product can be used either on or off site. Examples of waste recovery are waste-to-energy processes and composting. These processes convert waste into energy sources (e.g., heat, electricity, fuel) and useful materials, respectively.

Recovery also includes treatment of waste in order to remove, stabilize, or destroy pollutants or reduce the bulk volume of waste before final disposal.

Although final disposal of waste in a landfill is the least preferred strategy for waste management, waste recovery can also take place during this stage.

8.3 | SOLID WASTE SYSTEM DECODING

SOLID WASTE GENERATION

SOLID WASTE DIVERSION MODEL

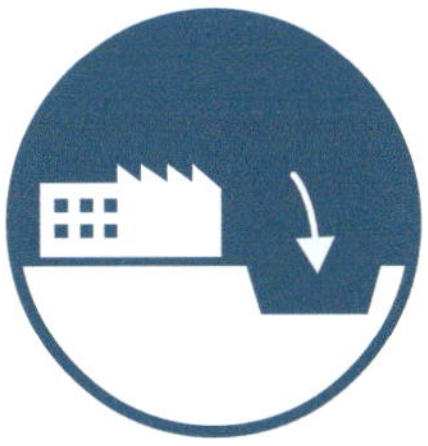

SOLID WASTE TREATMENT AND FINAL DISPOSAL

SOLID WASTE COLLECTION AND TRANSFER

Within the framework of the Zofnass Sustainable Planning Guidelines, Solid Waste infrastructure is decoded on the following four system levels: Waste Generation, Waste Diversion Model, Treatment and Final Disposal, and Collection and Transfer. Sustainability objectives are set for each of these system levels, which are then analyzed through the planning guidelines.

The system level of Waste Generation involves the analysis of data (statistics or estimations) that summarize the volume and composition of a community's or city's generated waste. Waste generation analysis is the crucial basis for the development of waste management strategies.

Next, the Diversion Model system level involves the strategic decisions regarding the management methods selected to divert waste from landfills. These are defined in correlation with the socioeconomic and environmental context. Options and scenarios for sustainable management are assessed and hierarchized, and diversion targets are set in order to determine the required waste streams. The long-term plans involve mostly goals toward zero waste going to landfill.

Consistent with the general framework, the third system level, Treatment and Final Disposal, is where the main facilities and essential installations and processes of the system are addressed. This level is related to two basic processes: waste processing for recycling and recovery, and final disposal of waste that is not diverted. All the strategies and targets defined in the Diversion Model should be implemented with the most appropriate methods and technologies in order to maximize the system's efficiency and minimize environmental and social impacts. Similarly, landfill sites and configurations should be properly planned and managed.

The fourth system level, Collection and Transfer, focuses on the transfer of solid waste from the points of generation to the points of processing or final disposal. All components serving the transfer of waste are examined, including collection vehicles, routes, and schedules as well as temporary storage facilities and collection points.

SOLID WASTE GENERATION

A product's useful life begins when raw materials enter the manufacturing process; it continues with the product's distribution to users and finally ends with the user's decision to discard it. Waste generation can occur either during manufacture/production or at the moment of disposal by the user. Sustainable waste management calls for reduction of waste generation rates and waste volumes. A first priority toward this aim is to extend the useful life of products, therefore reducing the volume of waste entering the waste management system. The composition of waste generated is another important factor that will determine the opportunities for waste diversion and recovery. All the above features of waste (quantity, generation rate, and composition) are closely related to consumers' behavior. That is why the system level of Generation focuses on the promotion of consumption habits and practices that contribute to sustainable waste management. In addition, the Waste Generation system level underlines the need for recording quantitative and qualitative data describing the particularities of an area or community. Waste characteristics vary depending on the location, the social patterns, the climate, the economy of each area, and the population. It is essential to regularly monitor waste generation rates and patterns in order to provide a trustworthy basis for assessing waste management options, proposing sustainable strategies, and setting realistic targets. Analyzing and understanding the composition of waste are therefore essential for the definition of the management plan.

DATA ON WASTE GENERATED

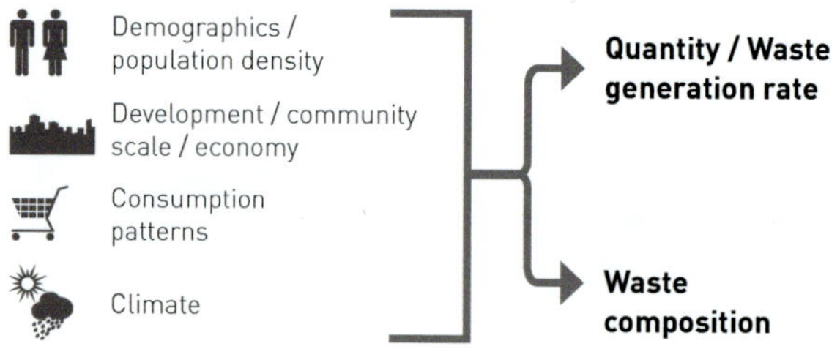

WASTE COMPOSITION

WASTE GENERATION SECTORS

SOLID WASTE DIVERSION MODEL

The Diversion Model system level is where waste generation data are integrated with the socioeconomic and environmental context in order to define requirements and potential for collection, treatment, and disposal, to set and reassess sustainability goals, and finally to select the most suitable strategy. The Diversion Model involves the core decisions that will affect the whole system's planning. Therefore the role of both authorities and experts is crucial.

A very basic step is to analyze the waste composition of a certain community in order to decide on the waste streams that will be managed. The decision on number of waste streams is affected by a number of social factors, environmental or economic. Having a single waste stream (dirty sorting) is the option in which the user does not participate in recycling, leaving the whole responsibility for waste diversion to the community. Waste recycling and waste recovery are still feasible under this option, but the economic cost of waste separation in material recovery facilities will be increased while amounts of waste treated will be rather low. If the community decides to have more than one waste stream (clean sorting), defining each waste stream depends on a number of social, economic, and environmental factors. Options regarding collection methods, vehicles, and receptacles should be assessed, along with the society's degree of engagement, as well as spatial and technical requirements of units such as treatment plants, transfer stations, and landfills. An essential step toward responsible decision making is setting up sustainable waste management goals. The more the goals set are specific and realistic, the higher the probability of achieving them. Measurable and time-scaled goals allow for monitoring whether the system performs as planned or not. The waste diversion ratio and the percentage of materials recovered from treatment within a time framework are examples of specific targets in waste treatment. It is also important to set long-term goals, as the impact of strategies toward high waste diversion or zero landfill is more evident in the long term.

The waste diversion model analyzes all waste types and their potential destinations (treatment plants, disposal, collection network) and conversion. Feasibility studies and expert analysis are required in order to produce waste management scenarios for elaboration. The quantities and types of waste streams that will be defined by the selected scenario will further determine waste collection, transfer, treatment, and disposal.

VARIOUS OPTIONS OF WASTE STREAMS

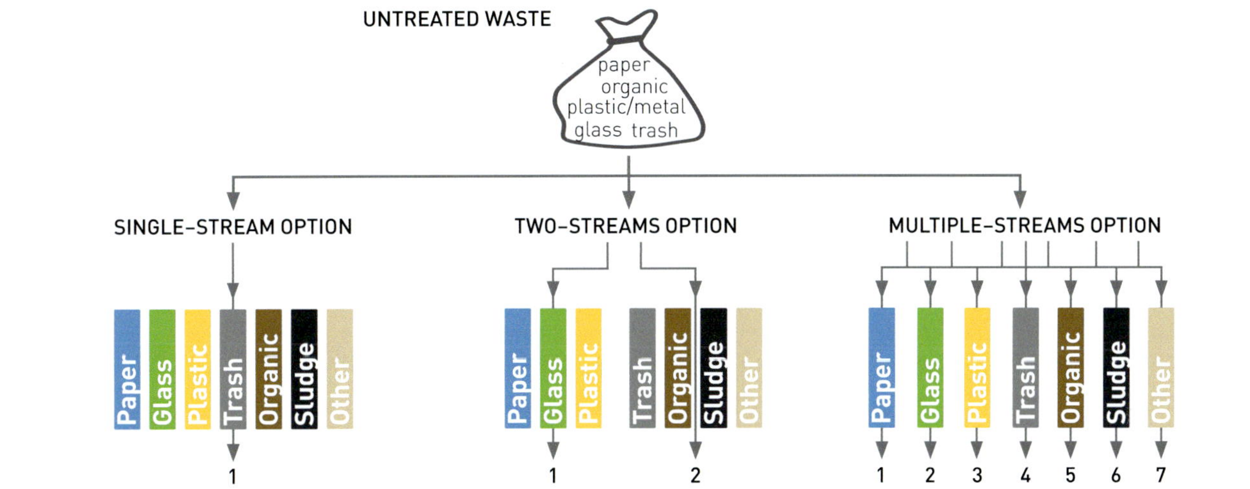

WASTE STREAMS MANAGEMENT PLAN

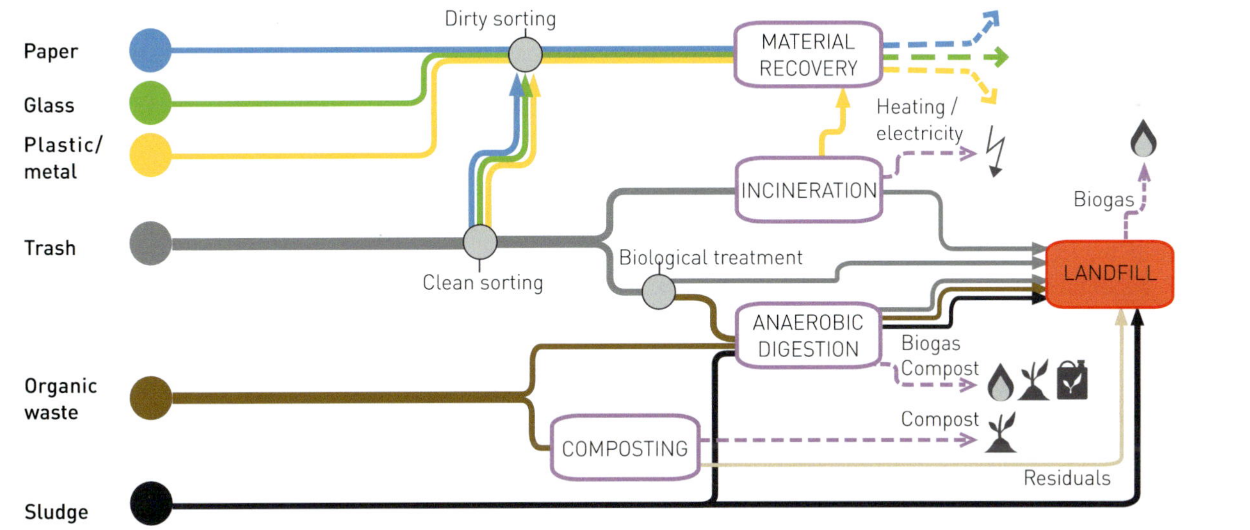

WASTE MANAGEMENT GOALS

DIVERSION — Target: To divert a minimum percentage of generated waste within a certain time framework

MATERIAL RECOVERY — Target: To recover a minimum percentage of materials through separate collection within a certain time framework

COMPOST — Target: To convert a minimum percentage of waste into compost through separate collection of organic waste within a certain time framework

ENERGY PRODUCTION — Target: To produce a minimum amount of energy through waste-to-energy plants

TARGETS TOWARD ZERO WASTE

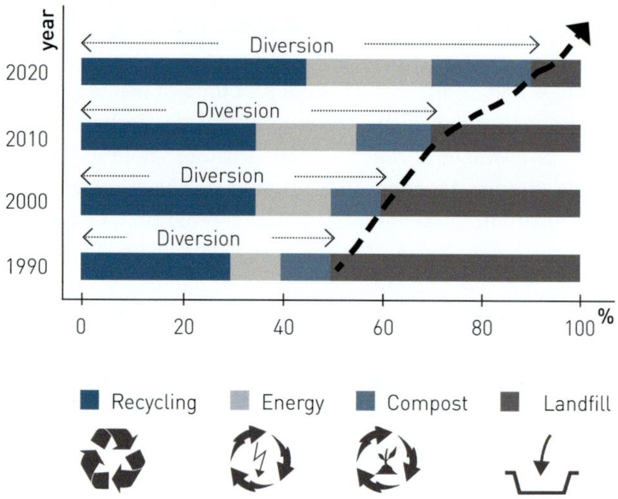

SOLID WASTE TREATMENT AND FINAL DISPOSAL

Treatment and Final Disposal is the third system level of Solid Waste infrastructure. It studies infrastructure units that form the nodes of sustainable waste management. These are the facilities where the processes of waste diversion, resource recovery, and safe disposal take place.

Depending on the decisions of the diversion model regarding waste stream types and sustainable targets, these units may include material recovery facilities, waste-to-energy facilities, anaerobic digestion plants, or composting plants and sanitary landfills.

The planning of the entities of this system level aims to minimize the environmental and social impact of treatment and final disposal. All waste streams should be treated before reaching a landfill. For every treatment type there are several options and technologies to be considered. Benefits and impacts should be examined before the final selection. It is understood that treatment facilities can be beneficial as, depending on their function, they either produce recycled materials, reduce waste, or recover resources in the form of compost, biogas, or energy. Apart from the benefits, however, they may also cause several impacts, such as noise pollution, odors, visual impact, and greenhouse gas emissions, which should be addressed during planning.

Although landfilling is an option of last resort, there are still ways to improve landfill sites and installations, as there are technologies that permit resource recovery even at the stage of final disposal and more specifically in sanitary landfills. In sustainable waste management, uncontrolled deposit should be avoided as it can cause soil and groundwater contamination, GHG emissions, odors, and visual impact. Controlled deposit implements significant measures to regulate and mitigate these negative impacts, particularly groundwater and soil contamination as well as GHG emissions. In addition, as previously mentioned, it offers the possibility of resource recovery through the capture of landfill gas and production of electricity, biogas, etc. In order to minimize impacts of final disposal, detailed analysis and planning of services is required. The aim is to optimize landfill sites by proper control of all phases of development, i.e., predevelopment, construction, annual operation, closure, and post-closure.

TREATMENT ENTITIES

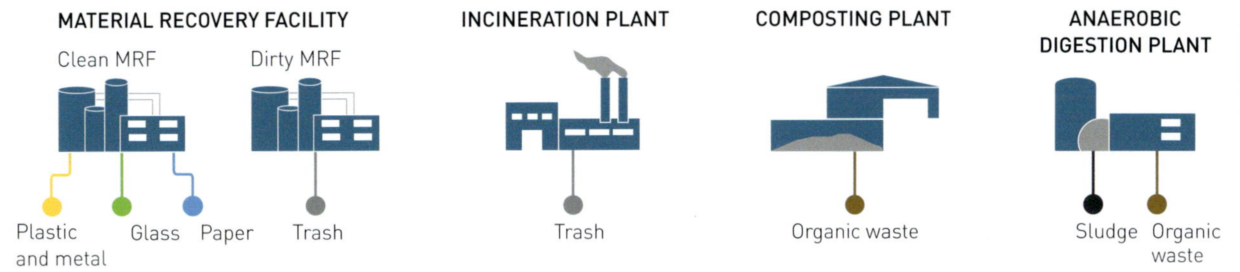

TREATMENT ENTITIES	PROCESSES	BENEFITS	LIMITATIONS
Materials recovery facility			
Clean MRF:	• Control of origin/weight • Manual or mechanical sorting • Compaction/size reduction • Packaging for recycling • Temporary storage	Recovered materials Waste reduction	Noise Odors
Dirty MRF:	• Control of origin/weight • Sorting/separation of recyclables • Compaction/size reduction • Biological treatment • Packaging in packets for final disposal • Temporary storage		
Incineration plant	• Waste storage and feed preparation • Combustion in a furnace • Production of hot gases • Production of bottom ash residue for disposal • Treatment of the cooled gas to remove air pollutants • Disposal of residuals from gas treatment process • Dispersion of treated gas to the atmosphere • Monitoring of emissions • Generation of energy	Electricity Waste reduction	Noise Emissions Odors Visual impact
Composting plant	• Pretreatment and mixture of different organic stocks • Decomposition phase in tunnels • Maturing phase in turned heaps • Post-treatment • Biofiltration • Storage of liquid extracts (leachates) • Production of compost • Production of digestate that either can be composted or can be dried to produce fertilizer	Compost Waste reduction	Noise Odors
Anaerobic digestion plant	• Pretreatment and/or mixture of residues • Digestor's alimentation/digestion • Dewatering of digested materials • Gasification • Safety flaring of gas • Biogas adaptation • Cogeneration motors for energy generation • Distribution of electric energy • Exploitation of thermal energy	Biogas Waste reduction	Noise Odors Visual impact

FINAL DISPOSAL ENTITIES	PROCESSES	BENEFITS	LIMITATIONS
Controlled landfill 	• Artificial impermeabilization of the deposit's vessel • Control of acceptable/unacceptable waste • Gradual covering of the disposed waste • System for the collection and evacuation of liquid outflows (leachates) • Treatment of leachates • Control and collection of landfill gas • Soil and groundwater monitoring	Soil/groundwater contamination	Emissions Odors Visual impact
Bullet controlled landfill 	• Artificial impermeabilization of the deposit's vessel • Control of acceptable/unacceptable waste • Gradual covering of the disposed waste • System for the collection and evacuation of liquid outflows (leachates) • Treatment of leachates • Control and collection of landfill gas • Soil and groundwater monitoring • Compaction and packing of waste into "bullets" prior to disposal	Soil/groundwater contamination Waste reduction Odors	Emissions Visual impact
Controlled deposit with landfill gas flaring 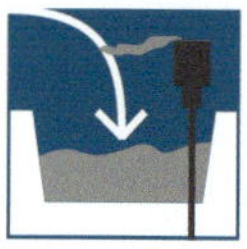	• Artificial impermeabilization of the deposit's vessel • Control of acceptable/unacceptable waste • Gradual covering of the disposed waste • System for the collection and evacuation of liquid outflows (leachates) • Treatment of leachates • Control and collection of landfill gas • Soil and groundwater monitoring • Safety flaring gas	Soil/groundwater contamination Reduced emissions	Odors Visual impact
Controlled deposit with energy production 	• Artificial impermeabilization of the deposit's vessel • Control of acceptable/unacceptable waste • Gradual covering of the disposed waste • System for the collection and evacuation of liquid outflows (leachates) • Treatment of leachates • Control and collection of landfill gas • Soil and groundwater monitoring • Treatment of biogas • Generation of thermal energy • Generation of electric energy	Soil/groundwater contamination Reduced emissions Electricity Biogas	Odors Visual impact

SOLID WASTE COLLECTION AND TRANSFER

The fourth system level of Solid Waste infrastructure studies the collection networks that serve to transfer waste from the point of first disposal, by consumers, to points of treatment or final disposal (landfill). Among all services of waste management, this is the most visible and noticeable to the public, not only because it starts from people's homes but also because it is a scheduled daily service. In this system level all the equipment used should be carefully analyzed, selected, and managed in order to comply with commitments developed on the Diversion Model and Treatment and Final Disposal levels. It includes receptacles and containers, collection vehicles, and transfer stations. This equipment is the part of waste management that citizens interact with most. It can support hygienic conditions and environmental protection as well as a sense of well-being for residents and visitors. An investment in modern equipment can be significantly effective in waste collection.

Waste streams are strongly connected to the planning of collection and transfer. Waste quantities, goals set, and assigned destinations for each stream will define the characteristics of the collection system entities, such as collection schedules and equipment. Separate waste collection (separation at source) is generally preferred to mixed collection, as higher percentages of recyclable waste can be recovered by the processing facilities. In order to minimize costs, reduce space requirements of the system, and minimize the visual impact of numerous collection points, alternative methods of waste collection can be introduced, such as door-to-door collection on demand (curbside collection) or collection at mobile collection points. The first method requires a higher degree of involvement by the users and special reserved space in buildings for short storage of waste, whereas the second method requires specialized collection trucks.

Apart from types, sizes, capacities, and number of receptacles and collection trucks, collection routes should also be carefully planned, focusing on minimizing and optimizing routes both for users moving toward receptacles and for vehicles moving from the receptacles to the assigned destinations. When disposal and treatment sites are located at a long distance, efficiency in waste transfer routes is very important. Long waste transfer routes and large numbers of shipments are unadvisable with regard to economic costs, time efficiency, and environmental protection. The optimization of collection routes can be achieved by properly distributing receptacles and by reducing the length and number of trips. The reduction of vehicular trips to and from the disposal site is usually achieved by introducing transfer stations. By combining loads of several individual collection trucks into one single shipment, communities can save money on labor and operating costs.

The distances between collection points, treatment points, and landfill, in combination with vehicle routes, should be examined before deciding on the need for and site of a transfer station. Routes of waste transfer can also be optimized by the spatial combination of facilities.

RECEPTACLES AND CONTAINERS

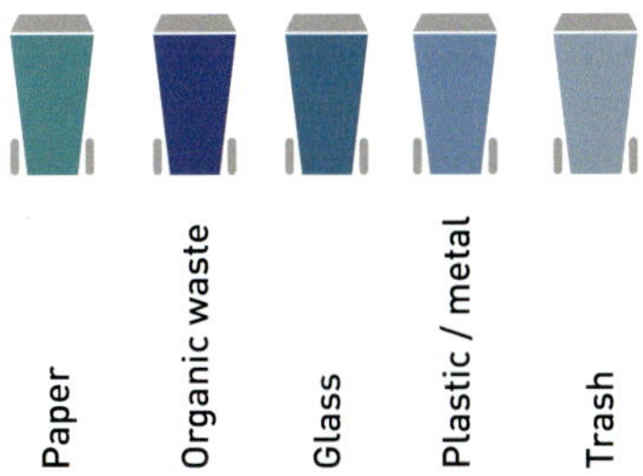

- Location/distribution
- Type
- Size
- Capacity
- Material
- Content
- Cost
- Standards

Receptacles for separate collection of waste streams are allocated throughout the city

ROUTES AND VEHICLES

- Location/distribution
- Type
- Size
- Capacity
- Material
- Fuel consumed
- Cost/operational cost
- Standards

Collection vehicles with multiple routes transfer all waste streams to the transfer station

TRANSFER STATION

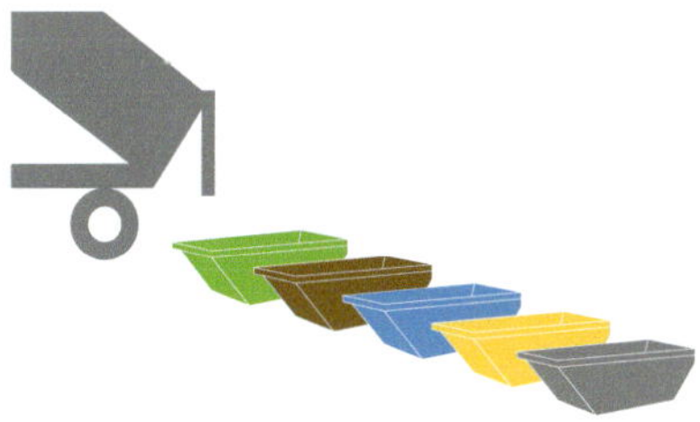

- Location/access
- Size/area
- Vehicle parking capacity
- Temporary waste storage capacity (intermodal containers capacity)
- Processes (compaction and/or baling into self-contained bales etc.)
- Immediate context
- Construction and operation cost

Waste transfer stations are facilities where MSW is unloaded from collection vehicles and briefly held while it is reloaded onto larger long-distance transport vehicles for shipment to landfills or other treatment or disposal facilities

OPTIONS FOR COLLECTION ACCORDING TO DEFINED WASTE STREAMS

Mixed collection of recyclables

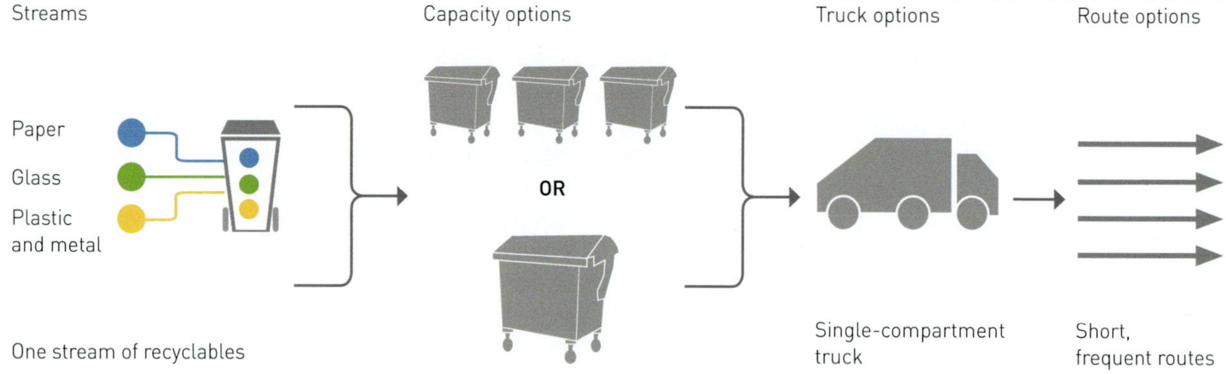

Separate collection of recyclables

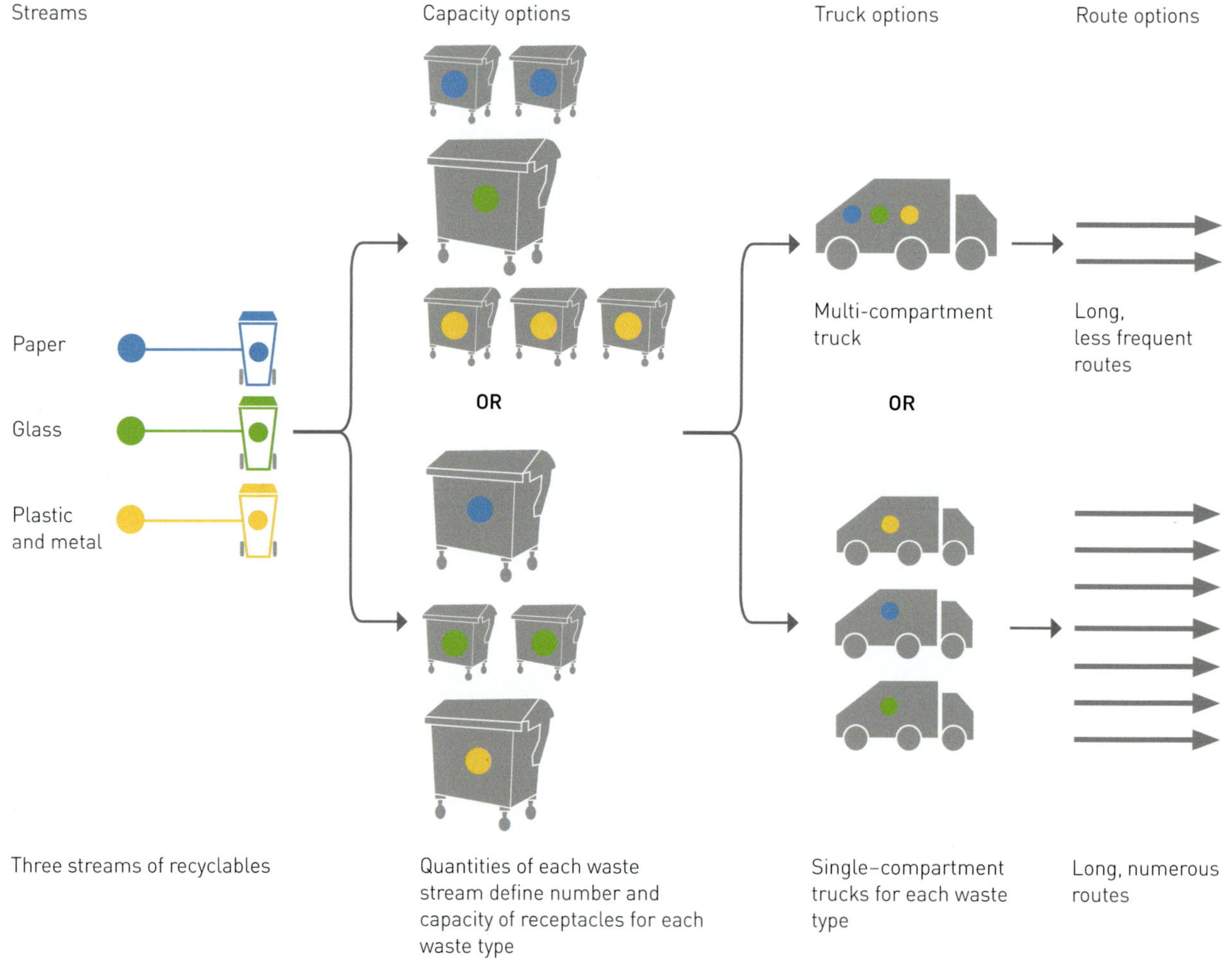

8.4 | OBJECTIVES AND GUIDELINES FOR A SUSTAINABLE SOLID WASTE SYSTEM

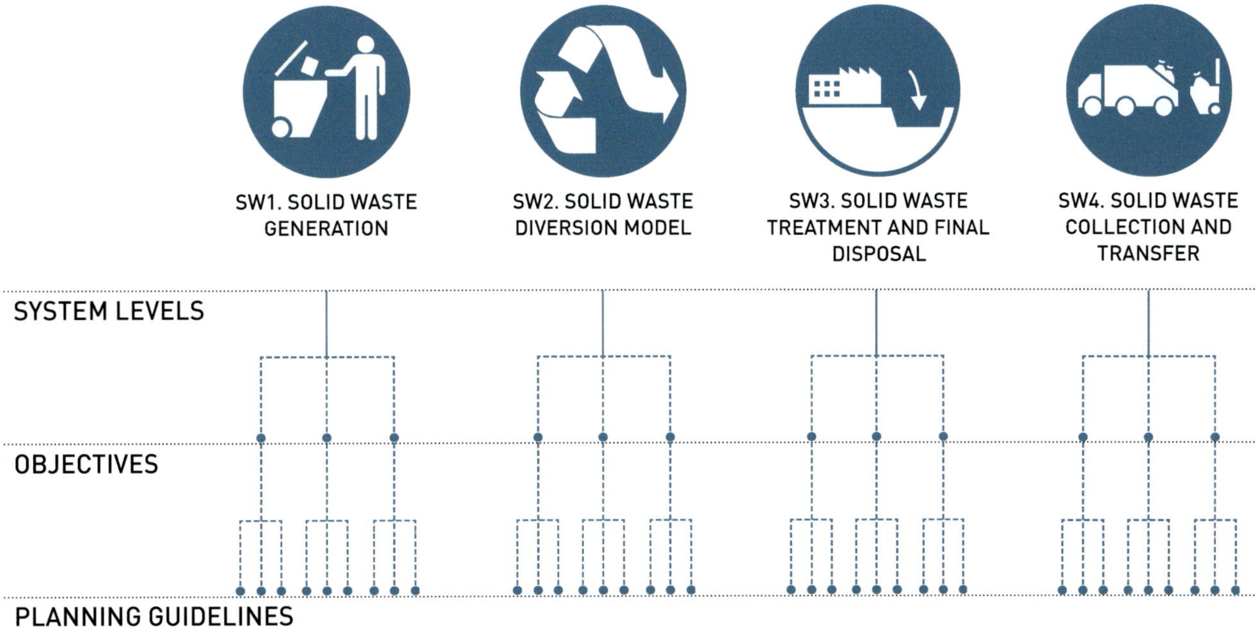

This section serves as an introduction to the planning guidelines for a sustainable solid waste system and traces the processes that lead to them. Following the decoding of Solid Waste through its four main system levels, objectives representing the main sustainability challenges for each level are defined. The objectives are organized through the five focus categories of the Zofnass Program, namely Leadership, Quality of Life, Natural World, Resource Allocation, and Climate and Risk. The Sustainable Planning Guidelines aim to lead toward these objectives, highlighting synergies and connections with the other six infrastructure categories of Energy, Water, Transportation, Landscape, Information, and Food.

OBJECTIVES AND GUIDELINES SYNOPSIS

The main objective of the Waste Generation level refers to the reduction of solid waste that enters the waste management system. The corresponding guidelines promote the participation of communities in waste reduction practices and suggest that authorities set long-term goals and standards for waste generation. Another objective is to monitor solid waste production rates and patterns in order to acquire useful data for all waste management stages. For this objective to be addressed, regular measurements of solid waste weight are suggested, combined with composition analysis with respect to seasonal alterations. Objectives of this level are related to the Leadership category and are to be addressed by decision-making planning bodies.

SW1. SOLID WASTE GENERATION

- SW1.1. Reduce Solid Waste entering the waste management system
- SW1.2. Monitor Solid Waste Generation trends

The objectives in the Diversion Model level lie within the areas of Leadership, Quality of Life, and Resource Allocation. A complete analysis of a community's waste mix is suggested, including classification of waste types and quantities generated, as well as investigation of potential for provisioning services. Guidelines for these objectives refer to information on the quantity and type of recyclable, recoverable, and energy-producing materials included in the solid waste generated. They also encourage defining all types of hazardous waste included in the waste system.

In addition, objectives of this level refer to the relation of waste management to economic and city growth. It should be ensured that the solid waste diversion strategy developed will provide a long-term viable system. The guidelines for this level suggest first that all correlations of alternative solid waste management products with city competitiveness should be thoroughly traced. By-products of selected waste management strategies must be absorbed by the market, and their connection with resource and feedstock needs of other infrastructure systems should be carefully studied. Economic viability studies of the system must take into consideration community size and waste generation patterns and composition in order to secure its long-term financial viability. At this stage, it is also important to coordinate administration actions on solid waste management such as technical, environmental, financial, regulatory, etc.

Together with providing a solid analysis of facts, an objective at this stage is to set clear goals and standards that are to be achieved through the solid waste diversion model. A regular monitoring of the system's performance, through benchmarks and goals, is necessary in order to allow for the required corrective and updated actions in planning. The goals should address long-term municipal growth and encourage efforts toward the evolution of waste management systems. Regular updates on the waste diversion plan are also important in order to keep up with community growth and development requirements. It is advised that strategies toward waste diversion be clearly set and hierarchized.

Objectives also include determining waste streams, treatment methods, and recovery strategies for the system at a strategic level, based on the suggested analysis.

Finally, the guidelines for this level highlight that decisions on selected waste management strategies should take into account social factors and ensure public consensus. Steps in this direction include allocating financial costs rationally, encouraging the public to participate in conversations, and ensuring social equity through the proper location and distribution of management units. For the public's consistent access to information on solid waste management strategies, appropriate promotional and educational programs should be developed. Also, the Information infrastructure system can be addressed in order to provide the necessary mechanisms.

SW2. SOLID WASTE DIVERSION MODEL

- SW2.1. Characterize and classify the Solid Waste generated
- SW2.2. Identify opportunities for Solid Waste system to offer provisioning services
- SW2.3. Secure a viable Solid Waste Diversion Model in the long term
- SW2.4. Set clear goals and standards for Solid Waste Diversion Model
- SW2.5. Establish specific processes for Solid Waste Diversion Model
- SW2.6. Achieve consensus on Solid Waste Diversion strategies
- SW2.7. Define basic Solid Waste streams
- SW2.8. Safely dispose of all Solid Waste management residues

The objectives of Treatment and Final Disposal are related to the achievement of waste diversion and resource recovery goals that were set on the previous level. Guidelines refer to providing necessary and appropriate infrastructure facilities, technologies, and procedures for managing the waste categories identified.

Securing the reliability and flexibility of solid waste treatment and disposal as well as careful distribution and placement of the respective facilities are also objectives of this system level. It is essential to optimize technologies and processes as well as to secure the long-term financial viability of treatment and disposal facilities. The objectives highlight the precautions that should be taken regarding the social and environmental impacts of these units. The integrity of natural-world elements and systems is addressed through synergic guidelines with other infrastructure systems such as Food and Landscape. Community well-being is another very important objective. Guidelines include minimizing the social and health impacts by addressing odors and noise pollution and in some cases by establishing compensatory measures for the communities that host facilities of treatment and disposal. With respect to Climate and Risk, the objectives include provision for system resilience against natural and man-made threats, and control of and compensation (e.g., carbon offsets) for greenhouse gas and air pollutant emissions. Objectives concerning Resource Allocation encourage the production of solid waste by-products that serve and are adapted to the requirements of other infrastructure systems for further processing (feedstock and maintenance). Another set refers to the promotion of energy- and water-efficient technologies for the operation of treatment and disposal facilities.

The fourth level of Solid Waste deals with optimizing the flows within the system. The main objectives for the Leadership category include: the provision for collection of all waste types across the city's jurisdiction, the improvement of the system's efficiency, and its integration with other infrastructure systems. The first objective is for all identified waste categories to be collected according to the defined streams. Particular emphasis is given to the collection of hazardous waste under the necessary safety conditions. The collection system should be planned based on projections of long-term city development. Locations and spaces should be identified and reserved for solid waste transfer facilities expected to be required in the future. Regarding the objective of system efficiency, the guidelines suggest optimization of transfer station locations in relation to final disposal and treatment units as well as the selection of appropriate equipment (receptacles and vehicles) according to the management methods applied. Efficiency also calls for coordination in frequency of collection and adequate administrative structures to optimize asset management.

SW3. SOLID WASTE TREATMENT AND FINAL DISPOSAL

- SW3.1. Achieve goals set for Solid Waste Diversion and resource recovery
- SW3.2. Secure the reliability and flexibility of Solid Waste Treatment and Disposal services
- SW3.3. Place and distribute Treatment and Disposal facilities for optimized Solid Waste system performance
- SW3.4. Optimize landfill facilities
- SW3.5. Integrate Solid Waste Treatment and Disposal with other infrastructure systems
- SW3.6. Promote community well-being
- SW3.7. Enhance quality of place
- SW3.8. Maintain the integrity of natural-world elements and systems
- SW3.9. Prepare for short-term hazards and projected future threats
- SW3.10. Control greenhouse gas emissions
- SW3.11. Control air pollutant emissions
- SW3.12. Optimize Energy use in Solid Waste Treatment and Disposal facilities
- SW3.13. Optimize Water use in Solid Waste Treatment and Disposal facilities

SW4. SOLID WASTE COLLECTION AND TRANSFER

- SW4.1. Secure Collection of all Solid Waste
- SW4.2. Improve Solid Waste Collection and Transfer efficiency
- SW4.3. Integrate Solid Waste Collection and Transfer with other infrastructure systems
- SW4.4. Promote community well-being
- SW4.5. Enhance quality of place
- SW4.6. Maintain the integrity of natural-world elements and systems
- SW4.7. Prepare for short-term hazards and projected future threats
- SW4.8. Reduce greenhouse gas emissions
- SW4.9. Optimize Energy use in Solid Waste Collection and Transfer
- SW4.10. Optimize Water use in Solid Waste Collection and Transfer

As for integration with other systems, it is suggested that Solid Waste should work in synergy with Water, Energy, and Food infrastructures in order to collect and transfer by-products of water treatment, energy generation, and food production. Coordination with Transportation infrastructure is also needed in order to secure adequate spaces along transportation networks for the placement of solid waste collection equipment. Objectives in the areas of Quality of Life, Natural World, Resource Allocation, and Climate and Risk are mainly addressed through guidelines connected to other infrastructure systems (synergic guidelines). Specifically regarding quality of life, it is important to mitigate odors and noise pollution during waste collection. Additionally, in order to enhance quality of place, reduction of the visual impact of solid waste collection is necessary, as well as mitigation of traffic congestion through appropriate scheduling of waste collection itineraries.

The Objectives of the respective Guidelines as well as the relevant Guidelines of other infrastructure systems are presented in the following tables. (TO) and (FROM) indicate that the guideline can assist or can be assisted by the other infrastructure system, where the arrow is placed. (WITH) indicates that the guideline might require planning for an Information or Landscape subsystem and should follow multiple Guidelines from these two systems.

GUIDELINES

OBJECTIVES	GUIDELINES	SYNERGIES

SW1 SOLID WASTE GENERATION

▲ TO
+ WITH
▼ FROM

OBJECTIVES	GUIDELINES	SYNERGIES
SW1.1. Reduce Solid Waste entering the waste management system	SW1.1.1. Guide, assist, and motivate communities to participate in reuse, donation, and recycling programs	
	SW1.1.2. Set long-term goals and standards for reducing Solid Waste Generation	
	SW1.1.3. Promote on-site processing to reduce incoming Solid Waste	
	SW1.1.4. Reduce Solid Waste Generation by other infrastructure systems	
	SW1.1.5. Regulate use of hazardous substances bound to enter the Solid Waste management system	
	SW1.1.6. Regulate use of difficult-to-process materials bound to enter the Solid Waste management system	
SW1.2. Monitor Solid Waste Generation trends	SW1.2.1. Regularly measure volume and weight of produced Solid Waste and record seasonal fluctuations	
	SW1.2.2. Regularly analyze composition of Solid Waste and record seasonal fluctuations	

SW2 SOLID WASTE DIVERSION MODEL

▲ TO
+ WITH
▼ FROM

OBJECTIVES	GUIDELINES	SYNERGIES
SW2.1. Characterize and classify the Solid Waste generated	SW2.1.1. Generate comprehensive information on the quantity and type of recyclable, recoverable, and energy-producing materials composing Solid Waste	
	SW2.1.2. Define the types of hazardous materials entering the Solid Waste system	
SW2.2. Identify opportunities for the Solid Waste system to offer provisioning services	SW2.2.1. Identify potential for recovery or recycling of materials entering the Solid Waste system	
	SW2.2.2. Identify potential for Energy recovery and Generation through Solid Waste management	
SW2.3. Secure a viable Solid Waste Diversion Model in the long term	SW2.3.1. Trace correlation of alternative Solid Waste Diversion options with city competitiveness and economic sectors	
	SW2.3.2. Ensure that Solid Waste Diversion Model residues and by-products can be absorbed by the market	
	SW2.3.3. Match Solid Waste Diversion Model residues and by-products with resource and feedstock needs of other infrastructure systems	
	SW2.3.4. Secure long-term financial viability of Solid Waste Diversion Model based on community size and Waste Generation patterns and composition	
	SW2.3.5. Coordinate administration actions such as technical, environmental, financial, regulatory, etc. toward implementation of Solid Waste Diversion Model	

OBJECTIVES	GUIDELINES	SYNERGIES
SW2.4. Set clear goals and standards for the Solid Waste Diversion Model	SW2.4.1. Set goals for Solid Waste Diversion ratio	
	SW2.4.2. Integrate long-term city growth and development goals with Solid Waste Diversion Model	
	SW2.4.3. Adopt quantitative indicators to benchmark and monitor Solid Waste Diversion Model performance	
	SW2.4.4. Regularly update Solid Waste Diversion Model according to community growth and development	
SW2.5. Establish specific processes for the Solid Waste Diversion Model	SW2.5.1. Define specific processes for the reuse of materials entering the Solid Waste system	
	SW2.5.2. Define specific processes for the recycling of materials entering the Solid Waste system	
	SW2.5.3. Define specific Solid Waste Treatment and Disposal processes for further material and energy recovery	
SW2.6. Achieve consensus on Solid Waste Diversion strategies	SW2.6.1. Develop strategies for fairly allocating the financial cost of Solid Waste management	
	SW2.6.2. Enhance social equity through the location and distribution of Solid Waste system facilities	
	SW2.6.3. Carry out public consultations for strategic decisions on Solid Waste Diversion Model	
	SW2.6.4. Provide consistent and organized access to public information on performance of Solid Waste Diversion Model	+
SW2.7. Define basic Solid Waste streams	SW2.7.1. Collect and transfer Solid Waste in streams according to the needs of selected management processes	
SW2.8. Safely dispose of all Solid Waste management residues	SW2.8.1. Resolve final disposal of residues from all Solid Waste Treatment processes	

SW3 SOLID WASTE TREATMENT AND FINAL DISPOSAL

▲ TO
+ WITH
▼ FROM

SW3.1. Achieve goals set for Solid Waste Diversion and resource recovery	SW3.1.1. Provide for the Treatment and Disposal of all Solid Waste categories identified	
	SW3.1.2. Select optimum material recovery technologies from Solid Waste if applicable	
	SW3.1.3. Select optimum energy recovery technologies from Solid Waste if applicable	
	SW3.1.4. Select optimum procedures for stabilizing and disposing all Solid Waste Treatment and Disposal residues	
	SW3.1.5. Ensure Solid Waste Treatment and Disposal methods selected can be supported by locally available know-how	
	SW3.1.6. Ensure consistency in quality and quantity of Solid Waste Treatment and Disposal by-products	

OBJECTIVES	GUIDELINES	SYNERGIES
SW3.2. Secure the reliability and flexibility of Solid Waste Treatment and Disposal services	SW3.2.1. Develop decentralized Solid Waste Treatment and Disposal facilities as necessary	
	SW3.2.2. Size and distribute Solid Waste Treatment and Disposal facilities to establish economies of scale	
	SW3.2.3. Provide for adaptability of Solid Waste Treatment and Disposal facilities to changes in Solid Waste composition	
	SW3.2.4. Provide thorough environmental regulations for Solid Waste Treatment and Disposal facilities	
	SW3.2.5. Develop smart, integrated management systems to optimize the operation of Solid Waste Treatment and Disposal facilities	+
	SW3.2.6. Secure long-term financial viability of Solid Waste Treatment and Disposal facilities	
	SW3.2.7. Optimize Solid Waste Treatment and Disposal asset management through adequate administrative structures	
SW3.3. Place and distribute Treatment and Disposal facilities for optimized Solid Waste system performance	SW3.3.1. Optimize placement of Solid Waste Treatment and Disposal facilities so as to minimize distances from Waste Generation areas	▲
	SW3.3.2. Reserve spaces for Solid Waste Treatment and Disposal facilities taking into consideration long-term city development	
SW3.4. Optimize landfill facilities	SW3.4.1. Establish strict monitoring mechanisms for landfill operation procedures	
	SW3.4.2. Provide all additional Treatment facilities required to support landfill operation	▼
SW3.5. Integrate Solid Waste Treatment and Disposal with other infrastructure systems	SW3.5.1. Provide required traffic connections to all Solid Waste Treatment and Disposal facilities	▼
	SW3.5.2. Integrate Solid Waste Treatment and Disposal with Energy Generation facilities if applicable	▲
	SW3.5.3. Solid Waste and Energy systems to work in synergy for the Treatment and Disposal of Energy Supply and Energy Generation residues	▲
	SW3.5.4. Solid Waste and Water systems to work in synergy for the Treatment and Disposal of Water Treatment residues	▲
SW3.6. Promote community well-being	SW3.6.1. Assess and resolve potential health impacts of Solid Waste Treatment and Disposal facilities	
	SW3.6.2. Avoid disproportionate social impact on communities hosting Solid Waste Treatment and Disposal facilities	
	SW3.6.3. Establish compensatory measures for communities hosting Solid Waste Treatment and Disposal as necessary	
	SW3.6.4. Assess and resolve potential noise pollution issues caused by Solid Waste Treatment and Disposal facilities	+
	SW3.6.5. Assess and resolve potential odor issues caused by Solid Waste Treatment and Disposal facilities	+

OBJECTIVES	GUIDELINES	SYNERGIES
SW3.7. Enhance quality of place	SW3.7.1. Visually integrate Solid Waste Treatment and Disposal facilities to the local context	+
	SW3.7.2. Minimize land use and fragmentation of urban fabric caused by Solid Waste Treatment and Disposal facilities	
SW3.8. Maintain the integrity of natural-world elements and systems	SW3.8.1. Minimize natural Landscape alteration caused by Solid Waste Treatment and Disposal facilities	
	SW3.8.2. Protect natural habitats and wildlife corridors and compensate for their fragmentation or disruption by Solid Waste Treatment and Disposal facilities	+
	SW3.8.3. Protect soil from contaminants found in Solid Waste Treatment and Disposal facilities	+ ▲ ▲
	SW3.8.4. Protect water elements and groundwater from contaminants found in Solid Waste Treatment and Disposal facilities	+ ▲
	SW3.8.5. Restore Solid Waste Treatment and Disposal sites after decommission and provide for long-term monitoring and maintenance where needed	
SW3.9. Prepare for short-term hazards and projected future threats	SW3.9.1. Assess and manage future threats from natural and man-made phenomena affecting Solid Waste Treatment and Disposal facilities	+
SW3.10. Control greenhouse gas emissions	SW3.10.1. Provide greenhouse gas emissions capture and storage technologies in Solid Waste Treatment and Disposal facilities	
	SW3.10.2. Compensate for Solid Waste Treatment and Disposal emissions through carbon offset and sequestration practices	+
SW3.11. Control air pollutant emissions	SW3.11.1. Promote clean technologies in Solid Waste Treatment and Disposal	
SW3.12. Optimize Energy use in Solid Waste Treatment and Disposal facilities	SW3.12.1. Use energy-efficient Solid Waste Treatment and Disposal technologies	
SW3.13. Optimize Water use in Solid Waste Treatment and Disposal facilities	SW3.13.1. Use water-efficient Solid Waste Treatment and Disposal technologies	
	SW3.13.2. Use gray or reclaimed water where possible for Solid Waste Treatment and Disposal processes	▲

SW4
SOLID WASTE COLLECTION AND TRANSFER

▲ TO
+ WITH
▼ FROM

SW4.1. Secure Collection of all Solid Waste	SW4.1.1. Organize Collection according to defined Solid Waste streams	
	SW4.1.2. Provide required coverage for Solid Waste Collection and Transfer networks	
	SW4.1.3. Provide for Collection of all identified Solid Waste categories	

OBJECTIVES	GUIDELINES		SYNERGIES
	SW4.1.4.	Provide safe Collection and Transfer of hazardous waste	
	SW4.1.5.	Reserve spaces for Solid Waste Collection and Transfer facilities taking into consideration long-term city development	
	SW4.1.6.	Size and distribute Solid Waste Collection and Transfer facilities to establish economies of scale	
	SW4.1.7.	Secure long-term financial viability of Solid Waste Collection and Transfer	
SW4.2. Improve Solid Waste Collection and Transfer efficiency	SW4.2.1.	Optimize placement of Solid Waste Transfer stations between Solid Waste Generation locations and Treatment and Disposal facilities	
	SW4.2.2.	Provide required traffic connections to all Solid Waste Transfer facilities	
	SW4.2.3.	Provide adequate type, location, and capacity of Solid Waste Collection points based on land use, population data, and population distribution	
	SW4.2.4.	Select Solid Waste Collection receptacles and vehicles compatible with the Treatment and Disposal methods selected	
	SW4.2.5.	Coordinate frequency of Solid Waste Collection based on projected or recorded Waste Generation patterns	
	SW4.2.6.	Provide flexible systems for Collection of Waste of non-constant Generation flow	
	SW4.2.7.	Optimize Solid Waste Collection and Transfer asset management through adequate administrative structures	
SW4.3. Integrate Solid Waste Collection and Transfer with other infrastructure systems	SW4.3.1.	Solid Waste and Water systems to work in synergy for the Collection and Transfer of Water Treatment residues	
	SW4.3.2.	Solid Waste and Transportation systems to work in synergy for the Collection and Transfer of solid waste	
	SW4.3.3.	Solid Waste and Energy systems to work in synergy for the Collection and Transfer of Energy Generation residues	
	SW4.3.4.	Solid Waste and Food systems to work in synergy for the Collection and Transfer of food waste	
	SW4.3.5.	Solid Waste and Information systems to work in synergy for the Collection and Transfer of discarded electronic equipment	
	SW4.3.6.	Solid Waste and Landscape systems to work in synergy for the Collection and Transfer of Landscape Maintenance residues	
SW4.4. Promote community well-being	SW4.4.1.	Assess and resolve potential health impacts of Solid Waste Collection and Transfer	
	SW4.4.2.	Assess and resolve potential noise pollution issues caused by Solid Waste Collection and Transfer	
	SW4.4.3.	Assess and resolve potential odor issues during Solid Waste Collection and Transfer	

OBJECTIVES	GUIDELINES		SYNERGIES
SW4.5. Enhance quality of place	SW4.5.1.	Visually integrate Solid Waste Collection and Transfer facilities and equipment to the local context	
	SW4.5.2.	Minimize land use and fragmentation of urban fabric by Solid Waste Transfer facilities	
	SW4.5.3.	Schedule Solid Waste Collection and Transfer itineraries so as to mitigate traffic congestion	
SW4.6. Maintain the integrity of natural-world elements and systems	SW4.6.1.	Protect natural habitats and wildlife corridors and compensate for their fragmentation or disruption by Solid Waste Collection and Transfer processes	
	SW4.6.2.	Protect soil from contaminants during Solid Waste Collection and Transfer	
	SW4.6.3.	Protect water elements and groundwater from contaminants during Solid Waste Collection and Transfer	
SW4.7. Prepare for short-term hazards and projected future threats	SW4.7.1.	Assess and manage future threats from natural and man-made phenomena affecting Solid Waste Collection and Transfer facilities	
SW4.8. Reduce greenhouse gas emissions	SW4.8.1.	Select low-emission and alternative-fuel Solid Waste Collection and Transfer vehicles	
SW4.9. Optimize Energy use in Solid Waste Collection and Transfer	SW4.9.1.	Optimize Solid Waste Collection and Transfer itineraries	
SW4.10. Optimize Water use in Solid Waste Collection and Transfer	SW4.10.1.	Use gray or reclaimed water where possible for Solid Waste Collection and Transfer processes	

RELEVANT GUIDELINES

FROM ▸ ▸ TO
+
WITH

SW1 SOLID WASTE GENERATION

GUIDELINES		RELEVANT GUIDELINES	
SW1.1.3.	Promote on-site processing to reduce incoming Solid Waste	F1.1.2.	Promote on-site processing of food waste by consumers
		F4.9.1.	Promote on-site processing of food waste in Food Processing facilities
		F3.6.1.	Promote on-site processing of food waste in Food Trade facilities
		W3.9.1.	Promote on-site processing of Water Treatment residues
SW1.1.4.	Reduce Solid Waste Generation by other infrastructure systems	I2.8.1.	Promote Information Communication methods and media with reduced Waste Generation
		I3.11.1.	Select Information Nodes equipment for reduced e-Waste Generation
		I4.11.1.	Select Information Networks equipment aiming for Solid Waste reduction
		F4.6.1.	Reduce unnecessary food packaging
		F4.6.2.	Promote use of recyclable materials for food packaging
SW1.1.5.	Regulate use of hazardous substances bound to enter the Solid Waste management system	E2.5.1.	Resolve Energy Supply residues handling and disposal requirements
		W3.9.4.	Resolve routes and methods for the collection, treatment, and disposal of Water Treatment residues and by-products
SW1.2.1.	Regularly measure volume and weight of produced Solid Waste and record seasonal fluctuations	I1.3.1.	Identify infrastructure systems' processes to be monitored/documented through Information infrastructure
SW1.2.2.	Regularly analyze composition of Solid Waste and record seasonal fluctuations	I1.3.1.	Identify infrastructure systems' processes to be monitored/documented through Information infrastructure

FROM ▸ ▸ TO
+
WITH

SW2 SOLID WASTE DIVERSION MODEL

GUIDELINES		RELEVANT GUIDELINES	
SW2.2.2.	Identify potential for Energy recovery and Generation through Solid Waste management	E2.1.1.	Identify all potential Energy sources
SW2.3.3.	Match Solid Waste Diversion Model residues and by-products with resource and feedstock needs of other infrastructure systems	L4.10.3.	Incorporate Solid Waste and wastewater Treatment residues and by-products in regular Landscape Maintenance if available
		E2.1.5.	If waste is selected as an alternative Energy source, integrate with overall Solid Waste management planning

GUIDELINES			RELEVANT GUIDELINES	
			F2.10.2.	Take advantage of Solid Waste system residues and by-products for Food Production processes
			T3.9.3.	Incorporate Solid Waste management residue and by-products in Transportation Nodes construction and maintenance
			T4.10.4.	Incorporate Solid Waste management residues and by-products in Transportation Networks construction and maintenance
			T2.4.3.	Use efficient and/or alternative-fuel vehicles
SW2.3.3.	Match Solid Waste Diversion Model residues and by-products with resource and feedstock needs of other infrastructure systems	+	I1.3.1.	Identify infrastructure systems' processes to be monitored/documented through Information infrastructure

SW3 SOLID WASTE TREATMENT AND FINAL DISPOSAL

FROM ▸ ▸ TO + WITH

SW3.2.5.	Develop smart, integrated management systems to optimize the operation of Solid Waste Treatment and Disposal facilities	+	I1.3.2.	Identify other infrastructure systems' processes to be managed through Information infrastructure
SW3.3.1.	Optimize placement of Solid Waste Treatment and Disposal facilities so as to minimize distances from Waste Generation areas		T1.2.4.	Optimize placement of infrastructure facilities that generate significant traffic
SW3.4.2.	Provide all additional Treatment facilities required to support landfill operation		W3.3.1.	Integrate Water Treatment facilities with Solid Waste infrastructure for leachate treatment
SW3.5.1.	Provide required traffic connections to all Solid Waste Treatment and Disposal facilities		T3.3.1.	Investigate needs of Transportation Nodes integration with Solid Waste Treatment and Disposal facilities
			T4.3.5.	Investigate needs of Solid Waste Treatment and Disposal facilities' connection to Transportation Networks
SW3.5.2.	Integrate Solid Waste Treatment and Disposal with Energy Generation facilities if applicable		E3.3.1.	Investigate potential of Energy Generation integration with Solid Waste Treatment and Disposal facilities
SW3.5.3.	Solid Waste and Energy systems to work in synergy for the Treatment and Disposal of Energy Supply and Energy Generation residues		E3.10.1.	Resolve routes and methods for the collection, treatment, and disposal of Energy Generation residues and by-products
SW3.5.4.	Solid Waste and Water systems to work in synergy for the Treatment and Disposal of Water Treatment residues		W3.9.4.	Resolve routes and methods for the collection, treatment, and disposal of Water Treatment residues and by-products

GUIDELINES			RELEVANT GUIDELINES
SW3.6.4.	Assess and resolve potential noise pollution issues caused by Solid Waste Treatment and Disposal facilities	L1.7.2.	Define needs for mitigating and compensating for other infrastructure systems' impact on community life through Landscape infrastructure
SW3.6.5.	Assess and resolve potential odor issues caused by Solid Waste Treatment and Disposal facilities	L1.7.2.	Define needs for mitigating and compensating for other infrastructure systems' impact on community life through Landscape infrastructure
SW3.7.1.	Visually integrate Solid Waste Treatment and Disposal facilities to the local context	L1.7.2.	Define needs for mitigating and compensating for other infrastructure systems' impact on community life through Landscape infrastructure
SW3.8.2.	Protect natural habitats and wildlife corridors and compensate for their fragmentation or disruption by Solid Waste Treatment and Disposal facilities	L1.7.1.	Define needs for mitigating and compensating for other infrastructure systems' impact on the natural world through Landscape infrastructure
SW3.8.3.	Protect soil from contaminants found in Solid Waste Treatment and Disposal facilities	L1.7.1.	Define needs for mitigating and compensating for other infrastructure systems' impact on the natural world through Landscape infrastructure
		W2.3.3.	Identify and manage basic sources of contaminants within the watershed
		F2.5.5.	Identify and manage basic sources of contaminants within the foodshed
SW3.8.4.	Protect water elements and groundwater from contaminants found in Solid Waste Treatment and Disposal facilities	L1.7.1.	Define needs for mitigating and compensating for other infrastructure systems' impact on the natural world through Landscape infrastructure
		W2.3.3.	Identify and manage basic sources of contaminants within the watershed
SW3.9.1.	Assess and manage future threats from natural and man-made phenomena affecting Solid Waste Treatment and Disposal facilities	L1.7.3.	Define needs for protection of infrastructure systems against projected climate change and extreme events through Landscape infrastructure
SW3.10.2.	Compensate for Solid Waste Treatment and Disposal emissions through carbon offset and sequestration practices	L1.7.1.	Define needs for mitigating and compensating for other infrastructure systems' impact on the natural world through Landscape infrastructure
SW3.13.2.	Use gray or reclaimed water where possible for Solid Waste Treatment and Disposal processes	W1.6.3.	Identify opportunities for use of different nonpotable water qualities

GUIDELINES		RELEVANT GUIDELINES	
SW4	**SOLID WASTE COLLECTION AND TRANSFER**	**FROM ▸ ▸ TO + WITH**	
SW4.2.1.	Optimize placement of Solid Waste Transfer stations between Solid Waste Generation locations and Treatment and Disposal facilities	T1.2.4.	Optimize placement of infrastructure facilities that generate significant traffic
SW4.2.2.	Provide required traffic connections to all Solid Waste Transfer facilities	T4.3.6.	Investigate needs of Solid Waste Collection and Transfer facilities' connection to Transportation Networks
SW4.3.1.	Solid Waste and Water systems to work in synergy for the Collection and Transfer of Water Treatment residues	W3.9.4.	Resolve routes and methods for the collection, treatment, and disposal of Water Treatment residues and by-products
SW4.3.2.	Solid Waste and Transportation systems to work in synergy for the Collection and Transfer of solid waste	T4.3.7.	Provide spaces for Solid Waste Collection equipment along Transportation Networks
		T3.10.1.	Conform waste management practices within Transportation Nodes to requirements of Solid Waste Collection and Transfer planning
SW4.3.3.	Solid Waste and Energy systems to work in synergy for the Collection and Transfer of Energy Generation residues	E3.10.1.	Resolve routes and methods for the collection, treatment, and disposal of Energy Generation residues and by-products
SW4.3.4.	Solid Waste and Food systems to work in synergy for the Collection and Transfer of food waste	F2.10.3.	Organize Waste recovery systems in Food Production facilities
		F3.6.2.	Organize Waste recovery systems in Food Trade areas
		F4.9.3.	Organize Waste recovery systems through all Food Processing and Distribution stages
SW4.3.5.	Solid Waste and Information systems to work in synergy for the Collection and Transfer of discarded electronic equipment	I3.11.2.	Provide plan for collection and diversion of Information Nodes' discarded equipment
SW4.3.6.	Solid Waste and Landscape systems to work in synergy for the Collection and Transfer of Landscape Maintenance residues	L4.10.1.	Conform waste management within Landscape entities to the requirements of Solid Waste Collection and Transfer planning
SW4.4.2.	Assess and resolve potential noise pollution issues caused by Solid Waste Collection and Transfer	L1.7.2.	Define needs for mitigating and compensating for other infrastructure systems' impact on community life through Landscape infrastructure
SW4.5.1.	Visually integrate Solid Waste Collection and Transfer facilities and equipment to the local context	L1.7.2.	Define needs for mitigating and compensating for other infrastructure systems' impact on community life through Landscape infrastructure

GUIDELINES		RELEVANT GUIDELINES	
SW4.5.3.	Schedule Solid Waste Collection and Transfer itineraries so as to mitigate traffic congestion	T1.4.2.	Optimize itineraries within other infrastructure systems
SW4.6.1.	Protect natural habitats and wildlife corridors and compensate for their fragmentation or disruption by Solid Waste Collection and Transfer processes	L1.7.1.	Define needs for mitigating and compensating for other infrastructure systems' impact on the natural world through Landscape infrastructure
SW4.6.2.	Protect soil from contaminants during Solid Waste Collection and Transfer	L1.7.1.	Define needs for mitigating and compensating for other infrastructure systems' impact on the natural world through Landscape infrastructure
		F2.5.5.	Identify and manage basic sources of contaminants within the foodshed
		W2.3.3.	Identify and manage basic sources of contaminants within the watershed
SW4.6.3.	Protect water elements and groundwater from contaminants during Solid Waste Collection and Transfer	L1.7.1.	Define needs for mitigating and compensating for other infrastructure systems' impact on the natural world through Landscape infrastructure
		W2.3.3.	Identify and manage basic sources of contaminants within the watershed
SW4.7.1.	Assess and manage future threats from natural and man-made phenomena affecting Solid Waste Collection and Transfer facilities	L1.7.3.	Define needs for protection of infrastructure systems against projected climate change and extreme events through Landscape infrastructure
SW4.9.1.	Optimize Solid Waste Collection and Transfer itineraries	I1.3.2.	Identify other infrastructure systems' processes to be managed through Information infrastructure
SW4.10.1.	Use gray or reclaimed water where possible for Solid Waste Collection and Transfer processes	W1.6.3.	Identify opportunities for use of different nonpotable water qualities

NOTES

1. Daniel Hoornweg and Perinaz Bhada-Tata, "What a Waste: A Global Review of Solid Waste Management", World Bank, March 2012, No. 15.
2. Forbes R. McDougall, Peter R. White, Marina Franke, and Peter Hindle, *Integrated Solid Waste Management: A Life Cycle Inventory* (Oxford: Blackwell Publishing, 2001).
3. Ibid.
4. United Nations Environment Programme, *Waste: Investing in Energy and Resource Efficiency* (2011).

IMAGE CREDITS

P. 180: Photograph by Bicanski, accessed on October 14, 2015, http://www.public-domain-image.com/free-images/miscellaneous/junkyards-pictures/waste-materials-in-waste/attachment/waste-materials-in-waste. **P. 191:** (Product's life cycle) Adapted from "Life Cycle Analysis for Shipping Crate," Clemson University, accessed on January 13, 2014, http://www.clemson.edu/ces/crb/ece495/past_projects/WasherMover/Spring2012/Group2_spring12_ece495_website/lifecycle.html.

Off
HDR Off
VIDEO
PHOTO
SQUARE

Information as Infrastructure

Chapter 9

i

Information

is the dynamic and vital resource of our century, a web of linkages continuously expanding across sectors to meet emerging needs.

9.1 | INDUSTRY PERSPECTIVES

Information as a New Form of Infrastructure

We all know that we are being inundated with data. Cities are in the same boat. But not all that data easily translates into useful information, and even less into actionable knowledge. Therefore, it's time to start thinking about information as a new form of infrastructure on which our new generation of smart cities must be built.

Data deluge

One of the characteristics of today's world is the abundance of information. Never before has society generated as much information as it does now. Scandinavian researcher SINTEF estimated that 90 percent of all data in the world was generated over the past 2 years.[1] Others have estimated that every day we create around 2.5 billion gigabytes of new information, and by 2020 the rate of adding new information is predicted to double every 72 hours.

A good part of all this data is geospatially based and relevant to infrastructure, with sources ranging from new sensors used to gather climate, pavement, bridge, or water system conditions, to building environments, to social media sites, cell phones, smart phones, and tablets generating geospatially located digital pictures, videos, and more. Is all this data going to allow us to make better sustainable planning decisions and help to address our forward-looking infrastructure problems?

Governments, whether federal, state, or municipal, have traditionally been among the most prolific generators and users of information. But now more traditional data sources such as census and land records are being constantly supplemented by legions of digitally enabled citizens and sensor networks.

Information generated by machines, for use by machines, is also on the rise, and just as this abundance of information has transformed traditional business, so it will transform cities as well.

Useful and usable

Over the next few years, cities will see an exponential increase in the supply of urban data, and increasingly novel and sophisticated ways of using that data. As the cost of technology such as sensors decreases, there will also be an increase in the types of indicators and the level of detail that can be measured.

So how do we convert all this data noise into a signal that can drive better planning of urban environments and ultimately a more sustainable future? People often mistake data for information, and information for knowledge or wisdom. Data consists of facts captured at a single moment in time—for example, sensors may capture levels of particle pollution in the air. Data is then processed into information—levels of particle pollution are compiled into an air quality report for the city.

From this information, knowledge can be extracted and decisions can be made: for example, the city knows high levels of particle pollution are bad for the environment and citizen health, and so makes a short-term decision to declare a "Spare the Air" day, and a long-term decision to address the sources of pollution.

The ability to process and analyze information in real time helps enable infrastructure planners, designers, owners, and investors to both maximize the values of their assets and create smarter cities. Two main types of information are needed:

- "Academic" information: more traditional forms of information, such as census data, land records, or the results of transportation models.
- "Real-time" information: information from a network of sensors embedded in buildings, or from citizen crowd-sourcing.

Both types of information are necessary to enable strategic and tactical decisions. Better decision making during planning is critical if we are truly to have smart cities. Therefore we need to start thinking about information as really another infrastructure type. Digital infrastructure is no less important than physical infrastructure assets such as roads, highways, or a water system, and its resilience is no less critical. Impacts to the information grid could have lasting and perilous consequences. But how do we leverage information as infrastructure and turn it into a smart system for smarter infrastructure?

The rise of smart cities

The first step is to define what we mean by smart systems and smart infrastructure. In the context of this book, "smart" means having the capability to adapt to varying infrastructure demands and conditions which impact the systems this information is running, including new technology currently available or in development. Information as infrastructure works when it creates the ability to make informed decisions, when the system

i

it drives can respond intelligently to changes in the surrounding environment (digital or physical) and interact with other infrastructure or user demands, all geared toward achieving an improved performance.

The smart infrastructure systems that drive smart cities need to facilitate informed decision making through feedback loops. The system must be able to monitor change, measure the degree of change, then analyze options for addressing it and communicate those options in order to drive human actions or in some cases take action without human interaction. In the longer term this decision-making information will help in the planning of newer, more efficient versions of the information infrastructure itself. It becomes a positive feedback loop that improves both the physical and digital infrastructure of the city. A special report on managing information by the *Economist*[2] estimates that only 5 percent of data generated is "structured"—that is, easily understood and digested by people, software, or computers. For data to have any value, it needs to be both usable and useful. The information received should allow cities to act strategically and tactically.

For example, cities want to be able to use data in order to plan infrastructure. Census data predicting population growth and job opportunities can be used along with spatial data to identify and prioritize areas for development within the city. Transportation modeling can predict vehicle miles traveled, and possible ridership on public transit. The city can make strategic decisions about what it will look like in the future, and how its infrastructure might perform.

Once a project has been designed and built, different information can be used to make sure that that infrastructure is optimized. If, to accommodate growing population and transit-oriented development, it was decided to implement a bus rapid transit system, then sensors along the route might monitor traffic in real time, and switch the direction of travel for some lanes in order to optimize traffic flow. Those same sensors could also check emissions and air quality, in order to meet GHG and health targets.

A mobile application developed by citizen hacktivists can also make use of city data to enable transit passengers to alert fellow passengers or the transit agency to delays in the schedule.

Applications

A smart city powered by information used as infrastructure is generally arranged around seven major pillars: energy, water, food, waste, transportation, landscape, and information.[3] Information systems can be embedded within each of the other six systems in order to enhance their performance.

For example, energy management today leverages information as a key infrastructure component through smart grids, smart meters, and intelligent energy storage. Buildings can have intelligent lighting and cooling systems and energy performance management systems. Key to success is the ability to monitor, remotely control, and adjust components, balancing interactions between consumers, generators, and the networks themselves. Increasingly these systems are being implemented across the globe in the planning of new cities or revitalization of developed ones.

These systems driven by information are more adaptive and can reconfigure themselves in response to demands and supply or in the case of an emergency. Over time this performance information allows for more predictive capability and optimization for more efficiency, since models of the systems can be created and used in their operations. This integrated system allows supply, consumption, and control at many stages or locations, helping with redundancy and resilience.

The same is also true on the water side. Smart water systems enabled through networks of sensors enhance water management, and today deliver more integrated and resilient wastewater and flood protection infrastructure, all driven by more intelligent information systems. While water infrastructure has traditionally been unintelligent, operated mostly by human interaction, the use of information as an infrastructure subsystem has created smart meters for consumers, allowing them to be better informed about usage and become more active participants. During the planning of infrastructure, the ability to leverage this usage information creates better predictability about what is required and reduces demand and waste in the long run. This information will also allow future water infrastructure to adapt more easily to supply and demands, which reduces energy needed for treatment, pumping, and wastewater collection, whether in a closed-loop wastewater system with energy recovery or a runoff/flood management system.

Smart transportation can encompass intelligent parking systems, predictive analysis, congestion management, and public portals. We know that making transportation systems smarter doesn't always solve all the problems we face, since we don't control when and where people want to use them. By leveraging information as an infrastructure subsystem, however, future systems could have the ability to communicate with individual users and help plan, predict, and provide alternative methods of transportation and routes to reach destinations.

Interconnected infrastructure requires information, especially as rapid advances in technology force local, national, and international governments to decide what future needs to accommodate. It requires a change in thinking, a concern not only with traffic lights, signs, and crosswalk lights but with how to add intelligence into pavement, utilities, and the like. This will require changes in how we plan future infrastructure, its business models and job functions, all with an information underpinning.

Our roadways and highways aren't inert objects—they're dynamic systems comprised of the interactions between vehicles, pedestrians, bikers, traffic signals, and the environment. The vehicles and drivers must play a role as real-time data contributors to the information infrastructure transportation network; this is where the intelligence of an autonomous system really shines. There will come a day when the vehicles of today, even the autonomous versions, are replaced by something even better, and the network and the information that drives it should be adaptable to those changes.

The amount of data being made available by government organizations is also being used in increasingly innovative ways to make cities work better. Citizens have used this data to create apps that track open bike parking spots in real time, find federally funded health centers, or figure out whether a car has been towed by the city or stolen. Cloud computing services, open data platforms, and emergency response systems all work toward making city governments smarter and more transparent.

What does a sustainable information system look like?

A sustainable information system is one that aids decisions in meeting the needs of the present without compromising future generations' meeting their own needs. It is one that:

- Protects the privacy of individuals, businesses, and government,
- Offers credible systems to help users manage the information overload, and affects distance, time, and the behavior of systems,
- During its life cycle is affordable, operates fairly and efficiently, offers choices of information services, and supports a competitive economy and balanced regional development,[4]
- Is designed, installed, operated, maintained, rehabilitated, and reused/recycled with an emphasis throughout its life cycle on using natural resources efficiently and preserving the global environment.

At the same time, an information infrastructure should enable planning strategies for reducing a city's impact on the natural environment by coordinating the rest of the city's infrastructure and services.

It's these kinds of smart infrastructure interconnections—at a community, metropolitan, or even national level—that will change the planning of sustainable cities and provide the foundation for even more holistic visions of smart cities where all aspects of infrastructure talk to each other to optimize performance, minimize energy use, and make life more enjoyable and productive for the people who live in them.

Terry D. Bennett
LS, LPF, MRICS, ENV SP, LEED AP, Senior Industry Strategist for Civil Infrastructure, Autodesk Inc.
Lynda Sharkey
Sustainable Cities and Transportation Product Manager, Autodesk Inc.

i

NOTES

1. "Big Data—for Better or Worse," http://www.sintef.no/home/Press-Room/Research-News/Big-Data--for-better-or-worse/, 2013 (last accessed May 2014).
2. "Data, Data Everywhere," http://www.economist.com/node/15557443, 2010 (last accessed May 2014).
3. http://www.navigantresearch.com/wp-assets/secure/SCIT-13+ Navigant+Research.pdf.
4. "From Green to Sustainable Information Communication Technology Systems (ICT)," http://www.bicsi.org/uploadedfiles/Conference_Websites/Winter_Conference/2012/presentations/From%20Green%20to%20Sustainable%20ITS.pdf, 2012 (last accessed May 2014).

9.2 | **SUSTAINABLE INFORMATION PLANNING**

COMMUNICATION AND DECISION-MAKING PROCESS

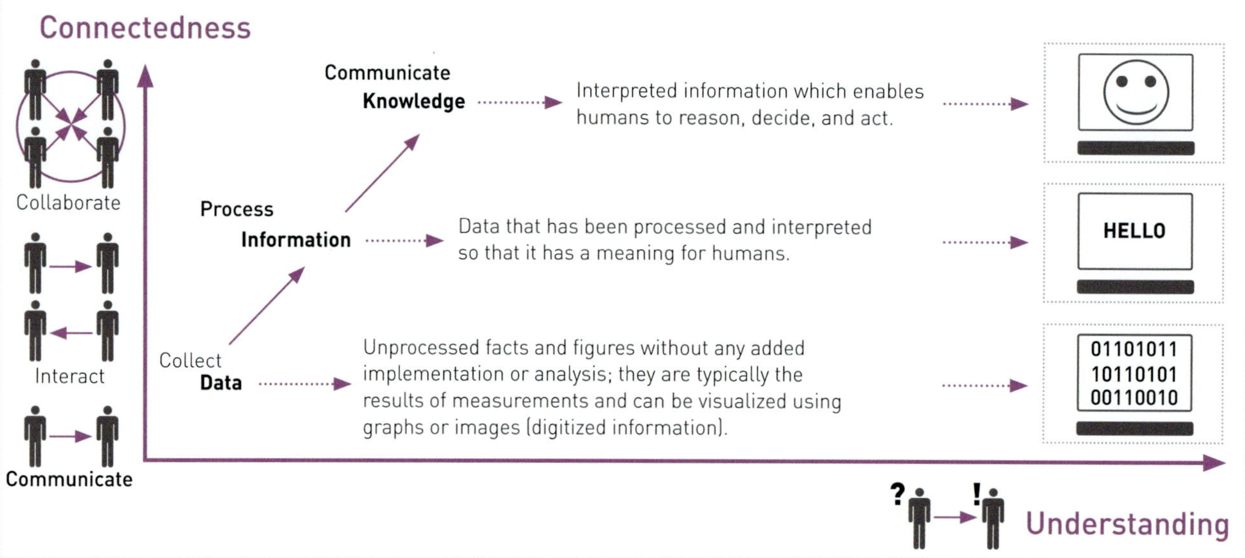

Information is the dynamic and vital resource of our century, a web of linkages continuously expanding across sectors to meet emerging needs.

Information as infrastructure constitutes the integrated framework that allows the collection, flow, and processing of information. It can be characterized as a next-generation infrastructure serving to support operations, decision making, and actions. It consists of infinite subsystems, public or private, created and used by individuals and organizations. While intangible, information can have a strong impact on physical space, determining how cities are studied, planned, and managed. It has the potential to affect the performance of all other infrastructure systems, their ability to respond to ever-evolving demands and challenges. As such, Information infrastructure is closely interrelated with global urban and rural economies, acting as a catalyst for growth, competitiveness, and resilience.

"Information Communication Technologies help reduce the minimum unit cost of a particular service by reducing transaction time and improving staff productivity; maximize the total benefit of services in the short run; or increase incentives for continuing investment, innovation, and improvement of service provision."[1]

Just as water infrastructure exists in order to offer reliable access to water, or transportation infrastructure to serve flows of people and freight without interruptions, so information infrastructure is responsible for making information available at the required quantity, quality, and format. It offers its users the ability to input, access, process, filter, store, and disseminate the information they need.

Data, information, and knowledge are three interrelated terms found at the core of the information system and the processes that it serves (see diagram above). Data are raw and unprocessed facts and figures that enter the system without having been subjected to any conversion or analysis. They digitalize the qualitative and quantitative characteristics and parameters of the real world, allowing them to be manipulated by machines. Information is processed and meaningfully organized data, available for consumption. Finally, knowledge is interpreted information which allows end users to reason, decide, and act.

Information infrastructure has transformed the concepts of physical distance and time, supporting the communication, interaction, and collaboration of communities. It allows information to extend everywhere and at any time, affecting economic and cultural dynamics. It includes the technologies and operations that are responsible for the canalization of information flow through fixed wired or wireless networks. At the same

UBIQUITOUS CITY DIAGRAM: Infrastructure with intelligence

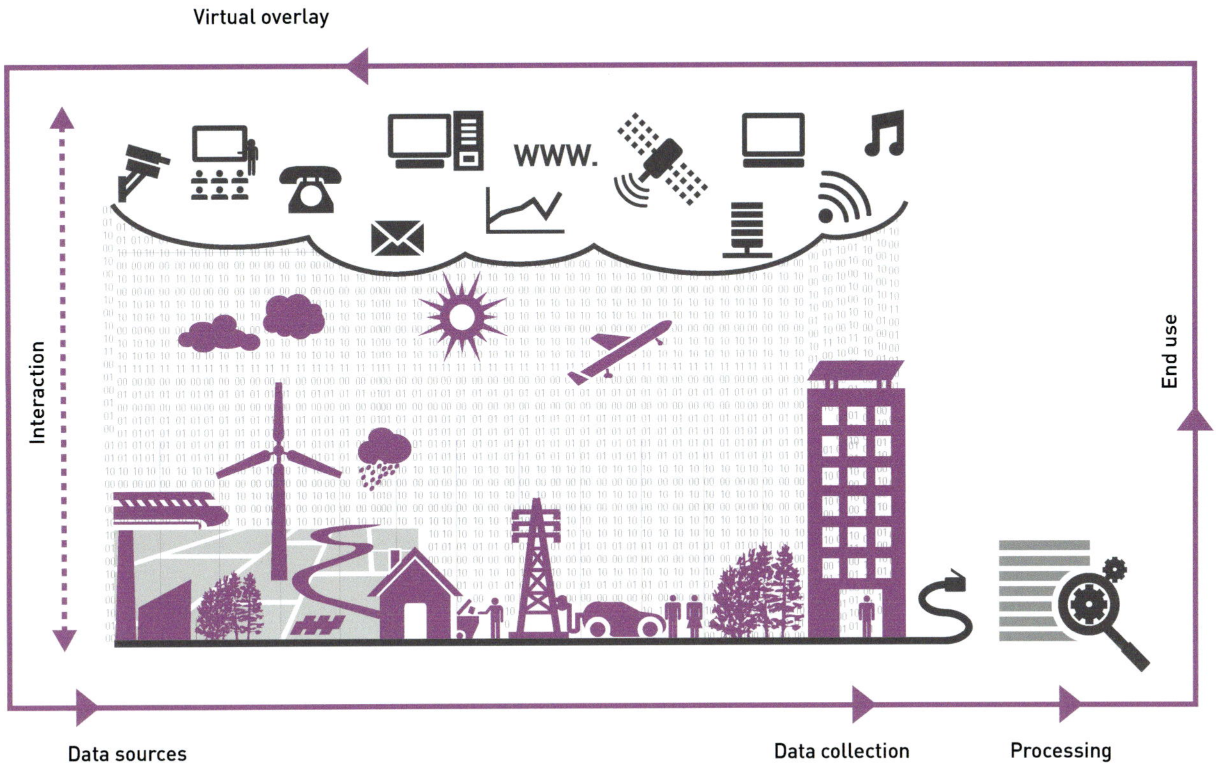

time, the increase of urban populations, built space, and new infrastructure together with the evolution of digital technologies has resulted in an accelerating increase in the production and utilization of information. It is an iterative process, as availability of connection opportunities leads to a demand for more interaction. More and more information needs to be communicated faster and to more stakeholders, often leading to an information overload that can prove difficult to manage. As a result, it becomes imperative to distinguish between useful or excess information.

This new digital era brings a shift in urban research and conventional development patterns across disciplines. Cities today require a seamless coordination of systems and processes in order to meet their targets for sustainability and resilience. Every day new technologies become available, increasing the speed and volume of transferred data and allowing for flexibility, adaptability, and responsiveness within the urban conditions. New possibilities are made available for operating city services and public infrastructure, such as electric power, transportation, and water, through the utilization of the large amount of information collected and the combination of different data fields. In order to establish and support the required connections, including human to human, human to machine, and machine to machine, a virtual overlay is being created within the physical urban context ready to monitor and manage a city's operation and performance. This new knowledge and the disappearance of geographic and time limitations allow for interdisciplinary decisions and integrated planning.

The integrated planning approach to information as infrastructure is based on a twofold analysis of the system. It focuses first on the telecommunication and internet services and secondly on information's function as a subsystem that can be incorporated in all other infrastructure systems (Energy, Water, Solid Waste, Transportation, Landscape, Food) to interconnect and assist them in achieving their performance goals. This makes information synergic at its core and able to contribute toward overall city sustainability. A high-performance system can effectively regulate, provision, and manage information, affecting the everyday life of citizens and the performance of all urban infrastructure and services. Information infrastructure therefore will play a crucial role in addressing the problems of 21st-century urbanization.

"It will sense the state of the world and different ecosystems in real time and alert humans and applications to emerging threats. It will also be able to allocate resources intelligently to compensate for natural disasters, storms, and environmental damage—much in the way that the

INTEGRATED INFORMATION PLANNING

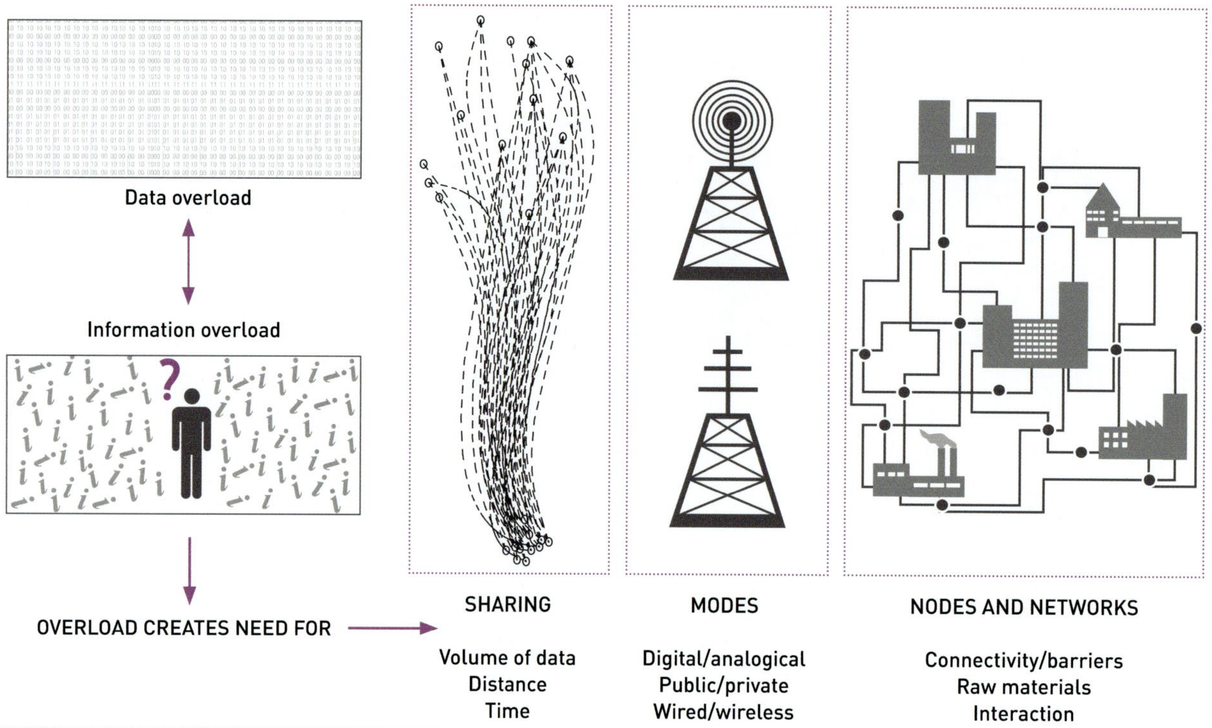

air traffic control systems allocate and manage airplane traffic."[2]

As a product of the digital revolution of the past few decades, Information infrastructure contains attributes different from the rest of the infrastructure systems, which have been serving humanity for much longer time periods. Water, Food, and Landscape infrastructure systems have been gradually established in parallel with the development of the first cities, going back to early urbanization in Mesopotamia during the fourth and third millennia BC. Energy, Transportation, and Solid Waste systems started to become essential to the operation of cities during the industrial revolution of the 18th and 19th centuries AD. The establishment of Information infrastructure as a critical system for cities is a phenomenon of our 21st century.

Thus the Information infrastructure system has a substantially different structural, scientific, technological, and sociopolitical starting point. Specifically, it is the product of networked structures, nonlinear processes, digital technologies, globalized economies, and interconnected societies. Understanding the information system has another dimension of difficulty, since its integration as a necessity of urbanization has not yet matured but is still in progress. The particularity of Information infrastructure is expressed in our framework through the twofold approach based on two basic strategic dimensions. In the first, encompassing telecommunications, the information infrastructure system is an infrastructure per se that supports the function of human societies and cities based on demand and supply. In the second, it is a synergistic subsystem that is networked with other infrastructure systems in order to enhance them.

Information infrastructure per se consists of a huge number of fiber-optic cables, data centers, and terminals that are spread throughout cities and across the globe. This infrastructure supports primarily telephone networks, computer networks, and especially the internet. The information circulated inside these networks is the actual commodity that is offered by the information infrastructure system. In this dimension, the information system promotes the economic and social development of cities by fostering productivity and efficiency in activities that regard the economy, services, governance, and social groups.

As a synergistic subsystem, Information infrastructure is being embedded in all other infrastructure systems (Energy, Water, Solid Waste, Transportation, Landscape, and Food) in order to enhance their performance. Recent years have seen numerous examples of the deployment of information systems in order to improve the performance of transportation, energy, water, and waste management, while significant steps have also begun in the case of landscape and food infrastructure systems. Infrastructure systems are becoming digitized, networked, and thus enhanced in order to address better the complexity of present and future urban challenges.

The particularity of Information infrastructure is also reflected in the twofold challenges of the system in terms of environmental, financial, and social sustainability. An information infrastructure system is sustainable not only when it is sustainable by itself, but also when it is integrated synergistically in other infrastructure systems in ways that enhance their sustainable performance.

The major sustainability challenge of Information infrastructure itself, as the carrier of telecommunications, is its capacity and reliability in supporting the functions of economy, services (e.g., healthcare), governance, and social groups (e.g., social networks). We can realize the importance of sustainability in information if we imagine the breakdown of the internet for a few days or hours in a city. Although information as a commodity is immaterial, Information infrastructure has a substantial material component. Networks, data centers, sensors, terminals, and personal devices dominate the urban space, while future projections describe an intensification of the phenomenon. Sustainability challenges of the physical component of Information infrastructure are observed in the increasing energy demand for its operation, in the exploitation of natural resources to cover material needs, and in electronic waste.

Simultaneously, the synergistic dimension of Information infrastructure implies another set of sustainability challenges. These concern enhancements of the other infrastructure systems that will be crucial for their sustainable performance. It is hard to imagine a future with secure and sustainable performance of energy systems without smart grids, of water systems without smart leakage monitoring, of waste management without dynamic pricing (pay-as-you-throw), or of transportation without real-time traffic control. The sustainability of all infrastructure systems partially depends on their convergence with Information infrastructure.

Information infrastructure doesn't generally consume direct resources from the environment in order to operate, in contrast to other infrastructure systems, though it has gradually become a significant consumer of electricity. This doesn't seem to be a serious issue yet, since demand increases simultaneously with the technological advancement of Information infrastructure, which becomes continuously more energy-efficient. An important environmental issue stems from the phenomenon of electronic waste with the increasing volume of discarded digital devices. Another such issue is the material-intensive manufacturing processes. On the other hand, the impact of Information infrastructure on the environment undoubtedly has a tremendous positive side. The integration of information in economy, services, governance, and social groups leads to a series of digital processes that replace activities and processes in the material world. Saving natural resources (e.g., through the reduction of paper use) and reducing waste generation are conspicuous benefits. It is estimated that a "15% reduction in emissions via ICT is possible by 2020."[3] The dematerialization impact of Information infrastructure rests also on its synergies with the other infrastructure systems. Specifically, the integration of information can lead to trip reduction in transportation, save energy through efficient grids, save water by minimizing leakages, reduce waste production through dynamic pricing, optimize food production through biometrics, and preserve natural areas through monitoring.

The information infrastructure system has a strong impact on society. Through the availability of the internet, computer, and telephone networks, knowledge is disseminated and access to information is provided in a way unprecedented in human history. Information infrastructure systems are positive catalysts for social equity, community development, and eventually quality of life. Information provides social capital in economy, promotes equal opportunities in services such as education or healthcare, fosters transparency and participation in governance, and finally augments dramatically the communication among social groups. Through its synergic dimension Information infrastructure contributes to social sustainability by facilitating the reliability of the other infrastructure systems. Moreover, through urban informatics it provides information on the performance of systems such as energy, water, or waste, so that it develops public awareness and social responsibility. On the other hand, Information infrastructure either as infrastructure per se or as synergic to other systems is vulnerable to data security and privacy issues. Data security is crucial for the maintenance of the proper functioning of information networks and other infrastructures. Cyberterrorism threats regarding energy power plants or water systems are an indicative example of the security issues. At the same time, the protection of the privacy of individuals, businesses, and government is essential to maintain the democratization and accessibility of information systems.

The integration of Information infrastructure systems

in modern societies has revolutionized the way the economy works through the real-time interconnection of markets and the globalization of commerce. Apart from interconnecting businesses and customers around the world, Information infrastructure brings a high increase in productivity, from the level of countries to that of individual work. Efficiency is improved and costs are correspondingly reduced. The emergence of e-commerce is illustrative. It is estimated that "with 10 percent increase in high speed Internet connections, economic growth increases by 1.3 percent."[4] Information has also generated a long list of new professions (e.g., software engineers) and disciplines (e.g., biotechnology, social media) that foster economic development, competitiveness, and innovation. In governance and in services such as education or healthcare, Information infrastructure facilitates the reduction of operating costs, so that customers and taxpayers receive better and less expensive services. In social groups the impact has been revolutionary as well, with communication costs so sharply reduced that international communication is now almost costless through online platforms (e.g., Skype and Facebook).

ENVIRONMENTAL

Challenges for environmental sustainability include:

- Minimizing water and energy consumption required for operation
- Minimizing e-waste generation
- Controlling contaminants of e-waste

SOCIAL

Challenges for social sustainability include:

- Avoiding social exclusion due to fast-paced technological developments
- Adapting interfaces to needs of different social groups
- Assisting users in adapting to changes
- Ensuring privacy and minimizing e-crime
- Maintaining affordability for all end users

ECONOMIC

Challenges for economic sustainability include:

- Being constantly updated
- Boosting cities' productivity and innovation
- Integrating different information systems

SYSTEM PURPOSE: RELIABLE COMMUNICATION AND OPTIMIZATION	
	HIGH PENETRATION AND CONNECTIVITY: Provide coverage to include all users
	ADAPTABILITY: Data quality, speed, filtering
	USER FRIENDLINESS: Selection of media, applications, and interfaces which communicate best information
	SMART AND INTEGRATED SYSTEMS: Interaction and integration between services and systems, enhancement of processes, responsiveness

The strategies listed here suggest basic steps that could be included in the planning of an information system. They do not aim to outline all the requirements for a sustainable system, but instead highlight approaches that if implemented could resolve multiple issues simultaneously.

Connectivity, as a strategy for sustainable information infrastructure in the city, concerns the minimum infrastructure needed to provide access to computer and telephone networks. An adequate information infrastructure has to be established across the city through a resilient network of fiber-optic cables that is supported by processing and storage nodes. Connectivity provides the essential means for a community to interact and exchange.

"Behind many apps on a smart phone is a complex network of servers, routers, databases, fiber-optic cables, satellites, mass storage arrays, and sophisticated software."[5]

Connectivity implies serious investments similar to the investments made in transportation infrastructure networks. Worldwide, network connections are becoming an essential infrastructure for innovation, economic growth, and competitiveness. Connectivity is the first step in enabling access to information among citizens, businesses, and government.

The telecommunications sector by itself is essential but not sufficient in the provision of information infrastructure. Social and economic sustainability can be achieved when the potential of the information systems is utilized. Experience from many countries has proved that even if information infrastructure is available, governments, businesses, and even individuals might not harness its potential (as when public institutions and governments don't adopt e-governance services to overhaul bureaucracy and avoid corruption). New policies have to accompany the information systems, along with incentives for changing common practices (e.g., education programs) and the right organizational frameworks to promote innovation. The penetration rate of information and telecommunication systems is critical for the sustainable development of communities. Another factor in fostering use and integration of information infrastructure is the reduction of costs.

After integrating the information infrastructure system extensively in the economic and social structures of the city's community, further strategies toward sustainability need to be employed. The efficiency and reliability of the system have to be addressed, securing and optimizing the performance of existing information and telecommunication systems. Since Information infrastructure concerns the major topics of urban life, governmental, business, and personal data need to be secured. Technological and legal frameworks can develop the system's reliability.

"It's 1 AM on a Saturday night in August 2011. The Dutch Minister for Homeland Security calls a press conference. People panic. Is it a flood? Is it an invasion by the Belgians? No, it's a cyber-attack: an obscure company called a Certificate Authority—DigiNotar—has been breached, paralyzing internet communications in an entire country for over a week."[6]

Big data management is also important, since globally every day we create around 2.5 billion gigabytes of new information. The systems should be reliable to deal efficiently with the continuously increasing volumes of data both in terms of computing power and storage capacity (more efficient software and hardware). The response and recovery of the information infrastructure under extreme conditions of urban disasters are also an issue for maximizing the system's efficiency and reliability.

A user-friendly information system offers an easy, efficient, and enjoyable interaction between the interfaces (media which communicate the information) and the end users. The development of user-friendly and affordable information communication media targeted toward citizens and local communities is critical for increasing information use and building an inclusive information infrastructure. In the last few decades there has been a revolution in computing and communications. The technological progress and increasing use of information will continue at a rapid pace. In a user-friendly system, users are involved in sourcing new systems or changing existing ones. They can guide the development of information services toward solutions that better suit their needs, skill levels, and working conditions. The whole information system will then work better and more efficiently.

Smart systems integration is the strategy for implementing sustainable information infrastructure in the city, in a way that leads to the realization of the concept of the smart city. Smart systems exploit in the highest degree the potential of information infrastructure systems, since they can upgrade existing linear operations to automated networked self-regulating processes. There is a big margin for smart systems integration in synergizing with the other infrastructure systems which can be eventually revolutionized.

"The smart city is so different in essence to the 20th century city that the governance models and organizational frameworks themselves must evolve."[7]

In the smart city, economy, services, governance, and social groups will be even more interconnected and adapted to a dynamic urban environment. While "cities have behaviors similar to other complex phenomena,"[8] only smart systems can decode and manage this complexity with maximum efficiency.

"At the very least, they will automate all of the physical systems of the city. At the very best, they will engineer entirely new ways for us to thrive. The infrastructure is being laid, but the ideas and software that will choreograph it will require years, if not decades, of research and development in test beds like Songdo."[9]

High penetration and connectivity

Monitor displays continuously adapt to information flows

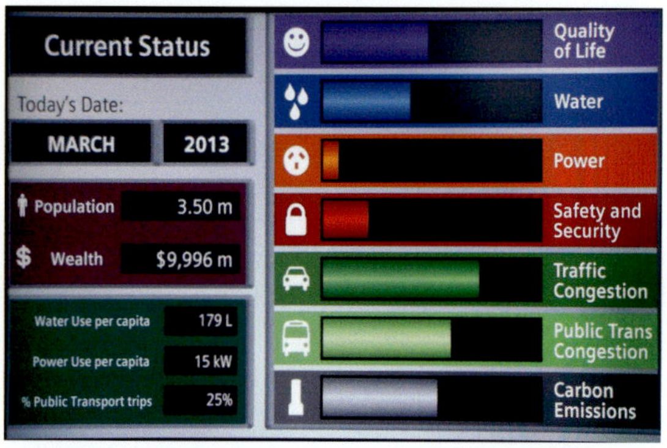

Application linking performance indicators of different urban systems

Upgrading existing infrastructure facilities with new smart grid applications

9.3 | INFORMATION SYSTEM DECODING

INFORMATION DEMAND

COLLECTION AND COMMUNICATION MODES

INFORMATION NODES

INFORMATION NETWORKS

Within the framework of the Zofnass Sustainable Planning Guidelines, Information as infrastructure is decoded on the following four system levels: Information Demand, Collection and Communication, Information Nodes, and Information Networks.

Information Demand is focused on defining the services that need to be provided by the system. As the first step toward a sustainable infrastructure system, this level is crucial as it identifies end users and their needs within the specific urban or regional context. Planners should investigate how telecommunications and information applications can be used to reach more users, improve operational performance in sectors like governance, health, and infrastructure, support city growth, and provide for resilience and adaptability.

The level of Collection and Communication represents the main strategic decisions that need to be made concerning the selection of media, technologies, and resources that will be used for optimum data acquisition and information output. Despite data being collected daily and information flowing ceaselessly, knowledge is not necessarily the end result. This level focuses on collecting reliable and adequate data but also on communicating and disseminating information in ways that allow its utilization to maximum effect. Developing clear plans regarding data measurement and evaluation or the selection of communication media can change behaviors, solve problems, improve functions, and prevent failures. Economic or social changes result from the administrative, commercial, or technological choices we make. Planners should therefore determine how to collect, filter, organize, and share information in an "intelligent" manner that is responsive both to context and to all stakeholders' needs.

The third and fourth system levels include the main infrastructure entities that constitute the physical presence of information infrastructure. Nodes are the points of the system where data processing and storage operations as well as the communication of information take place. The spatial configuration of information infrastructure is determined on the level of Networks. This system level must be able to support the successful implementation of the system's activities and secure the reliable flow of information through wired or wireless media.

INFORMATION DEMAND

The level of Information Demand focuses on the purpose that this infrastructure system serves: the "consumption" of information. The purpose of information demand analysis is to systematically identify opportunities for information to meet needs of individual users, communities, and organizations. It is structured around the twofold approach to information as an infrastructure itself and a subsystem enhancing other infrastructure. Planners must be aware that creating a sustainable city requires both making systems and procedures smarter and optimizing users' behaviors and engagement. Users not only use but also produce information. This bidirectionality, with people functioning as transmitters and receivers, generators and consumers, makes imperative the need to implement bidirectional interactions, real-time communications, and coordinated planning.

End users may include residents and visitors, businesses and commerce, public authorities, amenities, utility operators, and government. Taking into consideration demographics, land use statistics, and environmental context, this system level traces how socioeconomic needs like administration and security services, spatial data collection, and infrastructure systems optimization can be supported through information applications.

TWOFOLD INFORMATION DEMAND

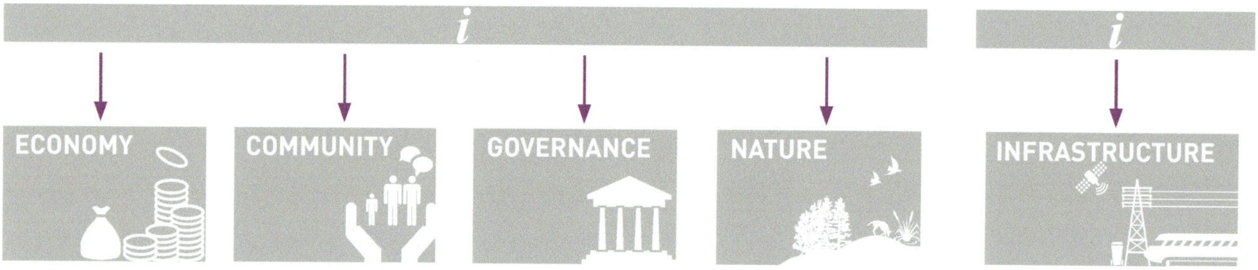

- Enhance productivity and economic growth
- Reinforce labor market and promote job creation

- Improve quality and reach of services to the public
- Increase equity and integration of marginalized groups

- Increase transparency, efficiency, and accountability in government function

- Nonstop monitoring and protection of natural world and resources

- Enhance and optimize other infrastructure systems
- Smart infrastructure leads to smart cities

INFORMATION NEEDS

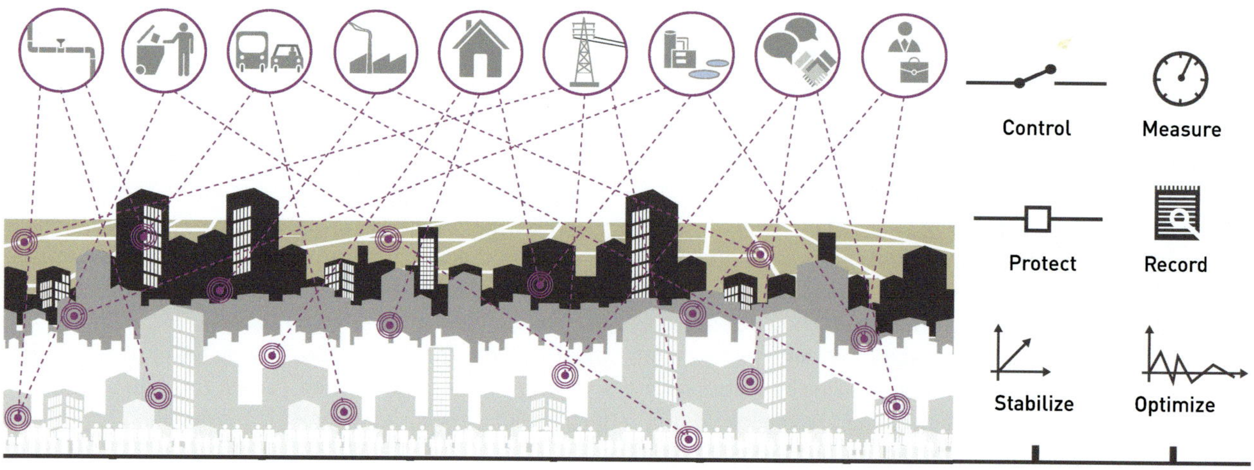

COLLECTION AND COMMUNICATION MODES

The level of Collection and Communication addresses the different modes of receiving data and transmitting information. It focuses on the strategic decisions concerning the selection of data input and information output methods and media in order to meet the needs defined on the Information Demand level and deliver a highly functional system. Notifying drivers of available parking spaces or a traffic accident ahead, selecting measurements to monitor air and water quality, informing residents in time about a water or electrical blackout are all indicative issues that should be addressed at the second level of the information system.

Data collection focuses on the input of signals and data that measure and trace real-world conditions and interactions. It supports decision making by aiming for objective and complete information, rather than subjective opinions. It recognizes data sources and determines how, where, and when data should be collected from the urban environment and entered into the information system. Data sources, as the following table indicates, can be very diverse. The acquisition methods must be compatible with the sources and must use adaptive, multifunctional media. The data collection methods can be categorized as manual or automatic. With manual methods of capturing, data are entered into the information system by users through input devices. Using automatic technologies, data are entered into the system without any outside human intervention. Signal input devices can capture physical phenomena and document behaviors and interactions, importing them digitally into the system.

Information communication refers to the decision-making process regarding the methods and media selected to communicate and disseminate information to end users. The optimum modes should be chosen depending on prevalent media culture, end user profile, economic feasibility, and desirable levels of openness of communication. The increased requirements imposed by the integration of modern devices in today's urban communities lead municipalities and information industries to continuously develop new user interface concepts. In order to give users a coherent experience, it is necessary to define and create sets of hardware and software media compatible with their capacity and knowledge.

Summarizing, data acquisition systems should be planned in parallel with the methods of communication. Both processes are highly dependent on the nature of the service that they intend to provide, the capabilities of the information receiver-user, and the performance targets of the system. Technologies and decisions integrated at a high level can predict, mitigate, or even prevent problems while encouraging more citizens' participation, interaction, and collaboration.

INDICATIVE DATA SOURCES

Climate

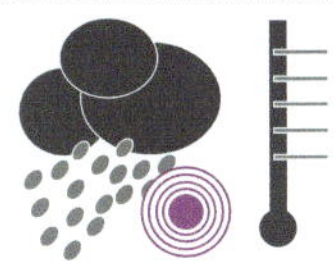

- Temperature
- Humidity
- Moisture
- Rainfall
- Air quality

Socioeconomic services and communication

- Health
- Education
- Administration
- Business
- Culture
- Sports
- Commerce
- Security

Transportation

- Traffic
- Congestion
- Frequency
- Schedule
- Accidents
- Network damages

Energy

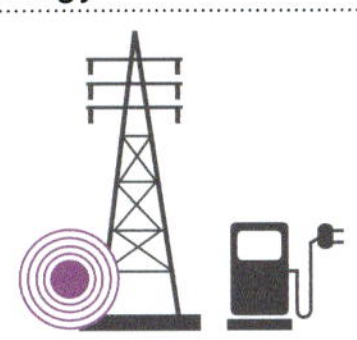

- Electricity
- Heat
- Gas/fuels
- Light intensity
- Peaks
- Network failures

Water

- Leaks
- Pressure
- Quality
- Stresses
- Amount

Solid waste

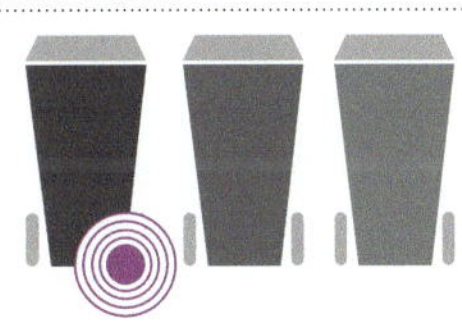

- Generation amount
- Pollution
- Streams

i

INDICATIVE DATA INPUT MODES

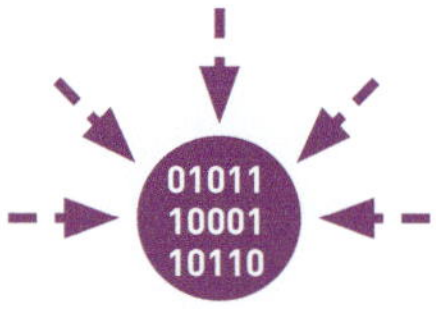

The data indicate the problem is ...

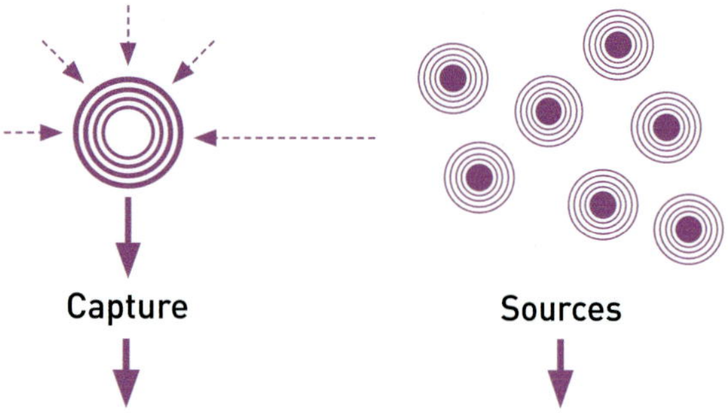

Automatically

- Sensors (converters of physical parameters to electrical signals)
- Signal conditioning circuitry
- Analog/time-to-digital converters
- Closed-circuit television (CCTV)
- Automatic block signaling (ABS)
- Microchips

Manually

- Personal computers (keyboard, mouse, touchpad)
- Speech/eye/thumbprint recognition technologies
- Automated teller machines (ATMs)
- Pen input devices
- Touch-sensitive screens
- Bar code scanners
- Telephones (mobile or landline)
- Documents
- Topographic instruments

INDICATIVE INFORMATION OUTPUT MODES

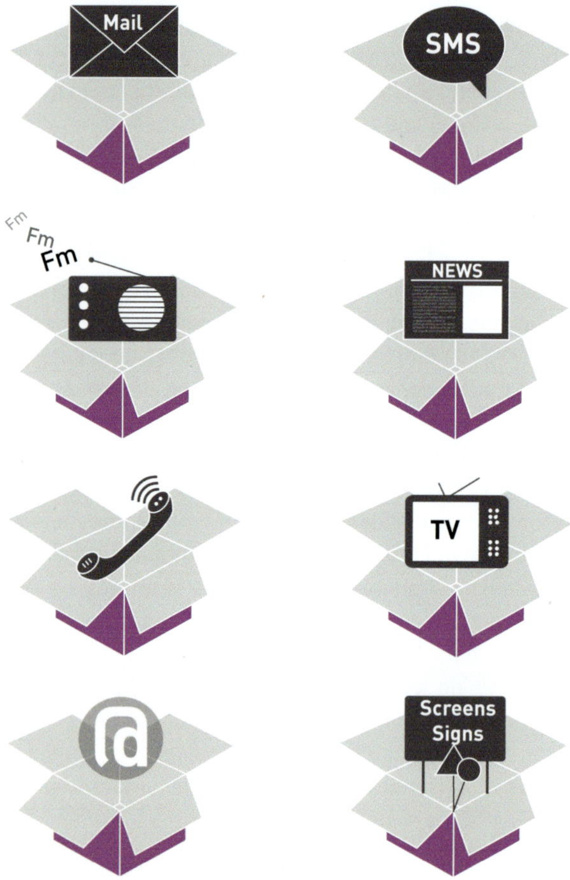

TRAFFIC MANAGEMENT EXAMPLE

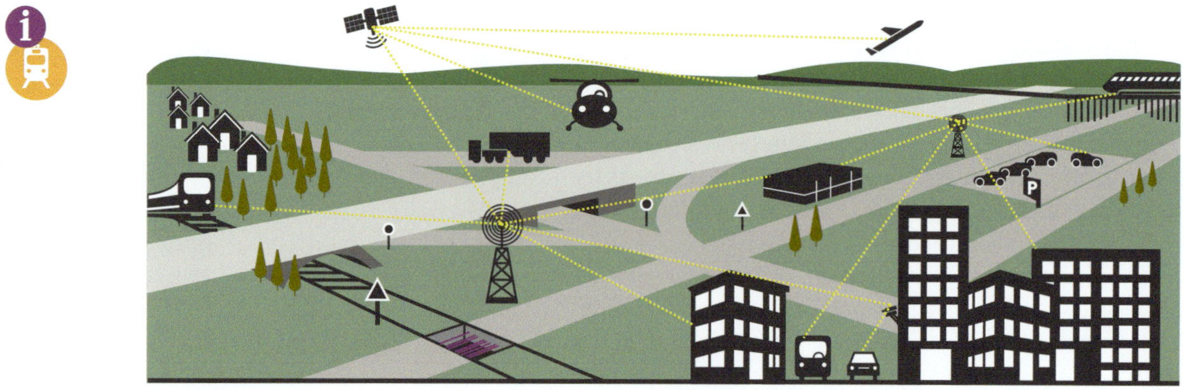

INFORMATION NODES

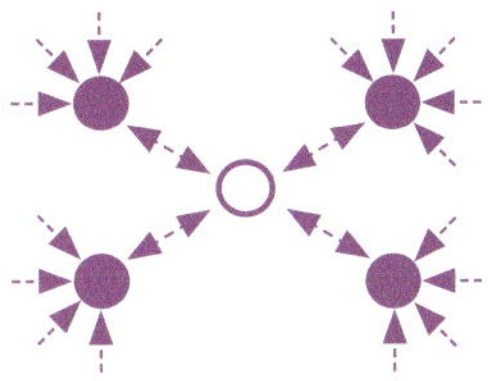

Information nodes originate–route–terminate data exchange ...

They include hardware and software technologies, the combination and interdependencies of which offer adaptation, minimization of outages and failures, optimized dissemination, reliability, and easy use.

Nodes refer to the connection and access points of information networks that are responsible for receiving, processing, storing, and redistributing data as well as communicating information. Based on their location and role, they can be categorized as intermediate or end nodes. Intermediate nodes function as connection and redistribution points. End nodes exist mainly to offer interaction opportunities to end users by entering information into the system or transmitting it. Info kiosks are examples of end nodes.

Nodes generally include both physical and virtual components. Their physical components include the facilities and equipment required for operation, differing in scale from a small device to a room or even a building. Virtual components consist of the software systems responsible for regulating the processes of filtering, storing, analyzing, and distributing. They can be seen as the "brain" of Information infrastructure as they often transform raw data to formats which allow easy comprehension and interpretation. The main virtual components are classified as operation systems and applications, both necessary in order to enable communication between humans and machines. Data centers and interfaces are the most representative type of node entities. Data centers receive, process, store, and forward data in the network. Interfaces are the points where data enter and/or leave the network. They exist in order to communicate information and provide the interaction between the information users and the system.

Planning of information nodes must always make provisions for the volume of data that will be generated, monitored, and transmitted. Nodes' organization, location, and capacity are critical to the efficiency and successful function of an urban information system. Nodes are usually located near major urban attractors and in areas where significant volumes of data are exchanged daily. They should be easily accessed (especially public information nodes) and interconnected, allowing for fast and secure information exchange. Arranging them in a decentralized configuration can significantly minimize the system's failures. When planning and managing a system consisting of thousands of nodes, it is crucial to provide high-level solutions and bylaws aiming to improve costs, performance, and resource use. As the consumers of information are multiplying rapidly, the demands for all nodes, particularly data centers, increase significantly. The adoption of new technologies and processes that will reduce negative impacts of energy use and carbon emissions is therefore imperative.

NODES ORGANIZATION AND INDICATIVE ENTITIES

Intermediate nodes: Redistributors

- Data circuit nodes (modems, hubs, switches)
- Computer LAN/MAN/WAN nodes
- Host computers/internet nodes with IP address
- Main and international telephone exchanges
- Remote concentrators
- Computers providing intelligent network services
- Base station telecommunication controllers
- Home location registers
- Gateway GPRS support nodes (GGSN)
- Serving GPRS support nodes (SGSN)
- Cellular base stations
- Fiber-optic nodes

End nodes: Receivers

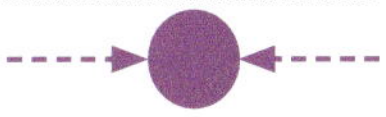

- Data terminal nodes (routers, workstations, servers, printers)
- Computer LAN/MAN/WAN nodes
- Public pay phones
- Private cell phones
- Private fixed phones
- Telephone poles

NODE COMPONENTS AND INDICATIVE ENTITIES

Physical

Data centers

- Transportation control center
- Airport terminal control center
- Telecommunication center
- Energy management center
- Water management center
- Electric power control center
- Grid control centers
- Industrial control center
- Geospatial and statistical data center
- Socioeconomic data center

Interfaces

- Personal electronic devices (telephones, tablets, smartphones)
- Public information kiosks
- Information boards
- Interactive touchscreens
- Network devices (fax machines, printers)
- Computer terminals

Virtual

Operating systems

- Algorithmic derivations
- Logical deductions
- Statistical calculations
- Distributed systems software

Software applications

- Enterprise
- Accounting
- Office suites
- Graphics
- Media
- Web
- Geographic information systems

INFORMATION NETWORKS

An information network is the complete system of links used for the transmission and reception of all types of information and data distributed in a particular area. They serve computer and telephone connections using either cable or wireless media. Indicative network components include copper and coaxial cables, optical fibers, wireless radio frequencies/waves, and communication satellites. Depending on their area of coverage, computer networks are divided into wide area networks (WANs), metropolitan area networks (MANs), and local area networks (LANs). The most popular computer network is the internet, potentially connecting all humans worldwide. Telephone networks allow for call exchanges through landline or mobile connections.

In this level the information system addresses the important challenges of resilience, security, and maintenance. Its entities have significant physical presence with important environmental, social, and economic impacts. Integrated information networks are mostly underground or wireless and have evolved today from silo-like transmission networks to multichannel routes that can be used for virtually everything. Consider a modern mobile device. Smart phones can be used to access an application, make a call, send an email, chat, receive an SMS text message, and browse the Web.

TYPES OF NETWORKS AND MEDIA

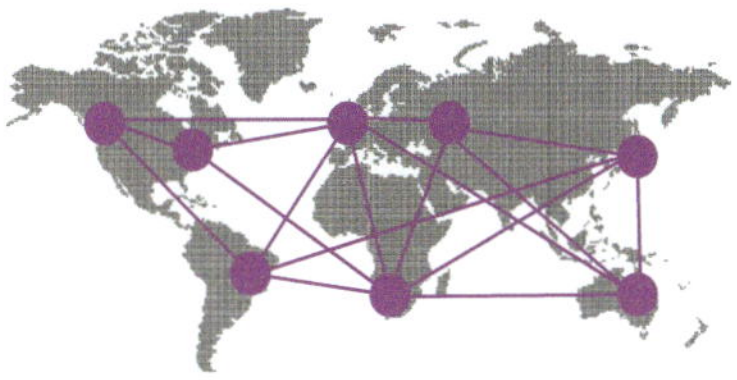

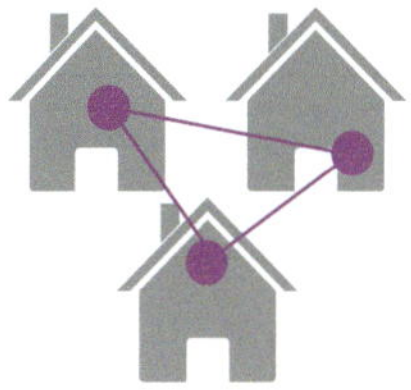

Computer networks

Wide area networks [WANs]
Span a large physical distance

Metropolitan area networks [MANs]
Connect nodes inside a city or a large campus

Local area networks [LANs]
Connect nodes inside a building or group of adjacent buildings, include higher data transfer rates

Telephone networks

Landline networks
Metal wire telephone lines for transmission, out of date

Mobile networks
Radio waves for transmission

Cable media

Twisted pair wires (TPW)
Copper wires, transmission of voice and data, relatively slow and sensitive to electrical interferences

Coaxial cables (CC)
Copper wires, 80 times transmission capacity of TPW, resist noise

Fiber-optic cables (FOC)
Thin strands of glass, transmission of pulsating beams of light, 26,000 times transmission capacity of TPW and CC, light, less expensive, more reliable

Wireless media

Radio frequencies, free-space optical, sonic, microwaves
Transmitted distance can be between a few meters and thousands of kilometers, increased risks of data loss and security, cheaper to install and maintain

SYSTEM LEVEL INTERCONNECTIONS

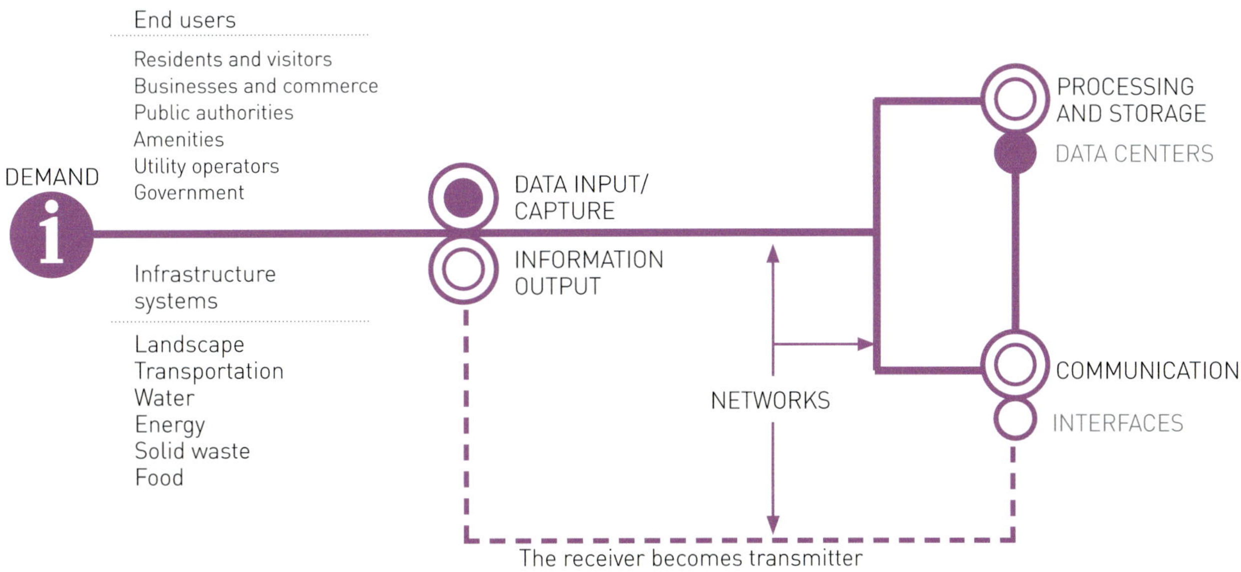

9.4 | OBJECTIVES AND GUIDELINES FOR A SUSTAINABLE INFORMATION SYSTEM

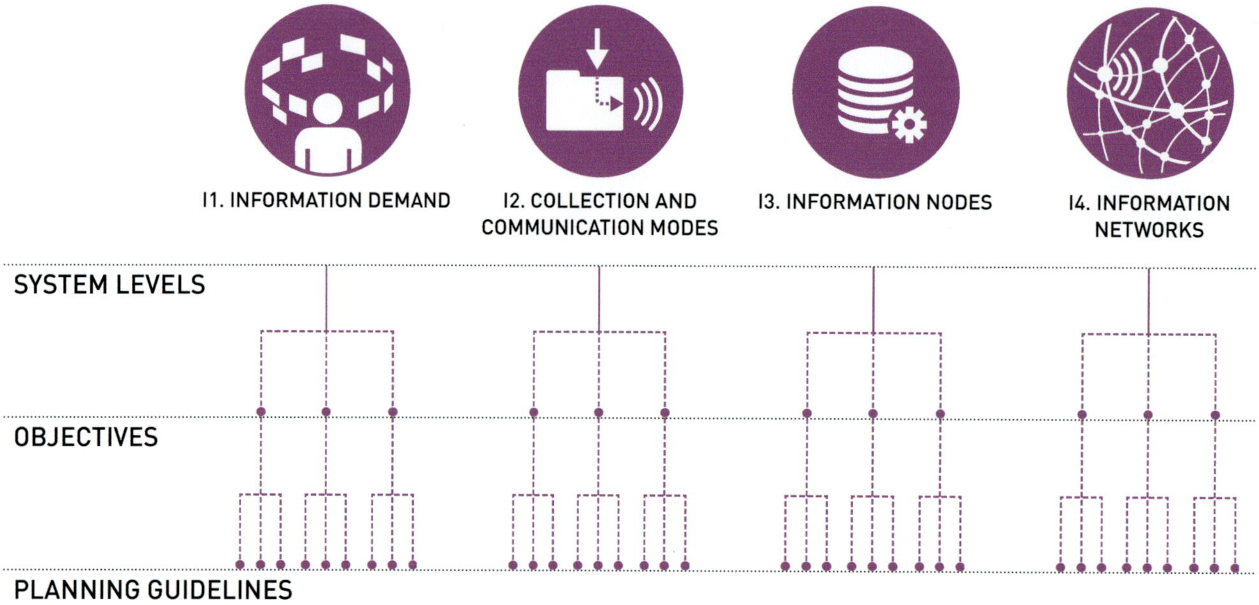

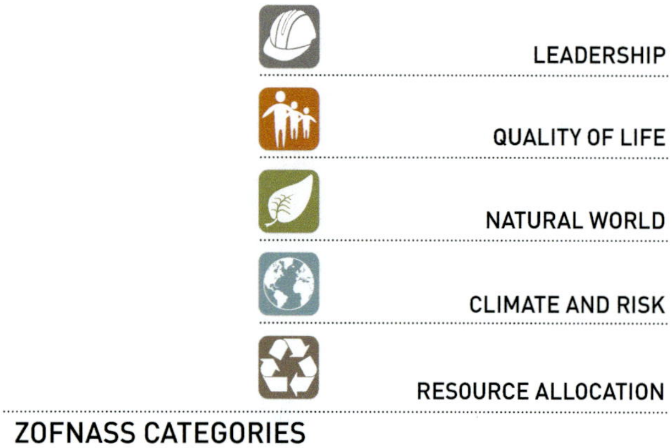

Following the decoding of the Information system on four system levels, planning objectives are defined for each level. They represent both challenges and strategies toward a sustainable system, expressing what should be the overall final outcome of Information planning.

The objectives are associated with the five focus categories of the Zofnass Program, namely Leadership, Quality of Life, Natural World, Resource Allocation, and Climate and Risk. The Sustainable Planning Guidelines aim to lead toward these objectives, highlighting synergies and connections with the other six infrastructure categories of Energy, Water, Solid Waste, Landscape, Transportation, and Food.

Planning an Information infrastructure system requires a good understanding of the needs for collecting, sharing, and using information and a high level of sensitivity toward the context in which the system will operate. Information can be a key tool in bridging the gap between theory and action, planning and operation. It allows for the observation of linkages and interrelations between stocks and flows of resources, the monitoring of consumption rates, and the mapping and management of obstacles toward the sustainable development of cities.

The guidelines presented in this section aim to highlight the important role of Information infrastructure and the broad and versatile field of its application through both sides of the twofold approach analyzed in previous sections. The guidelines express the needs for system reliability, extension of useful life, financial viability of operation and maintenance, and minimization of resource consumption throughout its life cycle.

The objectives and guidelines of the Demand system level do not refer directly to sustainable planning of the Information infrastructure system but instead aim to create a solid base for it. They are primarily associated with the Leadership category and aim to identify the potential regulating, provisioning, and cultural role of information. The guidelines of the other levels can then be used for meeting these needs.

Providing equal and reliable access to information is central at this level. The first objective is to define the potential for optimizing urban services and interactions. Assisted by information, citizens' skills and creativity can evolve, city services such as healthcare and education can function more efficiently, and the city's growth and development can be stimulated. The second objective is the performance optimization of all seven infrastructure systems, including telecommunication services. It represents the potential of information to act like a subsystem of Energy, Water, Solid Waste, Transportation, Food, and Landscape infrastructure systems, reducing the initial demand for their services and improving operation efficiency through real-time interaction, updates, and feedback. It represents a synergy between entire systems and not just individual guidelines. Specifically, through the main functions of monitoring, controlling, and managing information, urban infrastructure systems can improve their service delivery capacity by securing supply (e.g., energy or water) or adapting their functions to different user demands (e.g., trip generation rates). A thorough understanding of the demand for information services by other infrastructure makes it possible to mitigate systems' negative impacts, such as air pollutant emissions, to secure service during demand peaks or in periods of extreme natural phenomena, and to provide balance through the necessary bidirectional communication of supply and demand. All in all it creates the potential for smarter systems.

I1. INFORMATION DEMAND

- I1.1. Define Information demand for urban services and interactions
- I1.2. Define Information Demand for ecological services
- I1.3. Define Information Demand for the optimization of other infrastructure systems' performance

The level of Collection and Communication aims to guide policymakers through the strategic planning of the system's input and output modes. The objectives and guidelines are related to the Leadership, Quality of Life, and Resource Allocation categories. Overall, on this system level decisions are made to optimize both data acquisition and information communication processes. The needs traced on the first system level of Demand are translated into specific goals and benchmarks, setting up service delivery targets. Guidelines call for planning to be tailored to projections of the city's development, to be regularly updated and adaptable to the constantly changing information demand. Taking into consideration the diversity of available data sources, it is imperative to define the optimum data collection techniques, locations, and frequency in order to increase data quality and minimize failures.

The chosen information communication methods and media must be adapted to the particularities of each user group. They should facilitate perception and engagement, while remaining flexible to community changes and technological advancements. The access points allowing interactions with the system must be placed in optimum locations within the city based on activity, accessibility, and visibility. End user qualifications and preferences regarding communication media should be assessed so as to develop adequate alternatives. The communication options offered must be affordable and encourage the adoption of new applications and technologies. Protecting the privacy and confidentiality of data is also a strategic goal with strong social impact. Confidential and personal data should be protected during both collection and communication stages. The selection of communication methods and media can have a significant impact on resource consumption or conservation and subsequently on waste generation. The use of e-services and email in place of paper is a popular example.

I2. COLLECTION AND COMMUNICATION MODES

- I2.1. Optimize Information Communication
- I2.2. Optimize data Collection
- I2.3. Set clear goals and standards to be achieved through Information Collection and Communication
- I2.4. Secure a viable Information system in the long term
- I2.5. Ensure dissemination and openness of Information to end users
- I2.6. Protect privacy and confidentiality of data
- I2.7. Reduce resources consumed for Information Communication
- I2.8. Resolve Waste issues connected to the type of Collection and Communication technologies selected

The last two system levels of Nodes and Networks set objectives in all five categories of the Zofnass Program and establish multiple synergies with the other infrastructure systems. The configuration of nodes should ensure the reliability and flexibility of the overall system. Guidelines here emphasize the need for scalable organization and adequate processing and storage capacity in order to cover long-term city needs. Administrative structures should be developed to optimize the nodes' asset management and the financial viability of operations. Similarly, a plan for hardware and software updates must be put in place and be regularly revised in order to keep up with the pace of technologies' evolution. Another basic target for the sustainable function of nodes is their placement and distribution within the information network, in correlation with collection and communication needs. The need for integration with elements of Energy and Water infrastructure should be investigated, as nodes often demand significant amounts of these resources for their operation. The need for planning to minimize data losses is also highlighted. Procedures must be established to check incoming and transmitted data for accuracy and continuity, and offer data storage and energy backups. In the field of Quality of Life, the objectives promote community well-being and quality of place. The operation of information nodes must not pose threats to the health of surrounding communities, while larger facilities should be integrated into their local context, avoiding obstruction of views or urban flows with their physical presence. Natural-world elements and processes such as habitats, vegetation, and terrain must be protected and maintained during planning and operations. Conversely, the placement and configuration of nodes should provide resilience against risks like extreme natural events, data security threats, etc. Finally, planning for optimum Resource Allocation is once more emphasized at this level. Main objectives here refer to optimizing water and energy consumption and waste management. Guidelines call for water- and energy-efficient technologies, selection of environmentally friendly technological equipment, reduction of waste generation, and proper collection and diversion of discarded equipment.

I3. INFORMATION NODES

- I3.1. Secure reliable Information Node services and provide flexibility
- I3.2. Place and distribute Nodes for optimized Information system performance
- I3.3. Minimize data losses
- I3.4. Integrate Information Nodes with other infrastructure systems
- I3.5. Promote community well-being
- I3.6. Enhance quality of place
- I3.7. Maintain the integrity of natural-world elements and systems
- I3.8. Provide for city and Information system resilience
- I3.9. Optimize Water use in Information Nodes
- I3.10. Optimize Energy use in Information Nodes
- I3.11. Resolve Waste management issues of Information Nodes equipment

The objectives set for Network planning are very similar to those for Nodes. In the categories of Quality of Life, Natural World, Climate and Risk, and Resource Allocation, objectives focus again on promotion of the community's well-being, the enhancement of quality of place, the minimization of impacts on natural processes, city and system resilience, optimal resource use, and waste management. A feature specific to this system level is securing network stability and coverage across the city. The stability of information networks can be reinforced by promoting connectivity and compatibility of the different network branches as well as predicting and responding to failure or congestion events. Network routes and lengths define the system's performance, affecting flows and transmission speeds and the demand for resources for the network's construction and operation. Planning ahead for long-term network capacity and financial viability is also underlined here. Networks must be compatible with data collection and communication methods, and with the long-term requirements set by performance and capacity goals. Integrated planning should also extend to the administrative structures necessary to operate the networks. Similarly the potential to integrate information networks and the networks of other infrastructure systems must also be investigated. Transportation is a typical example, as information networks are often placed along roads for easier access.

The Objectives of the respective Guidelines as well as the relevant Guidelines of other infrastructure systems are presented in the following tables. (TO) and (FROM) indicate that the guideline can assist or can be assisted by the other infrastructure system, where the arrow is placed. (WITH) indicates that the guideline might require planning for an Information or Landscape subsystem and should follow multiple Guidelines from these two systems.

14. INFORMATION NETWORKS

- 14.1. Secure the reliability and flexibility of Information Networks
- 14.2. Secure Information Networks stability
- 14.3. Improve flow of Information
- 14.4. Integrate Information Networks with other infrastructure systems
- 14.5. Promote community well-being
- 14.6. Enhance quality of place
- 14.7. Maintain the integrity of natural-world elements and systems
- 14.8. Prepare for short-term hazards and projected future threats
- 14.9. Optimize material use in Information Networks
- 14.10. Optimize Energy use in Information Networks
- 14.11. Resolve waste management issues of Information Networks equipment

i

GUIDELINES

OBJECTIVES		GUIDELINES	SYNERGIES
I1 INFORMATION DEMAND			▲ TO + WITH ▼ FROM
I1.1. Define Information demand for urban services and interactions	I1.1.1.	Identify opportunities to improve the quality of life for the city's residents and communities through Information infrastructure	
	I1.1.2.	Identify opportunities to develop the city's economic sectors and industries through Information infrastructure	
	I1.1.3.	Identify opportunities to improve city administration and social amenities through Information infrastructure	▲ ▲
I1.2. Define Information Demand for ecological services	I1.2.1.	Identify opportunities to monitor and protect the natural context through Information infrastructure	▲ ▲ ▲
I1.3. Define Information Demand for the optimization of other infrastructure systems' performance	I1.3.1.	Identify infrastructure systems' processes to be monitored/documented through Information infrastructure	▲ ▼ ▲ ▲ ▲ ▲
	I1.3.2.	Identify other infrastructure systems' processes to be managed through Information infrastructure	▲ ▼▲ ▲ ▲ ▲
I2 COLLECTION AND COMMUNICATION MODES			▲ TO + WITH ▼ FROM
I2.1. Optimize Information Communication	I2.1.1.	Assess end users' capacity and preferences for different Information Communication media	
	I2.1.2.	Develop alternative Communication methods to cover the Information Demand identified	
	I2.1.3.	Define optimum spatial distribution of Information Communication points	
	I2.1.4.	Ensure end users' easy access to Information Communication points	
	I2.1.5.	Secure Information Communication points against threats and natural hazards	
	I2.1.6.	Optimize schedule and frequency of Information Communication	
I2.2. Optimize data Collection	I2.2.1.	Select adequate Collection methods and techniques for high-quality and reliable data	
	I2.2.2.	Define optimum data Collection locations	
	I2.2.3.	Secure data Collection points against threats and natural hazards	
	I2.2.4.	Optimize the schedule and frequency of data Collection	
	I2.2.5.	Respond to and manage peak flows in incoming data	

OBJECTIVES	GUIDELINES		SYNERGIES
I2.3. Set clear goals and standards to be achieved through Information Collection and Communication	I2.3.1.	Set goals for Information technologies penetration	
	I2.3.2.	Adopt quantitative indicators to benchmark and monitor Information system performance	
	I2.3.3.	Integrate long-term growth and development goals with Information system planning	
	I2.3.4.	Regularly update Information system planning according to community growth and development	
I2.4. Secure a viable Information system in the long term	I2.4.1.	Secure long-term financial viability of Information system	
I2.5. Ensure dissemination and openness of Information to end users	I2.5.1.	Encourage adoption of new applications and technologies by Information system end users	
	I2.5.2.	Ensure affordable Information Communication options for all members of the community	
	I2.5.3.	Disseminate Information through public space installations integrated to the local context	
I2.6. Protect privacy and confidentiality of data	I2.6.1.	Regulate personal data collection and use	
	I2.6.2.	Protect privacy during Information Collection and Communication	
I2.7. Reduce resources consumed for Information Communication	I2.7.1.	Promote Information Communication methods and media with reduced energy needs	
	I2.7.2.	Promote Information Communication methods and media with reduced material resources needs	
I2.8. Resolve Waste issues connected to the type of Collection and Communication technologies selected	I2.8.1.	Promote Information Communication methods and media with reduced Waste Generation	▲

I3 INFORMATION NODES

▲ TO
+ WITH
▼ FROM

OBJECTIVES	GUIDELINES		SYNERGIES
I3.1. Secure reliable Information Node services and provide flexibility	I3.1.1.	Provide Information Nodes with adequate processing and storage capacity to cover long-term city needs	
	I3.1.2.	Provide for Information Nodes scalability taking into consideration long-term city development	
	I3.1.3.	Reserve spaces for Information Nodes facilities taking into consideration long-term city development	
	I3.1.4.	Develop and regularly update a plan for Information Nodes hardware and software updates based on changing needs	
	I3.1.5.	Secure long-term financial viability of Information Nodes operation	
	I3.1.6.	Optimize Information Nodes asset management through adequate administrative structures	

OBJECTIVES	GUIDELINES		SYNERGIES
I3.2. Place and distribute Nodes for optimized Information system performance	I3.2.1.	Place Information Nodes facilities in correlation to data Collection and Communication needs	
	I3.2.2.	Optimize placement of Information Nodes with regard to Network routes	
I3.3. Minimize data losses	I3.3.1.	Check and filter incoming and transmitted data for accuracy and continuity	
	I3.3.2.	Develop plan for maintaining information and data storage backups	
	I3.3.3.	Integrate add-on security packages in software systems	
	I3.3.4.	Provide backup Energy Generation for Information Nodes	
I3.4. Integrate Information Nodes with other infrastructure systems	I3.4.1.	Integrate Information Nodes with Energy Distribution and Transmission facilities	▼
	I3.4.2.	Integrate Information Nodes with Water Networks	▼
I3.5. Promote community well-being	I3.5.1.	Assess and resolve potential health impacts caused by Information Nodes	
I3.6. Enhance quality of place	I3.6.1.	Visually integrate Information Nodes to the local context	
I3.7. Maintain the integrity of natural-world elements and systems	I3.7.1.	Protect natural habitats and wildlife corridors and compensate for their fragmentation or disruption by Information Nodes	+
I3.8. Provide for city and Information system resilience	I3.8.1.	Assess and manage risks from natural and man-made phenomena affecting Information Nodes	
	I3.8.2.	Plan response and recovery in case of emergency affecting Information Nodes facilities	
I3.9. Optimize Water use in Information Nodes	I3.9.1.	Use water-efficient technologies for cooling of Information Nodes where applicable	
	I3.9.2.	Use gray or reclaimed water where possible for the operation of Information Nodes facilities	▲
I3.10. Optimize Energy use in Information Nodes	I3.10.1.	Set requirements for energy-efficient technologies and equipment in Information Nodes	
I3.11. Resolve Waste management issues of Information Nodes equipment	I3.11.1.	Select Information Nodes equipment for reduced e-Waste Generation	▲
	I3.11.2.	Provide plan for collection and diversion of Information Nodes' discarded equipment	▲

I4 INFORMATION NETWORKS

▲ TO
+ WITH
▼ FROM

I4.1. Secure the reliability and flexibility of Information Networks	I4.1.1.	Provide required coverage for Information Networks for all types of data required	
	I4.1.2.	Provide adequate Information Networks capacity to cover long-term needs	

i

OBJECTIVES		GUIDELINES	SYNERGIES
	14.1.3.	Reserve spaces for Information Networks facilities taking into consideration long-term city development	
	14.1.4.	Secure long-term financial viability of Information Networks	
	14.1.5.	Optimize Information Networks asset management through adequate administrative structures	
14.2. Secure Information Networks stability	14.2.1.	Promote connectivity and integration within different Information grids	
	14.2.2.	Predict and respond to Information Networks failure and congestion events	
14.3. Improve flow of Information	14.3.1.	Define and provide required data transmission speed	
	14.3.2.	Optimize Information Networks routes and distances	
	14.3.3.	Maintain Information Networks compatibility with data Collection and Information Communication	
	14.3.4.	Maintain Information Networks compatibility with schedule and frequency of Collection and Communication requirements	
14.4. Integrate Information Networks with other infrastructure systems	14.4.1.	Integrate Information Networks with Transportation Networks	
	14.4.2.	Integrate Information Networks with Landscape entities	
14.5. Promote community well-being	14.5.1.	Assess and resolve potential health impacts of Information Networks	
14.6. Enhance quality of place	14.6.1.	Visually integrate Information Networks to the local context	
14.7. Maintain the integrity of natural-world elements and systems	14.7.1.	Minimize Landscape alteration caused by Information Networks	
	14.7.2.	Protect natural habitats and wildlife corridors and compensate for their fragmentation or disruption by Information Networks	+
14.8. Prepare for short-term hazards and projected future threats	14.8.1.	Assess and manage future threats from natural and man-made phenomena affecting Information Networks	+
	14.8.2.	Plan response and recovery in case of emergency affecting Information Networks	
14.9. Optimize material use in Information Networks	14.9.1.	Optimize length of Information Networks	
	14.9.2.	Source materials required for Information Networks locally when possible	
14.10. Optimize Energy use in Information Networks	14.10.1.	Use energy-efficient technologies for Information Networks operation	
14.11. Resolve waste management issues of Information Networks equipment	14.11.1.	Select Information Networks equipment aiming for Solid Waste reduction	
	14.11.2.	Provide plan for the collection and diversion of Information Networks' discarded equipment	

i

RELEVANT GUIDELINES

GUIDELINES		RELEVANT GUIDELINES	
I1	**INFORMATION DEMAND**		FROM ▶ i ▶ TO + WITH
I1.1.3.	Identify opportunities to improve city administration and social amenities through Information Infrastructure	T1.1.1.	Provide and promote remote access and e-services
		L1.1.4.	Collate local and regional social/economic data
		L1.1.5.	Collate local and regional data on cultural aspects
I1.2.1.	Identify opportunities to monitor and protect the natural context through Information infrastructure	L1.1.2.	Collate local and regional environmental data
		W2.3.1.	Map and monitor watershed and hydrological connections
		F2.5.1.	Map and monitor the city's foodshed and highly productive farmland
I1.3.1.	Identify infrastructure systems' processes to be monitored/ documented through Information infrastructure	T1.3.2.	Collect and organize data on Transportation needs of goods and passengers
		T1.3.3.	Monitor daily and seasonal Trip Generation fluctuations across the city
		T3.2.3.	Provide Node traffic operators with real-time information on traffic conditions
		T4.2.5.	Provide Network traffic operators with real-time information on traffic conditions
		T3.2.4.	Provide users of Transportation Nodes with real-time information on itineraries and traffic conditions
		SW1.2.1.	Regularly measure volume and weight of produced Solid Waste and record seasonal fluctuations
		SW1.2.2.	Regularly analyze composition of Solid Waste and record seasonal fluctuations
		SW2.6.4.	Provide consistent and organized access to public information on performance of Solid Waste Diversion Model
		W1.6.1.	Regularly collect and organize quantitative and qualitative data on Water Consumption
		W1.6.2.	Monitor daily and seasonal Water Consumption fluctuations
		W4.2.4.	Provide operators with real-time monitoring of Water Networks
		E1.5.1.	Monitor daily and seasonal Energy Consumption fluctuations
		E1.5.2.	Regularly collect and organize data on Energy Consumption
		E4.2.5.	Provide real-time monitoring of electricity grids
		F1.6.1.	Collect and organize quantitative and qualitative data on food consumption
		L4.2.1.	Identify and assess threats and problems in Landscape Maintenance through a robust and integrated monitoring system

i

GUIDELINES		RELEVANT GUIDELINES	
		L4.8.3.	Construct and operate efficient Landscape irrigation systems
I1.3.2.	Identify other infrastructure systems' processes to be managed through Information infrastructure	T3.2.3.	Provide Node traffic operators with real-time information on traffic conditions
		T4.2.5.	Provide Network traffic operators with real-time information on traffic conditions
		W3.1.7.	Develop smart, integrated management systems to optimize Water Treatment efficiency
		W4.2.1.	Provide for operational flexibility of Water Networks
		W4.2.2.	Optimize flow pressure in Water Networks
		SW3.2.4.	Develop smart, integrated management systems to optimize the operation of Solid Waste Treatment and Disposal facilities
		SW4.9.1.	Optimize Solid Waste Collection and Transfer itineraries
		E3.1.5.	Develop smart, integrated management systems to optimize Energy Generation efficiency
		E4.2.1.	Develop smart, integrated management systems to optimize balance of electricity generation and demand
		E4.3.2.	Develop smart, integrated management systems to optimize Energy Distribution and Transmission performance
		L4.2.1.	Identify and assess threats and problems in Landscape Maintenance through a robust and integrated monitoring system
		L4.8.3.	Construct and operate efficient Landscape irrigation systems

I2
COLLECTION AND COMMUNICATION MODES

FROM ▸ ▸ TO
+
WITH

I2.8.1.	Promote Information Communication methods and media with reduced Waste Generation	SW1.1.4.	Reduce Solid Waste Generation by other infrastructure systems

I3
INFORMATION NODES

FROM ▸ ▸ TO
+
WITH

I3.4.1.	Integrate Information Nodes with Energy Distribution and Transmission facilities	E4.4.2.	Plan Energy Distribution and Transmission networks to support the operation of Information Nodes
I3.4.2.	Integrate Information Nodes with Water Networks	W4.3.3.	Plan Water Networks to support the operation of Information Nodes facilities
I3.6.1.	Visually integrate Information Nodes to the local context	L1.7.2.	Define needs for mitigating and compensating for other infrastructure systems' impact on community life through Landscape infrastructure

GUIDELINES			RELEVANT GUIDELINES	
I3.8.1.	Assess and manage risks from natural and man-made phenomena affecting Information Nodes	I + L	L1.7.3.	Define needs for protection of infrastructure systems against projected climate change and extreme events through Landscape infrastructure
I3.9.2.	Use gray or reclaimed water where possible for the operation of Information Nodes facilities	I ▸ W	W1.6.3.	Identify opportunities for use of different nonpotable water qualities
I3.11.1.	Select Information Nodes equipment for reduced e-Waste Generation	I ▸ SW	SW1.1.4.	Reduce Solid Waste Generation by other infrastructure systems
I3.11.2.	Provide plan for collection and diversion of Information Nodes' discarded equipment	I ▸ SW	SW4.3.5.	Solid Waste and Information systems to work in synergy for the Collection and Transfer of discarded electronic equipment

I4 INFORMATION NETWORKS

FROM ▸ I ▸ TO + WITH

I4.4.1.	Integrate Information Networks with Transportation Networks	T ▸ I	T4.3.4.	Configure Transportation Networks to provide for needs of Information Networks and facilitate their maintenance
I4.4.2.	Integrate Information Networks with Landscape entities	L ▸ I	L3.5.1.	Configure Landscape entities' vegetation and paving to provide for other infrastructure networks' needs and facilitate access and maintenance
I4.7.2.	Protect natural habitats and wildlife corridors and compensate for their fragmentation or disruption by Information Networks	I + L	L1.7.1.	Define needs for mitigating and compensating for other infrastructure systems' impact on the natural world through Landscape infrastructure
I4.8.1.	Assess and manage future threats from natural and man-made phenomena affecting Information Networks	I + L	L1.7.3.	Define needs for protection of infrastructure systems against projected climate change and extreme events through Landscape infrastructure
I4.11.1.	Select Information Networks equipment aiming for Solid Waste reduction	I ▸ SW	SW1.1.4.	Reduce Solid Waste Generation by other infrastructure systems

NOTES

1. "Rationale for World Bank Group Involvement in ICT," chapter 2, p. 7, accessed April 15, 2014, https://ieg.worldbankgroup.org/Data/reports/chapters/chap2.pdf.
2. "Minding the Planet: The Meaning and Future of the Semantic Web," Lifeboat Foundation Scientific Advisory Board member Nova Spivack, accessed April 15, 2014, https://lifeboat.com/ex/minding.the.planet.
3. "The Smart Solution for Cities," Arup, accessed April 15, 2014, http://www.arup.com/~/media/Files/PDF/Publications/Research_and_whitepapers/UrbanLife_SmartSolutionForCities_July11.ashx.
4. Christine Zhen-Wei Qiang, "Information Communications Technology for Development," World Bank, accessed April 15, 2014, http://live.worldbank.org/information-communications-technology-development.
5. Paul Ceruzzi, *Computing: A Concise History* (Cambridge, MA: MIT Press, 2012), 157.
6. "Policymakers in Cyber Security: War Gaming HTTPS and the DigiNotar Attack," Harvard Leadership Conference 2014, accessed April 15, 2014, http://hgc.harvard.edu/hlc/2014/agenda.
7. Anthony Townsend, *Smart Cities* (New York: W. W. Norton, 2013), 32; see Arup, "Smart Cities: Transforming the 21st Century City via the Creative Use of Technology," last modified September 1, 2010, http://www.arup.com/Publications/Smart_Cities.aspx.
8. Tom Verebes, ed., *Masterplanning the Adaptive City: Computational Urbanism in the Twenty-First Century* (London: Routledge, 2014), 100.
9. Townsend, *Smart Cities*, 29; Songdo International Business District website, accessed April 15, 2014, www.songdo.com.

IMAGE CREDITS

P. 218: Photograph by Jordan McQueen, accessed on August 18, 2015, https://images.unsplash.com/photo-1425315283416-2acc503 23ee6?q=80&fm=jpg&s=94c3e17a79354c33dbd59ced862b1460. **P. 230:** (Center left) Photograph by Timothy Muza, accessed on August 18, 2015, https://images.unsplash.com/reserve/wi9yf7kTQxCNeY72cCY6_Images%20of%20Jenny%20Lace%20Plasticity%20Publish%20(4%20of%2025).jpg?q=80&fm=jpg&s=1dbe50cae7a6b5d620207d9b90a7ace5. **P. 230:** (Center right) Photograph by Eleonora Marinou.
P. 230: (Bottom left) Photograph by Eleonora Marinou from the Siemens Urban Development Center at the Crystal exhibition in London, March 2013. **P. 230:** (Bottom right) Photograph by Eleonora Marinou from the Siemens Urban Development Center at the Crystal exhibition in London, March 2013. **P. 232:** (Bottom) Adapted from "Overview of the Transmission Domain," Figure 10, http://www.sgip.org/Domains, accessed on September 15, 2015.

Sustainable Infrastructure Planning Guidelines

Food

as Infrastructure

Chapter 10

Food

is a basic human need, and its organized distribution is essential for the sustainable development of modern societies and contemporary cities.

10.1 | INDUSTRY PERSPECTIVES

Sustaining Cities, Sustaining Ourselves

Cities have often been compared to living systems—they live, grow, breathe, consume, evolve, and produce waste. We can describe urban metabolic networks built from the stocks and flows of energy, food, labor, and land that shape the health, resiliency, and sustainability of a city. Food, however, occupies a unique place among these resources, since it is necessary for daily survival and strongly linked to resident health across a range of time scales. In designing sustainable food systems for cities, we must first consider how we sustain ourselves. Diet and physical activity are fundamental determinants of both individual health and the health of an urban citizenry and economy. In the matter of diet, good planning, policy, and design can alleviate the deficiencies imposed by the market, deficiencies that currently often lead to food insecurity and significant nutritional gaps for low-income families.

The flows of food and food waste are key drivers of the environmental footprints of cities. When we think about climate change or environmental degradation, we immediately conjure up images of factories, industrial production, and traffic-choked cities. Our global food systems, however, are associated with a significant proportion of global carbon and water budgets and impart significant pollution burdens. The net effect of food consumption on environmental outcomes is driven by three factors: population trends, per capita consumption rates, and the energy intensity of the food we consume. Globally, all of these measures have increased in recent years. We also know that a changing climate is putting increased stress on our global food supply. Shifting from global to local or reducing food miles is not the only change needed. You can often make the most significant gains by shifting what is being consumed rather than shifting where it is being produced. Much attention has been given to the disproportionate environmental impact of livestock-derived food production for human consumption, a key element of the "energy intensity" factor. The demand and supply sides of this sector will need to be addressed in the coming years as the global economy continues to develop.

In policy debates, some of these important considerations are frequently oversimplified, with niche seasonal markets selling high-end produce or small-scale community gardens being viewed as the answers to a broken food system. It is true that intraurban agriculture has benefits. However, it's important to recognize that the food footprint of modern cities and megacities is large in scale, global in reach, and not likely to be completely offset by even the most ambitious and forward-thinking urban agriculture initiative. We need to understand the footprint of the urban organism and how it relates to the quantity and quality of food entering its borders. There is no single solution; good policy needs to be comprised of a basket of food programs. This chapter highlights many of the innovative models that are being used to increase urban food production, such as vertical farms, roof gardens, and communal garden allotments. Married to these creative structural solutions, we need to look for policies that connect families to healthy and affordable food where they live. Consumer demand, economic barriers, and zoning ordinances in urban and peri-urban areas are all relevant in this discussion.

It is clear that market forces do not necessarily distribute access to healthy and affordable foods equitably. Food access and food insecurity can be shaped and improved by municipal policy, and policymakers are taking notice. The city of Baltimore's Food Policy Initiative[1] has made health the driving goal of its portfolio of programs, through partnerships between government, community groups, business owners, schools, and university researchers. Their data-driven efforts include expansion of farmer's markets, urban agriculture, healthy food access, and healthy school meals. New York City's FRESH (Food Retail Expansion to Support Health) program[2] was established to respond to the lack of grocery stores in many of the city's low-income neighborhoods. The program uses a combination of zoning and tax incentives to incentivize the establishment and retention of local grocery stores. This is only one element of the solution; the availability of healthier food options affects the supply side of the equation. Consumer demand, driven by price, preference, competition, and shopping behavior, must also change.

Globally, planners and policymakers have good examples to emulate and adapt. According to the Food and Agriculture Organization of the United Nations, Brazil has become "the international benchmark for measuring national commitment to food security."[3] This achievement has its roots in the city of Belo Horizonte, which in 1993 made a choice to guarantee access to adequate and healthy food for all of its citizens. Since then, the city has become a model for creative solutions to food insecurity problems through 19 initiatives. For example, the city established public restaurants that offer subsidized meals for low-income families. Nutritionally fortified flour was distributed through health clinics and school meal programs. Another program allows licensed vendors to set up shop throughout the city on the condition of selling more than 20 produce items at reduced prices set by the government. Through the Direto da Roça (Straight from the Field) program, small-scale farmers are also allowed to sell their products directly to consumers at sites established by the city. The result? Striking public health gains have been realized at a cost equal to less than 2 percent of the city's annual budget.[4] Many of the successful programs in Belo were adapted for the nationwide Fome Zero (Zero Hunger) program launched in 2003. These models are not all directly portable to other settings, but they provide building blocks for policymakers and stakeholders.

Supporting a healthy food economy is consistent with supporting a healthy urban economy. As with most complex issues, there's no single answer. Cities need to adopt comprehensive policies that recognize the complexity of urban food systems and the effect of market forces, local and national policy, and human behavior. It is a failure of systems and policies that all residents of our cities do not have access to healthy and affordable food. Special attention should be given to the most vulnerable members of our communities such as pregnant women, children, and the elderly. In most cases, we have enough food and we have enough knowledge to achieve equity. The missing ingredient is a sustained commitment to put these to work.

Gary Adamkiewicz
Assistant Professor of Environmental Health and Exposure Disparities, Department of Environmental Health, Harvard T. H. Chan School of Public Health

NOTES

1. Archive.baltimorecity, "Planning / Baltimore Food Policy Initiative." Accessed November 20, 2015, http://archive.baltimorecity.gov/Government/AgenciesDepartments/Planning/BaltimoreFoodPolicyInitiative.aspx.
2. NYCEDC, "Food Retail Expansion to Support Health (FRESH)." Accessed November 20, 2015, http://www.nycedc.com/program/food-retail-expansion-support-health-fresh.
3. FAO, "Belo Horizonte: Growing Greener Cities In Latin America And The Caribbean." Accessed November 20, 2015, http://www.fao.org/ag/agp/greenercities/en/GGCLAC/belo_horizonte.html.
4. Frances Lappe, 2014. "The City That Ended Hunger." YES! Magazine, spring 2009 issue. Accessed November 20, 2015, http://www.yesmagazine.org/issues/food-for-everyone/the-city-that-ended-hunger.

10.2 | SUSTAINABLE FOOD PLANNING

Food is a basic human need, and its organized distribution is essential for the sustainable development of modern societies and contemporary cities. Thus over time, based on the systemization of food production, complex food distribution systems have been developed in order to cover the food demand of communities. Agriculture was the first food revolution that allowed the creation of the first urban centers. Since this beginning in the Fertile Crescent around 6000 BC, agriculture and urbanism have evolved hand in hand by aiming for the food security of organized communities. Another milestone in the nexus of human development, food systems, and cities was the industrialization of food that took place in parallel with the industrial revolution but was intensified over the 20th century. Food systems started becoming part of a global industry that gave a tremendous boost to sustaining larger communities, but this didn't come without a cost. "Dramatic food production gains over the past two centuries allowed the human population to grow from 1 billion up to 7 billion without any increased frequency of premature death, but these gains may not be environmentally sustainable."[1] Undoubtedly, the infrastructure that supports contemporary food systems has a significant spatial, environmental, and social impact.

Globally, agriculture accounts for 70 percent of water consumption, covers 34.3 percent of land area, is responsible for 17 percent of greenhouse gas emissions (or 30 percent if we include the associated deforestation), and occupies 37.3 percent of the labor force.[2] These data describe contemporary food demand; it is estimated that global food production must increase by 70 percent by 2050 to meet the expected increase in food demand, mainly caused by global population growth.[3] Increased food demand puts pressure on the spatial requirements for arable land, thus requiring an intensification of production through higher crop yields. In summary, the main issues of food infrastructure are the increasing food demand and the environmental impact of food systems. Increasing demand stems from rapid population growth in combination with increasing urbanization and the associated changes in consumption and dietary patterns. Environmental impact stems from increased production that puts a strain on the planet's natural resources and escalates climate change, while conversely climate change effects are already putting at risk stable food production. Apart from being a basic human need, food has been a critical factor for human economy, health, culture, and social life. Thus, every approach to food systems as infrastructure has to acknowledge the high degree of complexity and the multiple interrelated dimensions.

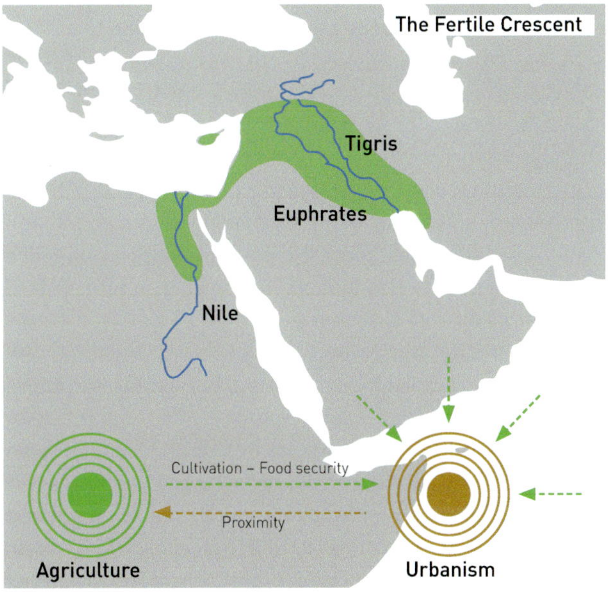

WORLDWIDE AGRICULTURE IMPACT

Food infrastructure has a significant spatial, environmental, and social impact. Global agriculture accounts for:

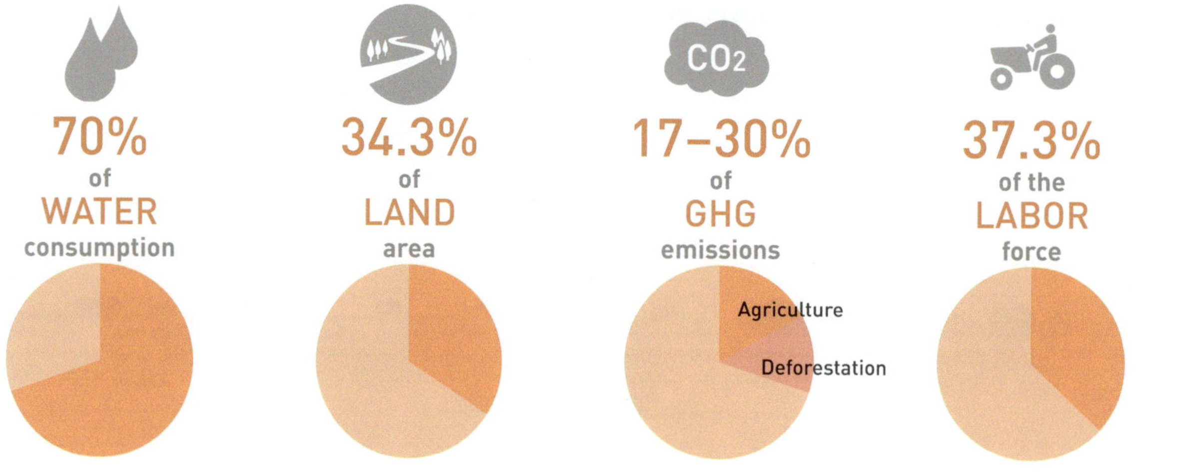

Data by: water IWMI, land Foresight, GHG emissions WRI, labor CIA, World Bank

CHALLENGES TOWARD A SUSTAINABLE FOOD SYSTEM

Food becomes an infrastructure system for cities when we start thinking beyond its status as a commodity. The industrialization of food has detached consumers from food's origin and of course the intermediate stages of its processing and delivery. The food we consume is one link in a complex combination of processes, from production, transportation, processing, and trade to consumption and partially disposal. The infrastructure that supports this series of actions is directly and indirectly related to cities.

Cities put pressure on food systems by increasing urban population in combination with limited global resources. At the same time, climate change that is caused partially by food system processes (high energy demand, greenhouse gas emissions, and intensive water use) brings risks to the operation of cities. Therefore, increasing food demand and climate change risks compel cities to enhance resilience by planning a more sustainable food infrastructure system and increasing self-sufficiency.

Much more than other infrastructure systems, the complex structures of food systems and their infrastructure highlight the issue of scale. For instance, one estimate shows "the average grocery store item in the United States travels 1,500 miles between production and consumption."[4] In such cases food production is completely detached from the urban environment and the consumers. Most commonly, food infrastructure is part of regional or global systems, whereas, on a much more local scale, in urban food production the food infrastructure is contained in the urban system. The sustainability of cities is connected with both scales, regional and urban. In parallel, regional and global food infrastructure will be in dialogue with urban and local food infrastructure. In the coming years, cities could improve their food security by lowering their unnecessary dependence on global and regional systems and promoting local or urban production. By itself, that development wouldn't be enough to secure food for a skyrocketing urban population that is under threat of climate change risks. The processes of both global and local food infrastructure systems should be revolutionized so that they provide higher crop yields and mitigate impacts to the environment. The role of food as infrastructure would be critical in such developments, while its synergies with other infrastructure systems have great potential. The integration of both natural and advanced technological solutions would be crucial.

A sustainable food system is one that covers the basic human need for food, maximizes the social, economic, and environmental benefits, and reduces the risks of negative impacts. For city planning, a sustainable food infrastructure system has to relink cities to surrounding food-productive landscape, integrate ways of urban food production into planning, and redesign food processing, distribution, and retail nodes to increase efficiency and reduce nuisance.

More specifically, the main environmental challenge for a sustainable food system is to protect the foodshed and ensure long-term stable production. This means exploiting food-productive resources within their regenerative capacity by adopting the most sustainable methods of production. Overexploitation and mass industrial food production lead to foodshed degradation and put at risk future production. Diminished soil fertility, erosion, land scarcity, and resource depletion, as for example through overfishing, are among the main impacts. Another challenge for an environmentally sustainable food system is the minimization of water and energy resources consumed for food supply. Water and energy are essential for food production and transportation. Actually, there is a nexus or interlocking relationship between food, water, and energy, with the management of each resource affecting the other two. Adapting cultivation to local climate conditions, minimizing resource-intensive meat production, and promoting local production can significantly reduce water and energy consumption. Natural ecosystems are also under increasing pressure from the food system. Unsustainable food production methods lead to pollution of soil and water bodies from chemicals and pesticides, while other negative impacts include deforestation, loss of biodiversity, water scarcity, and desertification.

In terms of economic sustainability, the main challenge of the food system is to provide healthy food at affordable prices. It should be able to minimize cost during production, processing, and distribution without compromising quality or fair trade conditions. Accessibility to adequate, nutritious food should be ensured for low-income families and other vulnerable social groups. Furthermore, financial incentives for local, urban, and organic production should be provided.

In terms of social sustainability, the main challenge for a socially sustainable food system is to provide food security by eliminating hunger and malnutrition. Protecting public health is an equally important challenge: a healthy diet should be promoted, and high standards of quality control should be set for all food products. In terms of accessibility it should ensure that all residents can walk, cycle, or take public transport to a full grocery store. Furthermore, the support and encouragement of urban production in private gardens, rooftops, communal space, or allotments can increase resilience and social cohesion while it provides opportunities for local employment and economic growth. The fair conditions and wages of food infrastructure workers should be protected.

ENVIRONMENTAL

Challenges for environmental sustainability include:

- Protecting the foodshed
- Minimizing greenhouse gas and air pollutant emissions
- Minimizing the water footprint (blue, green, and gray)
- Minimizing energy consumption for production, processing, and distribution
- Protecting the ecosystem (against deforestation, desertification, etc.)
- Regulating the use of chemical fertilizers and pesticides
- Resolving soil and groundwater contamination issues

SOCIAL

Challenges for social sustainability include:

- Ensuring food security
- Eliminating hunger and malnutrition
- Implementing high quality control of food products
- Promoting a healthy diet and ensuring accessibility of healthy food
- Supporting urban food production
- Creating local employment opportunities
- Preventing exposure to health risks for workers in agriculture
- Minimizing food waste

ECONOMIC

Challenges for economic sustainability include:

- Providing healthy food at affordable prices
- Minimizing of cost during production, processing, and distribution, without compromising quality of food
- Establishing fair trade, working conditions, and wages
- Securing accessibility of adequate and nutritious food for low-income households
- Providing financial incentives for local and organic production

STRATEGIES TOWARD A SUSTAINABLE FOOD SYSTEM

SYSTEM PURPOSE: FOOD SECURITY	
	FOODSHED MANAGEMENT: Secure city's food supply—support sustainable ways of production
	FOOD WASTE REDUCTION: Minimize food waste—divert food waste from landfills
	FOOD LOSS REDUCTION: Minimize food loss between production and consumption
	LOCAL AND URBAN PRODUCTION: Support local decentralized or urban production—minimize imports

The primary purpose and service of the Food infrastructure system is to ensure equal and reliable access to fresh nutritious food, or in other words to provide food security. The structures and operations put in place create several sustainability challenges, as discussed in the previous section. The strategies listed above suggest basic steps that could be included in the planning of a food infrastructure system. They do not aim to outline all the requirements for a sustainable system but instead highlight approaches that if implemented could resolve multiple issues simultaneously.

Sustainable foodshed management should be the first and most essential strategy of a food infrastructure system. Constant and uninterrupted food supply can only be secured in the long term by supporting sustainable ways of production and by protecting the natural resources needed for food production. The foodshed is inextricably linked with the natural landscape and water resources. Consequently, protecting the natural ecosystem and water cycle is of vital importance. The foodshed should be protected from pollution and from any human activity that may degrade its productive capacity. A close link occurs

with Energy infrastructure as well. Energy is necessary for food production and transportation, but in some cases, especially with biofuels, energy generation can be competitive and in conflict with food production.

Food waste reduction should also be a clear strategy for a sustainable food infrastructure system. The main target here has a dual purpose: to prevent food waste generation at the source and to divert generated food waste from landfills. In many developed countries the percentage of food ending in the waste bin is shockingly high, while the annual global food waste is estimated at up to a staggering one-third.[5] The main effort for prevention should focus on changing habits by educating and raising consumers' awareness of the sustainability impacts of the issue and the best practices for waste reduction. On-site recycling should be promoted, while separate organic waste collection for treatment in aerobic or anaerobic plants should be organized on a municipality scale. Redistributing unconsumed food through charitable or other organizations is also an optimum solution. To minimize other forms of waste, food product packaging should also be kept to a minimum, and the use of recyclable materials should be promoted.

Apart from food ending in the waste bin during retail and final consumption, a large amount of edible food is lost between production and pre-consumption, a factor commonly referred to as food loss. Food loss is also estimated to come to a staggering nearly 30 percent of total production.[6] Minimizing food loss can significantly increase the efficiency of the system. During the first stages of production and harvesting, the most efficient and adaptable methods should be implemented. The processing stage is responsible for the largest proportion of food loss. Avoiding unnecessary processing would not only reduce food loss but also save energy and water resources consumed in most commonly used processing methods and would protect food from nutrient loss occurring in highly processed products. Extreme weather events connected with climate change are also increasingly responsible for loss of production. Climate change is a much more general problem that demands comprehensive well-known solutions and actions; however, updating production and increasing resilience can be a way of responding.

The efficiency and resilience of the Food infrastructure system can be further reinforced by supporting local and urban agricultural production. Local products have minimum transportation and preservation needs, thus consume fewer resources and produce less carbon emissions, while they are more likely to reach consumers fresh and at lower cost. Incentives for local and urban agriculture should be provided, and prime farmland should be protected from urban sprawl and pollution related to industrial activities. City planning should foresee land use for private or communal gardens and urban farms. Infrastructure for farmer's shops and markets should be provided, and direct sales from producers to consumers should be encouraged. Consumers should be aware of the environmental and community benefits of buying local food, while the place of origin and place of processing of each agricultural product should be clearly labeled.

10.3 | FOOD SYSTEM DECODING

FOOD DEMAND

FOOD SUPPLY AND PRODUCTION

FOOD TRADE

FOOD PROCESSING AND DISTRIBUTION

The four system levels used to decode the food infrastructure system are Demand, Supply and Production, Trade, and Processing and Distribution.

Food Demand, the first system level, provides an analysis of the demand to be served by the infrastructure system. It examines and monitors the main factors that shape food consumption and the way changes in modern lifestyle combined with rising urbanization affect dietary preferences and consumption patterns. It underlines the importance of providing information and raising consumers' awareness on food sustainability, since their behavior and the choices they make are critical for the sustainability of the whole food system. The second level, Supply and Production, represents the basic decisions concerning the main strategies, goals, and standards set for the food supply and production system. It pinpoints ways to ensure a sustainable food supply system and promotes the most sustainable and efficient methods of production. Food security, uninterrupted production, the provision of nutritious and healthy products, and the support of local and organic food are among the main focus points of this level. It also explores ways in which cities can protect their foodshed and increase resilience by developing and supporting urban agriculture. The third system level, Trade, refers to every point inside a city where consumers can acquire food. It examines ways to increase accessibility of healthy, affordable food for all consumers. It supports local, seasonal, and fair trade products and provides solutions for food waste reduction. Finally, the fourth system level, Processing and Distribution, deals with all the processing facilities and transportation of food products between production and final retail. It protects quality of life and ensures high quality and sanitation at all stages. It explores and optimizes the spatial distribution of processing and food transportation facilities and provides ways to increase the efficiency of the system while reducing food loss and greenhouse gas emissions.

FOOD DEMAND

The first level of the food infrastructure system examines the demand for food per person and the way it is formed and determined by dietary preferences and consumption patterns. The Demand level has essentially different targets in the developing and the developed world. In the developing world the primary concern is food security, while in the developed world the main concerns are reduction of excess consumption and of food waste. It is estimated that 1.5 billion people worldwide overconsume while 1 billion are undernourished, with the majority of both categories living in cities.[7] In the last half-century important steps have been taken toward tackling malnutrition, increasing the global average daily food consumption from 2,193 calories in 1961 to 2,868 in 2011.[8] This progress, though, can be reversed. As already mentioned, rapid population growth in combination with increasing urbanism and changing consumption and diet patterns will lead to an unprecedented demand for food. Increasing production puts a strain on the planet's natural resources and escalates climate change; conversely, climate change effects are already putting at risk stable food production. Urbanization exacerbates the problem as it is linked with changing dietary habits and specifically an increase in the consumption of meat and processed food. Increasing meat consumption has a high ecological impact and is responsible for high greenhouse gas emissions, air and water pollution, freshwater scarcity, deforestation, and biodiversity loss. Similarly, processed food is connected with higher energy consumption and greenhouse gas emissions, higher waste generation, higher cost, and nutrient loss.

It is evident that raising consumers' awareness and promoting low-environmental-impact products are the key to the sustainability of the food production system and the reinforcement of food security in the long term. A balanced diet based on fresh healthy products should be the model for improved public health and low obesity rates. Efforts should be made to increase consumption of local and organic products, with incentives and targeted consumer information campaigns focusing on their environmental benefits. Efforts should also be made to inform consumers about the social benefits of fair trade products. Awareness should also be raised on the growing issue of food waste. Consumers should be encouraged to plan their food purchases appropriately, and mechanisms should be developed for the redistribution of excess food to those in need. Food ending in the waste bin should be diverted from landfills through municipal and (even better) on-site processing of organic waste, producing compost or other useful by-products.

FOOD INEQUALITY (WWF 2012)

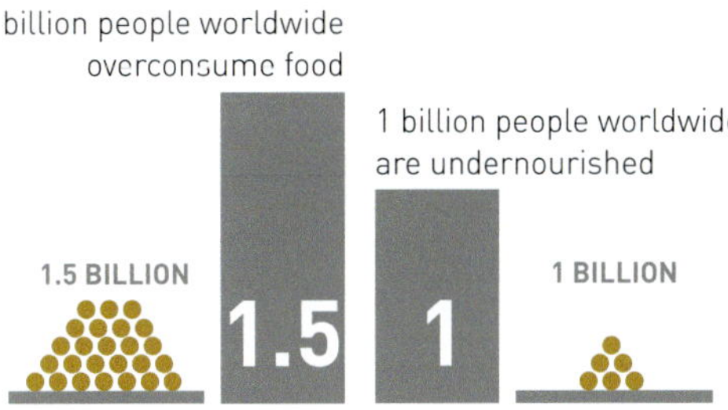

FOOD SUPPLY AND PRODUCTION

The Supply and Production level refers to the amount of food needed to cover a city's demand and the methods used to acquire it. The main aims of this level are to secure an adequate and high-quality food supply and to do so by promoting the most sustainable products and methods of production. At present, cities' food supply needs are mostly covered by national rural agriculture and fishery production or by imports from other countries. Local or urban agriculture production can also make a significant contribution.

The provision of a healthy balanced diet of fresh nutritious products at affordable prices should be the first priority of the supply system. This will help protect consumers from the negative effects of either obesity or malnutrition. In developed countries, obesity's increasing prevalence constitutes a serious threat to public health. The numbers are shocking: "more than 2.1 billion people—nearly 30 percent of the global population—are overweight or obese. That's nearly two and a half times the number who are undernourished. Obesity, which should be preventable, is now responsible for about 5 percent of all deaths worldwide. If its prevalence continues on its current trajectory, almost half of the world's adult population will be overweight or obese by 2030."[9]

The promotion of local agricultural products and the support of local farms are equally important. As already mentioned, it is estimated that the average grocery store item in the United States travels 1,500 miles between production and consumption.[10] Local food supply can reduce cost and energy resources needed for transportation and the consequent greenhouse gas emissions and pollution. Besides travel distance, it is also important for a food supply system to promote the products of organic and sustainable agriculture. The goals of sustainable agriculture are: "Enhance environmental quality and the natural resource base upon which the agriculture economy depends. Make the most efficient use of non-renewable resources and on-farm resources and integrate, where appropriate, natural biological cycles and controls. Sustain the economic viability of farm operations. Enhance the quality of life for farmers and society as a whole."[11]

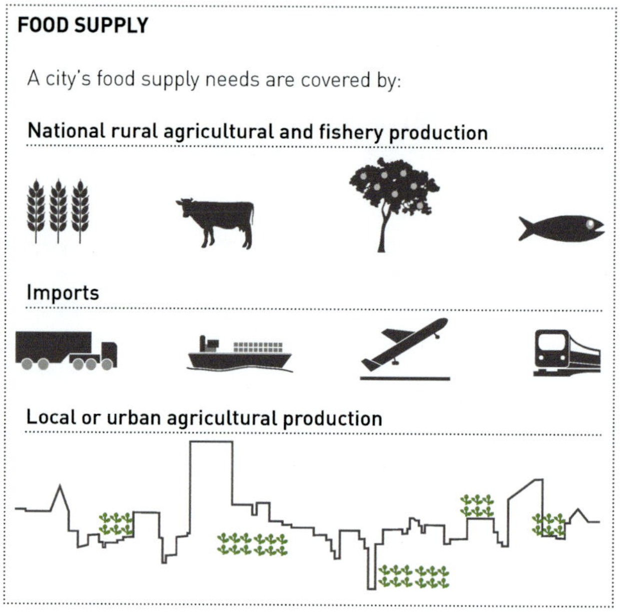

Finally, a sustainable food supply system should provide for food waste reduction and treatment. It should support home compost or/and organize municipal-scale collection of organic waste for aerobic or anaerobic digestion that can produce soil fertilizer or useful gaseous products for energy recovery.

FOOD PRODUCTION IS SPATIALLY ANALYZED, PLANNED, AND CONFIGURATED THROUGH THE LANDSCAPE INFRASTRUCTURE LEVELS

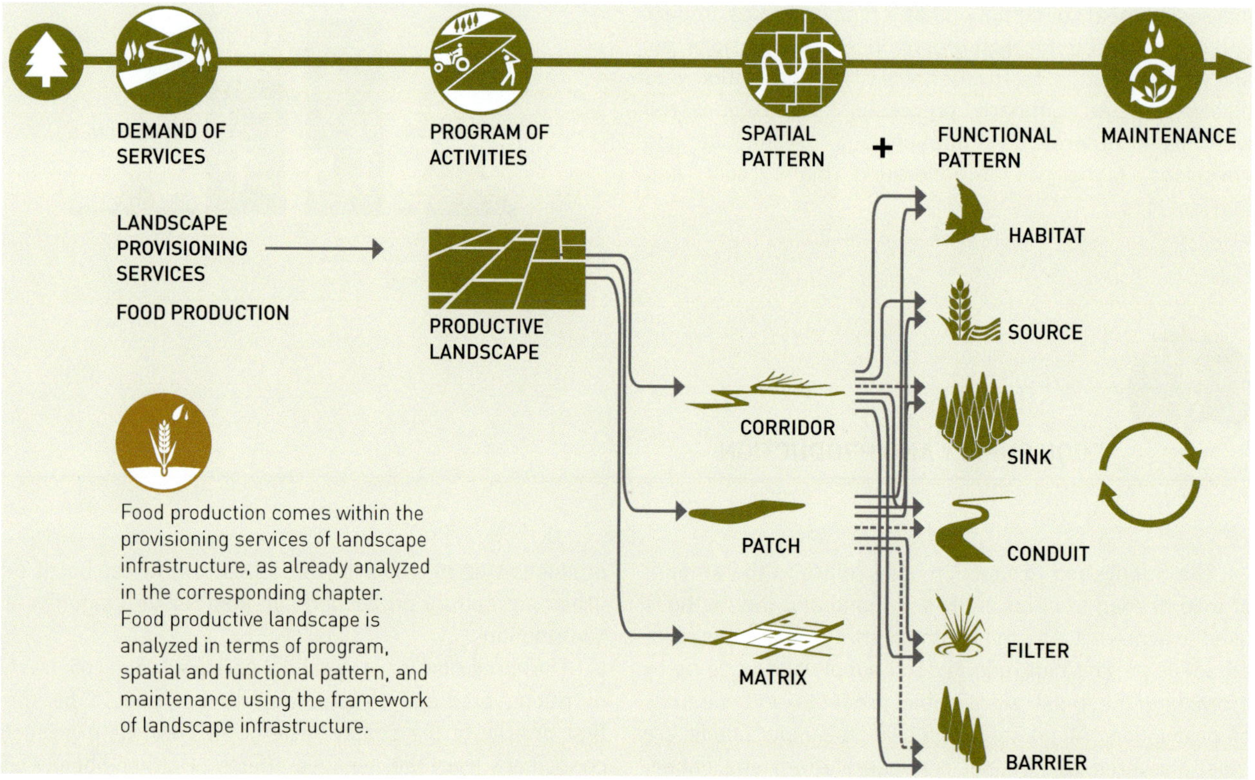

A sustainable food production system should ensure constant and undisturbed production, support environmentally friendly production methods, and minimize use of chemical fertilizers and pesticides. In terms of land use planning it should protect and preserve a city's foodshed while promoting urban production. A city's foodshed is defined as the geographic region in proximity to the urban center that produces part or the majority of the city's food supply. Most major cities' foodsheds are under constant pressure from urban expansion or contamination from adjacent industrial activities and transportation networks. Preserving foodshed increases cities' resilience and provides locally produced food, thus reducing energy and resources consumed by long-distance transportation.

As global and urban populations increase, an unprecedented demand for food is created. The natural world is constantly under threat due to pressure for larger and more intensive food production. Urban agriculture has the potential to cover a large portion of food demand, thus increasing cities' self-sufficiency and minimizing their impact on the environment. Urban agriculture can be implemented on various scales, from small individual production in private gardens and roof gardens to medium-scale production in communal gardens or municipal distributed allotments and up to intensive production in large-scale vertical farms, hydroponic greenhouses, or indoor fisheries.

Urban agriculture has significant environmental, social, and economic benefits. The environmental benefits include the preservation of natural resources and especially energy conservation since transportation needs are minimized, as well as the protection of wildlife habitats from the expansion of agricultural land. Socially, urban agriculture increases food security and improves health and nutrition by offering easier access to fresh products. In addition, small-scale and communal urban agriculture has proved to strengthen social cohesion and increase the feeling of well-being as a creative and fulfilling activity. On the economic level it can ignite entrepreneurial spirit, providing employment opportunities and a source of income especially for the most vulnerable and socially excluded groups.

Implementation of urban agriculture can be challenging, though, and in many cases compatibility issues with urban life arise. Nuisance, odors, or pest infestations can affect urban quality of life, while urban air pollution or contaminated soil can affect the quality of agricultural products. Therefore, comprehensive planning, clear strategy, and a rigid framework of bylaws are essential for the successful and beneficial application of urban agriculture. Furthermore, one should not overlook the very limited and usually very expensive land availability in most densely populated urban centers. In these cases, the most efficient and intensive methods of urban agriculture production should be promoted.

HOW CITIES CAN PRODUCE FOOD

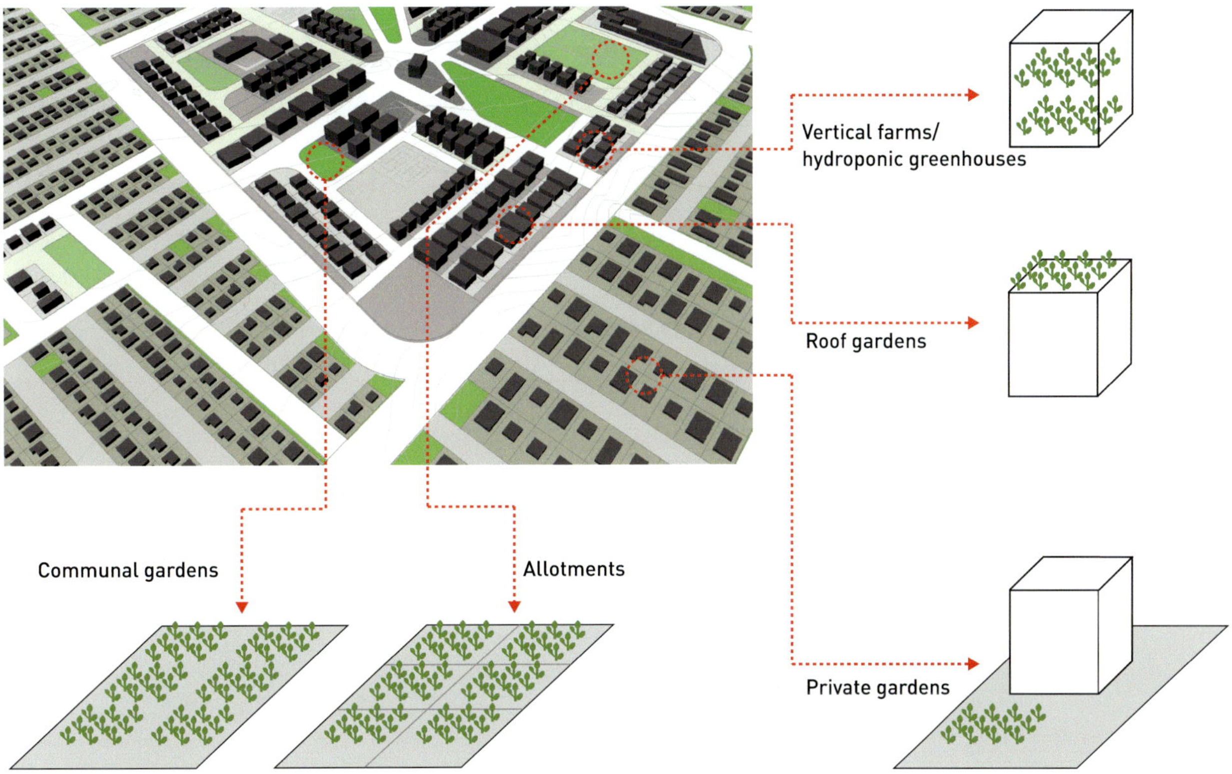

FOOD TRADE

The trade level examines every point inside a city at which consumers can acquire food. They include the key nodes of the food system, the nodes where production meets demand. They are part of everyday life, shaping cities' structure and visited almost daily by the majority of the population. The main entities of this level are grocery stores, butcher's shops, fish shops, bakeries, patisseries, supermarkets and convenience stores, farmer's markets, food markets, and food stands, as well as restaurants and street food spots.

A sustainable food trade system should ensure accessibility and affordability of healthy food, promote local, seasonal, and fair trade products, enhance quality of space, and minimize resources consumed and waste generation.

Accessibility of food retail facilities offering a variety of fresh food products is a key factor on this level. The distribution of trade nodes should ensure that food deserts (areas without ready access to fresh, healthy, and affordable food) are avoided. Whether an urban district can be classified as a food desert or not is determined by a number of quantitative and qualitative factors such as the actual average distance from a food node, the availability of public transportation, and the percentage of car ownership or the average income of the community. On the whole, a food trade system should ensure that all residents can walk, cycle, or use public transportation to reach a full grocery store. Residents with no direct access to fresh food markets tend to consume more processed food obtained from local and convenience stores, increasing the risk of obesity and other health hazards related to poor nutrition. Apart from accessibility, another essential factor is affordability. Instability in the global financial system, together with extreme weather phenomena associated with climate change, increase food price volatility. In many industrial world cities, low-income households cannot afford healthy fresh food. In these conditions, subsidies on basic food products combined with price regulations against market distortions and food price speculation are necessary.

It is also important for a sustainable food trade system to promote local and seasonal products. Energy

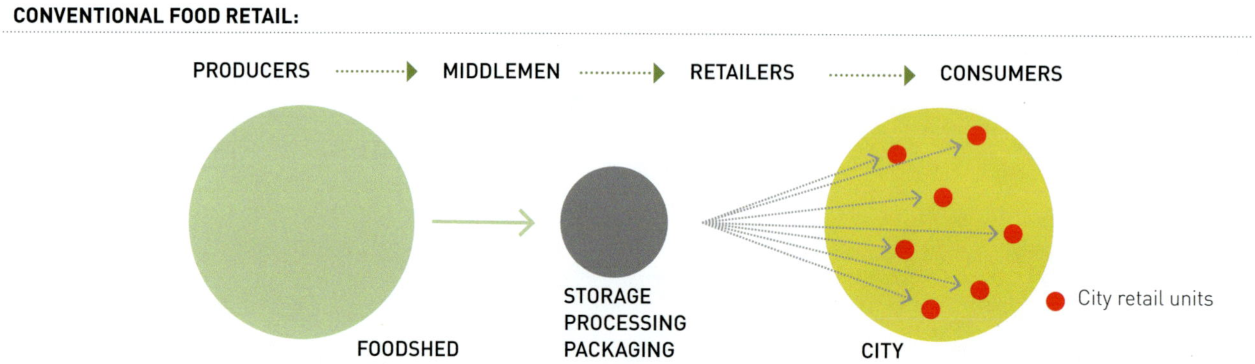

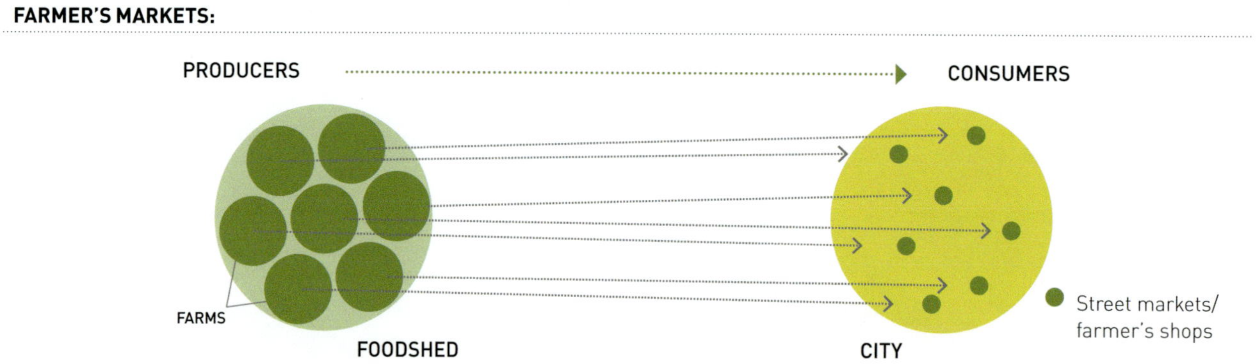

resources needed for transportation and corresponding greenhouse gas emissions are notably reduced, while local communities benefit from employment opportunities and economic growth. The provision for street markets and farmer's shops at which producers directly sell their products to consumers is an excellent way to support local high-quality products and preserve lower prices, due to fewer intermediaries being involved. It also spares much of the resources that would be used for processing and packaging. Cities can promote farmer's markets by designating suitable areas and providing permanent or temporary supporting facilities.

Another way in which food trade nodes can affect social and economic sustainability is by supporting fair trade practices. Fair trade ensures producers' and food production workers' well-being, especially for products exported from developing to developed countries. Fair trade products should be clearly designated and easily recognized, accompanied by corresponding information enhancing the consumer's ability to make informed and responsible choices.

The food trade system significantly contributes to food waste generation. This is due not only to food products remaining unsold on shelves past their expiration date but also to the higher standards imposed by many food retail chains on the aesthetic appearance of food, with the result that tons of absolutely edible food go to the waste bin every year. In the first case, coordination with social services and charities for redistributing food and restocking food banks offers the most suitable solution. In the second case, it is obvious that a relaxation in standards concerning the appearance (not the quality) of food products should be implemented, combined with efforts to raise consumers' awareness on the issue. Finally, in cases where food waste cannot be prevented, on-site processing and recovery should be promoted.

FOOD PROCESSING AND DISTRIBUTION

The Processing and Distribution level refers to the transportation, processing, and storage procedures that intervene between food production and final retail. The main entities on this level include processing facilities, food factories, slaughterhouses, quality control centers, packaging facilities, storage facilities and warehouses, logistic centers, and wholesale markets.

A sustainable food processing and distribution system should ensure high food quality and sanitation at all stages while minimizing processing and nutrient loss, minimizing transportation cost and thus greenhouse gas emissions, providing adequate storage and logistics facilities, and avoiding unnecessary packaging.

Ensuring high quality and sanitation at all stages should be the first priority of this level. Food scandals have broken out with increasing frequency in the last few decades following the gradual worldwide industrialization of food production and processing. A great risk is posed to human health, with unforeseeable consequences in the future. Frequent quality control checks combined with strict health regulations should be the answer to this threat.

Another important scope of this level is to increase the efficiency of the system while minimizing food processing needs. It is estimated that nearly 30 percent of food is lost between production and pre-consumption.[12] As food

FOOD LOSS

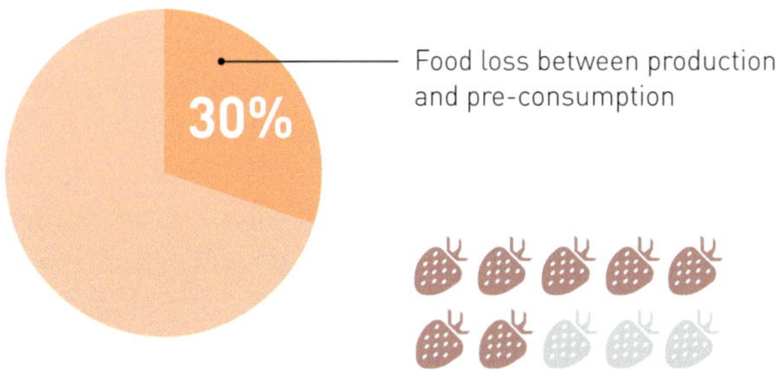

FOOD MILES

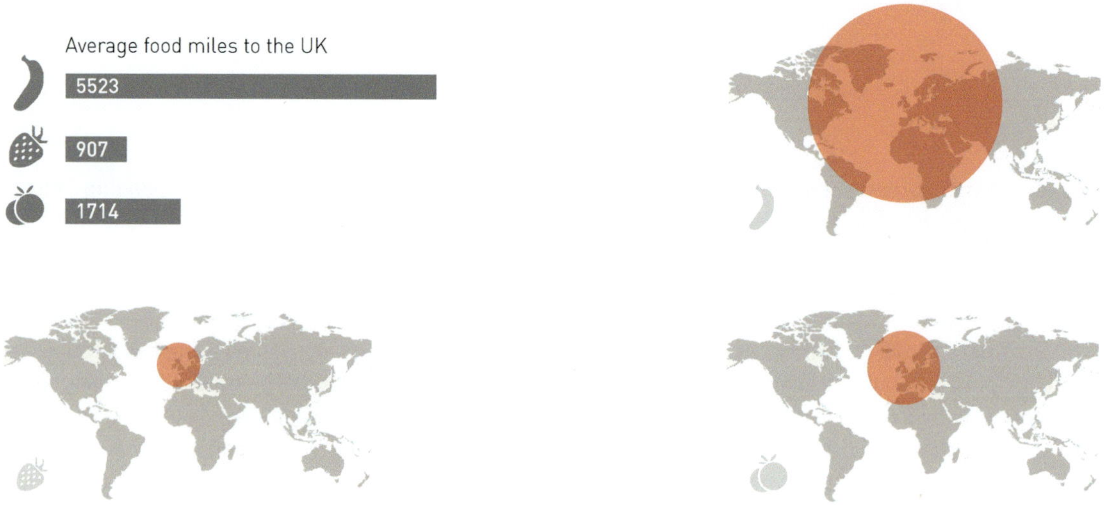

demand increases and resources for food production come under constant pressure, such a high percentage of food loss becomes unsustainable. More efficient methods of processing need to be adopted. Furthermore, minimizing unnecessary processing not only reduces the consumption of vital resources like energy and water, but also protects food quality from the nutrient loss observed in intensely processed products. Decisions made on this level can also significantly reduce the waste generation of the system, primarily by avoiding unnecessary packaging and promoting the use of recyclable materials.

Finally, the spatial organization and distribution of the entities of this level significantly affect the transportation cost and greenhouse gas emissions related to food products. Optimum placement of food processing and distribution facilities, combined with the use of low-emission vehicles and modes of transportation for food distribution, can substantially reduce the system's environmental impact. A commonly used method to assess the environmental footprint of food products is "food miles," "the distance food is transported from the time of its production until it reaches the consumer."[13] Although "food miles" is an important factor, it is not always enough for a complete assessment of the environmental impact of food products. Local options are not necessarily the most energy-efficient options. For example, some heated greenhouses can use excessive amounts of energy, which exceed the combined carbon footprint of production and the food miles associated with imported products that are less energy-intensive. A life cycle analysis is the optimum procedure. "Lifecycle analysis, a technique that meshes together a wide range of different environmental criteria including emissions and waste, is a more holistic way of assessing the real environmental impact of the food products. The technique accounts for energy input and output involved in the production, processing, packaging and transport of food. It also factors in resource depletion, air pollution and water pollution and waste generation/municipal solid waste."[14]

LIFE CYCLE ANALYSIS

Stages of food life cycle analysis:
land and water – production – processing – distribution – retail – consumption – waste recovery

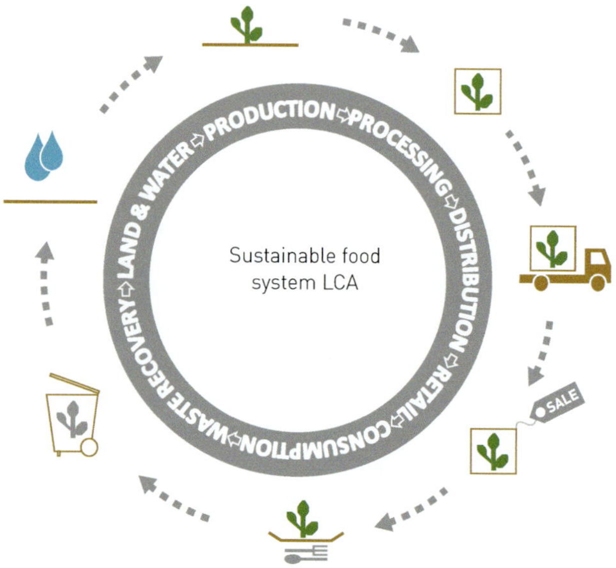

10.4 | OBJECTIVES AND GUIDELINES FOR A SUSTAINABLE FOOD SYSTEM

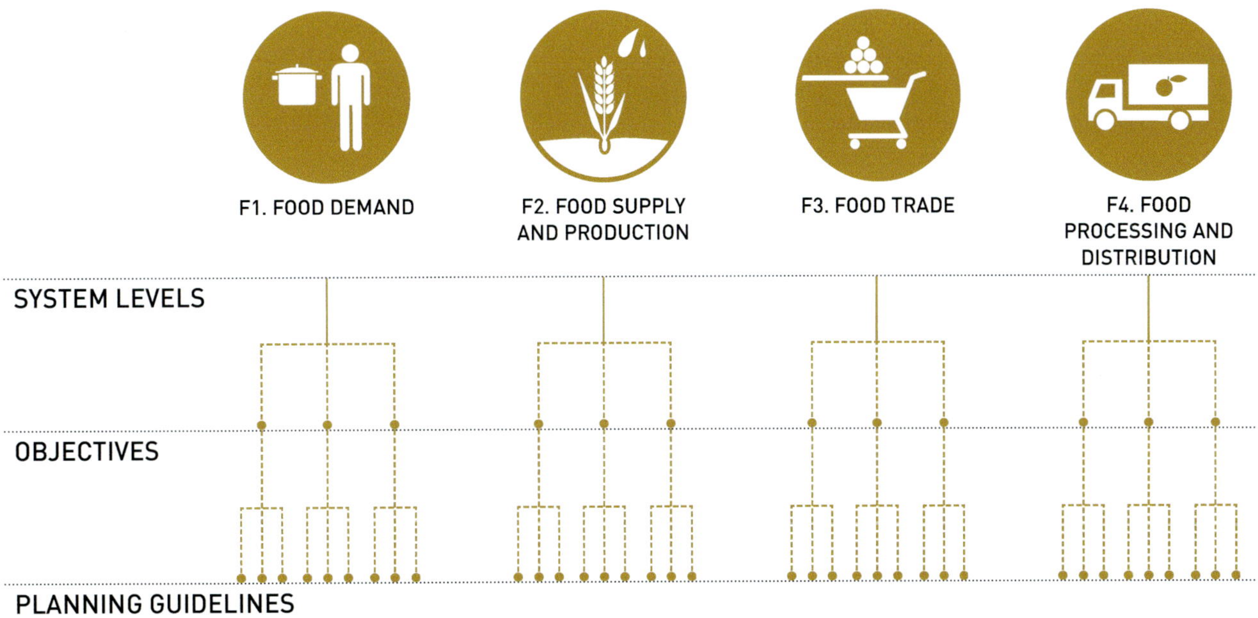

Following the decoding of the Food system on four system levels, planning objectives are defined for each level. They represent both challenges and strategies toward a sustainable system, expressing what should be the overall final outcome of food planning.

The objectives are associated with the five focus categories of the Zofnass Program, namely Leadership, Quality of Life, Natural World, Resource Allocation, and Climate and Risk. The Sustainable Planning Guidelines aim to lead toward these objectives, highlighting synergies and connections between Food and the other six infrastructure categories of Energy, Information, Solid Waste, Landscape, Transportation, and Water.

OBJECTIVES AND GUIDELINES SYNOPSIS

The objectives and guidelines of the Demand level are related to the Leadership category. The first objective deals with the increasing issue of food waste. Guidelines here call for educating consumers on rational planning of food purchases and promoting on-site processing of organic waste in synergy with Solid Waste infrastructure. A balance should be established between food consumption and demand patterns. Demand that is not met and cases of food wastefulness should be identified, and mechanisms for the redistribution of excess food should be developed and implemented. Other objectives focus on raising consumers' awareness and guiding them, through information campaigns and incentives, in making the most sustainable and environmentally friendly choices. A healthy balanced diet should be promoted based on fresh, local, and organic products. Fair trade products should also be supported for their social benefits. Finally, a constant monitoring of food consumption patterns and trends should be implemented and organized in synergy with Information infrastructure.

F1. FOOD DEMAND

- F1.1. Reduce food waste
- F1.2. Balance food consumption and demand patterns
- F1.3. Promote healthy food consumption habits
- F1.4. Increase demand for organic and local food products
- F1.5. Increase demand for fair trade products
- F1.6. Monitor food consumption trends

The Supply and Production level primarily deals with the important strategic decisions that define a city's food supply and the methods of production these require. The objectives of this level refer to the categories of Leadership, Natural World, and Resource Allocation. In terms of Leadership, the most important target of this level is to secure an adequate and uninterrupted food supply for the long term. Clear goals to be achieved through food supply and production planning should be defined and implemented. Food production should be protected from climate and man-made threats, and the city's food resilience should be enhanced by promoting local and urban production. In the Natural World category there is a clear synergy with Landscape infrastructure. Objectives here call for a sustainable management of the city's foodshed along with measures to preserve and enhance biodiversity and to protect natural-world elements and systems from degradation, contamination, and habitat loss caused by intensive, industrialized agriculture. Objectives connected to the Resource Allocation category provide guidelines for rational use of water resources and optimization of energy use for food production. Guidelines for solid waste reduction and resource recovery at this stage are also provided. It is evident that a strong synergy with Water infrastructure occurs at this level, together with synergies with the Energy and Waste infrastructure systems.

F2. FOOD SUPPLY AND PRODUCTION

- F2.1. Secure adequate and uninterrupted Food Supply flow in the long term
- F2.2. Define clear goals to be achieved through Food Supply and Production planning
- F2.3. Promote local Food Production and increase productivity
- F2.4. Promote urban agriculture
- F2.5. Manage city's foodshed
- F2.6. Preserve and enhance biodiversity through Food Production practices
- F2.7. Maintain the integrity of natural-world elements and systems
- F2.8. Optimize Water use in Food Production
- F2.9. Optimize Energy use in Food Production
- F2.10. Contribute to Waste reduction and resource recovery

Trade-level objectives refer to the Leadership, Quality of Life, and Resource Allocation categories. In the Leadership category the main objectives point to placing and distributing food retail spaces to ensure accessibility as well as to facilitate the trade of local and seasonal products and to minimize food waste. In terms of accessibility, it should be ensured that food deserts are avoided and that food stores are easily accessible from all residential areas by foot, cycling, or public transport. Local and seasonal products can be promoted by regulating and designating spaces for farmer's markets across the city, combined with incentives for higher penetration of these products in food markets and assisting local producers to establish direct connection with buyers. Food waste can be prevented by giving incentives for food donations and by addressing strict food quality standards while rejecting food supply based on aesthetic appearance. In the Quality of Life category, objectives focus on ensuring affordability of healthy food and enhancing quality of place, since food retail facilities are usually an important element of public space and social life. In the Resource Allocation category, objectives mainly concern the optimization of energy consumption by promoting the use of energy-efficient equipment in food trade facilities, and the contribution to waste reduction and resource recovery by promoting on-site processing of food waste and by organizing organic waste recovery systems in food trade areas.

F3. FOOD TRADE

- F3.1. Place and distribute food retail spaces to ensure accessibility
- F3.2. Facilitate the trade of local and seasonal products
- F3.3. Minimize food waste
- F3.4. Ensure affordability of healthy food
- F3.5. Enhance quality of place
- F3.6. Contribute to waste reduction and resource recovery

On the Processing and Distribution level, objectives refer to all five of the Zofnass Program focus categories. In the Leadership category, the importance of ensuring food quality during processing and distribution is pinpointed. It is essential to establish high quality and sanitation standards during these stages, along with strict control mechanisms for their implementation. It is equally important to increase efficiency and minimize the large percentage of food loss occurring during processing, storage, and distribution by promoting the most appropriate practices and technologies. The main objective of the Quality of Life category focuses on protecting community well-being from any health impacts, odors, or noise pollution usually caused by processing facilities or food transportation. In the Natural World category, maintaining the integrity of natural-world elements and systems is key, more specifically protecting soil, surface water, and groundwater sources from contaminating substances found in food processing facilities. In the Climate and Risk category, the main objective concerns the minimization of emissions caused by food distribution. Guidelines call for optimizing the placement of food processing and distribution facilities, regulating distribution itineraries, and promoting the use of low-emission transportation modes and vehicles. Objectives connected to the Resource Allocation category involve optimizing material use by reducing unnecessary food processing and packaging and promoting the use of recyclable materials. Guidelines should also be provided for optimizing water use in food processing, and waste reduction through on-site processing and resource recovery of food waste.

The Objectives of the respective Guidelines as well as the relevant Guidelines of other infrastructure systems are presented in the following tables. (TO) and (FROM) indicate that the guideline can assist or can be assisted by the other infrastructure system, where the arrow is placed. (WITH) indicates that the guideline might require planning for an Information or Landscape subsystem and should follow multiple Guidelines from these two systems.

F4. FOOD PROCESSING AND DISTRIBUTION

- F4.1. Ensure food quality is maintained during Food Processing and Distribution
- F4.2. Minimize food losses
- F4.3. Promote community well-being
- F4.4. Maintain the integrity of natural-world elements and systems
- F4.5. Minimize emissions caused by Food Distribution
- F4.6. Optimize material use during Food Processing and Distribution
- F4.7. Optimize Water Consumption in Food Processing
- F4.8. Optimize Energy Consumption in Food Processing
- F4.9. Contribute to Waste reduction and resource recovery during Food Processing and Distribution

GUIDELINES

OBJECTIVES	GUIDELINES		SYNERGIES
F1 FOOD DEMAND			▲ TO + WITH ▼ FROM
F1.1. Reduce food waste	F1.1.1.	Educate consumers on planning food purchases	
	F1.1.2.	Promote on-site processing of food waste by consumers	▲
F1.2. Balance food consumption and demand patterns	F1.2.1.	Identify Food Demand that is not met	
	F1.2.2.	Identify cases of food wastefulness	
	F1.2.3.	Develop mechanisms for the redistribution of excess food	
F1.3. Promote healthy food consumption habits	F1.3.1.	Raise consumers' awareness on balanced and healthy diet options	
F1.4. Increase demand for organic and local food products	F1.4.1.	Provide incentives for the consumption of local and organic food products	
	F1.4.2.	Provide public information about benefits of local food products	
	F1.4.3.	Provide public information on benefits of organic food products	
	F1.4.4.	Raise awareness on the environmental impact of food products	
F1.5. Increase demand for fair trade products	F1.5.1.	Provide public information on fair trade benefits	
F1.6. Monitor food consumption trends	F1.6.1.	Collect and organize quantitative and qualitative data on food consumption	+
F2 FOOD SUPPLY AND PRODUCTION			▲ TO + WITH ▼ FROM
F2.1. Secure adequate and uninterrupted Food Supply flow in the long term	F2.1.1.	Diversify Food Supply options	
	F2.1.2.	Plan for Food Supply in case of an emergency	
F2.2. Define clear goals to be achieved through Food Supply and Production planning	F2.2.1.	Set goals for Food Supply mix	
	F2.2.2.	Adopt indicators to benchmark and monitor Food Supply and Production planning	
	F2.2.3.	Integrate long-term city growth and evolution with Food Supply planning	
	F2.2.4.	Regularly update Food Supply planning according to community development	
F2.3. Promote local Food Production and increase productivity	F2.3.1.	Identify and assess climate and man-made threats on local Food Production	
	F2.3.2.	Prepare plans for short-term and long-term hazards affecting local Food Production	+

OBJECTIVES	GUIDELINES		SYNERGIES
	F2.3.3.	Avoid Food Supply dependence on monocultures	
	F2.3.4.	Provide know-how and assistance to local farmers on more productive methods and techniques	+ ▼
F2.4. Promote urban agriculture	F2.4.1.	Provide incentives for intensive urban agricultural farms	
	F2.4.2.	Provide space for communal gardens or allotments	▼
	F2.4.3.	Provide know-how and assistance on private gardens agricultural production	
F2.5. Manage city's foodshed	F2.5.1.	Map and monitor the city's foodshed and highly productive farmland	+ +
	F2.5.2.	Develop a foodshed management program	
	F2.5.3.	Protect foodshed from urban expansion	+ ▼
	F2.5.4.	Protect the foodshed's Water Supply	▼
	F2.5.5.	Identify and manage basic sources of contaminants within the foodshed	▼ + ▼ ▼
F2.6. Preserve and enhance biodiversity through Food Production practices	F2.6.1.	Preserve plant genetic resources	
	F2.6.2.	Preserve biodiversity avoiding alien monocultures	
F2.7. Maintain the integrity of natural-world elements and systems	F2.7.1.	Prevent soil degradation from intensified Food Production	
	F2.7.2.	Regulate use of fertilizers and pesticides in Food Production	
	F2.7.3.	Prevent habitat loss from agricultural and livestock expansion	
	F2.7.4.	Protect Water elements and groundwater from contaminants used in Food Supply and Production activities	+ ▲
	F2.7.5.	Provide know-how and assistance to local farmers on organic methods and techniques	
	F2.7.6.	Provide financial incentives and support to farmers using organic methods	
F2.8. Optimize Water use in Food Production	F2.8.1.	Select alternative Water Supply sources of adequate quality for Food Production activities	▲
	F2.8.2.	Promote efficient agriculture irrigation systems	▲
	F2.8.3.	Promote crops suitable to local climate and Water resources	▲
F2.9. Optimize Energy use in Food Production	F2.9.1.	Provide producers with know-how on more efficient and low-energy methods of production	
F2.10. Contribute to Waste reduction and resource recovery	F2.10.1.	Identify potential for using Food Supply and Production by-products as feedstock for other infrastructure systems	▲
	F2.10.2.	Take advantage of Solid Waste system residues and by-products for Food Production processes	▼
	F2.10.3.	Organize Waste recovery systems in Food Production facilities	▲

F3 FOOD TRADE

▲ TO
+ WITH
▼ FROM

OBJECTIVES		GUIDELINES	SYNERGIES
F3.1. Place and distribute food retail spaces to ensure accessibility	F3.1.1.	Promote the equal and dense distribution of food retail facilities across the city	▲
	F3.1.2.	Ensure that food stores are easily accessible from all residential areas by foot, cycling, or public transport	
F3.2. Facilitate the trade of local and seasonal products	F3.2.1.	Regulate and provide spaces for farmer's markets across the city	
	F3.2.2.	Provide incentives for higher penetration of local and seasonal products in food markets	
	F3.2.3.	Assist local producers to establish direct connection with buyers	
F3.3. Minimize food waste	F3.3.1.	Provide incentives for food donations	
	F3.3.2.	Address food quality standards that reject food supply based on appearance	
F3.4. Ensure affordability of healthy food	F3.4.1.	Regulate and protect basic food prices from volatility and speculation	
	F3.4.2.	Resolve potential conflicts with food resources used as supply for other systems	▼
	F3.4.3.	Promote healthy food products in food retail spaces	
F3.5. Enhance quality of place	F3.5.1.	Design food retail facilities to extend and enhance public space	
F3.6. Contribute to waste reduction and resource recovery	F3.6.1.	Promote on-site processing of food waste in Food Trade facilities	▲
	F3.6.2.	Organize Waste recovery systems in Food Trade areas	▲

F4 FOOD PROCESSING AND DISTRIBUTION

▲ TO
+ WITH
▼ FROM

OBJECTIVES		GUIDELINES	SYNERGIES
F4.1. Ensure food quality is maintained during Food Processing and Distribution	F4.1.1.	Establish quality and sanitation standards for Food Processing and Distribution	
	F4.1.2.	Establish control mechanisms for the implementation of quality and sanitation standards in Food Processing and Distribution	
F4.2. Minimize food losses	F4.2.1.	Promote practices and technologies to avoid food losses during storage, processing, and distribution	
F4.3. Promote community well-being	F4.3.1.	Assess and resolve potential health issues caused by Food Processing facilities	
	F4.3.2.	Assess and resolve potential noise pollution issues caused by Food Processing facilities	
	F4.3.3.	Minimize odor issues caused by Food Processing facilities	

OBJECTIVES	GUIDELINES		SYNERGIES
F4.4. Maintain the integrity of natural-world elements and systems	F4.4.1.	Protect Water elements and groundwater from contaminants found in Food Processing facilities	
	F4.4.2.	Protect soil from contaminants found in Food Processing facilities	
F4.5. Minimize emissions caused by Food Distribution	F4.5.1.	Promote use of low-emission Transportation modes and vehicles for Food Distribution	
	F4.5.2.	Optimize placement of Food Processing and Distribution facilities to reduce number and length of food distribution trips across the city	
	F4.5.3.	Regulate Food Distribution itineraries so as to mitigate traffic congestion	
F4.6. Optimize material use during Food Processing and Distribution	F4.6.1.	Reduce unnecessary food packaging	
	F4.6.2.	Promote use of recyclable materials for food packaging	
F4.7. Optimize Water Consumption in Food Processing	F4.7.1.	Promote water-efficient technologies in Food Processing (Provide incentives for Water recycling in Food Processing)	
F4.8. Optimize Energy Consumption in Food Processing	F4.8.1.	Promote energy-efficient technologies in Food Processing	
F4.9. Contribute to Waste reduction and resource recovery during Food Processing and Distribution	F4.9.1.	Promote on-site processing of food waste in Food Processing facilities	
	F4.9.2.	Identify potential for using Food Processing and Distribution residues and by-products as feedstock for other infrastructure systems	
	F4.9.3.	Organize Waste recovery systems through all Food Processing and Distribution stages	

RELEVANT GUIDELINES

GUIDELINES			RELEVANT GUIDELINES	
F1 FOOD DEMAND			FROM ▸ ▸ TO + WITH	
F1.1.2.	Promote on-site processing of food waste by consumers		SW1.1.3.	Promote on-site processing to reduce incoming Solid Waste
F1.6.1.	Collect and organize quantitative and qualitative data on food consumption		I1.3.1.	Identify infrastructure systems' processes to be monitored/documented through Information infrastructure
F2 FOOD SUPPLY AND PRODUCTION			FROM ▸ ▸ TO + WITH	
F2.3.2.	Prepare plans for short-term and long-term hazards affecting local Food Production		L1.7.3.	Define needs for protection of infrastructure systems against projected climate change and extreme events through Landscape infrastructure
F2.3.4.	Provide know-how and assistance to local farmers on more productive methods and techniques		L1.5.2.	Define strategies to enhance (benefit from) the productivity of landscape through Landscape infrastructure
F2.4.2.	Provide space for communal gardens or allotments		L2.2.1.	Translate identified Landscape services to specific activities
F2.5.1.	Map and monitor the city's foodshed and highly productive farmland		I1.2.1.	Identify opportunities to monitor and protect the natural context through Information infrastructure
			L1.5.1.	Identify opportunities for the provisioning of food, fiber, and feedstock provided by the natural context
F2.5.3.	Protect foodshed from urban expansion		T3.7.1.	Avoid construction of Transportation Nodes on undeveloped land
			T4.8.1.	Avoid the construction of Transportation Networks on undeveloped land
			T3.7.7.	Save, preserve, and reuse topsoil before the construction of Transportation Nodes where possible
			T4.8.12.	Save, preserve, and reuse topsoil before the construction of Transportation Networks where possible
			L1.2.5.	Define needs for regulating city growth and expansion patterns through Landscape infrastructure
F2.5.4.	Protect the foodshed's Water Supply		W2.3.2.	Develop a watershed management program
F2.5.5.	Identify and manage basic sources of contaminants within the foodshed		SW3.8.3.	Protect soil from contaminants found in Solid Waste Treatment and Disposal facilities
			SW4.6.2.	Protect soil from contaminants during Solid Waste Collection and Transfer
			E3.6.3.	Protect soil from contaminants found in Energy Generation facilities

GUIDELINES		RELEVANT GUIDELINES	
		W3.6.4.	Protect soil from contaminants found in Water Treatment facilities
		L1.2.4.	Define needs for restoration of contaminated areas through Landscape infrastructure
		L4.4.6.	Control flows of fertilizer- and pesticide-laden runoff water from Landscape elements
F2.7.4.	Protect Water elements and groundwater from contaminants used in Food Supply and Production activities	W2.3.3.	Identify and manage basic sources of contaminants within the watershed
		L1.7.1.	Define needs for mitigating and compensating for other infrastructure systems' impact on the natural world through Landscape infrastructure
F2.8.1.	Select alternative Water Supply sources of adequate quality for Food Production activities	W1.6.3.	Identify opportunities for use of different nonpotable water qualities
F2.8.2.	Promote efficient agriculture irrigation systems	W1.4.1.	Reduce water required for Food Production and Food Processing
F2.8.3.	Promote crops suitable to local climate and Water resources	W1.4.1.	Reduce water required for Food Production and Food Processing
F2.10.1.	Identify potential for using Food Supply and Production by-products as feedstock for other infrastructure systems	E2.1.1.	Identify all potential Energy sources
F2.10.2.	Take advantage of Solid Waste system residues and by-products for Food Production processes	SW2.3.3.	Match Solid Waste Diversion Model residues and by-products with resource and feedstock needs of other infrastructure systems
F2.10.3.	Organize Waste recovery systems in Food Production facilities	SW4.3.4.	Solid Waste and Food systems to work in synergy for the Collection and Transfer of food waste

F3 FOOD TRADE

FROM ▸ ▸ TO
+
WITH

F3.1.1.	Promote the equal and dense distribution of food retail facilities across the city	T1.2.2.	Include basic amenities within each residential area
F3.4.2.	Resolve potential conflicts with food resources used as supply for other systems	E2.4.1.	Avoid conflict between crops used both for Food Supply and Energy Supply
F3.6.1.	Promote on-site processing of food waste in Food Trade facilities	SW1.1.3.	Promote on-site processing to reduce incoming Solid Waste
F3.6.2.	Organize Waste recovery systems in Food Trade areas	SW4.3.4.	Solid Waste and Food systems to work in synergy for the Collection and Transfer of food waste

GUIDELINES		RELEVANT GUIDELINES	
F4 FOOD PROCESSING AND DISTRIBUTION		FROM ▶ ▶ TO + WITH	
F4.5.2.	Optimize placement of Food Processing and Distribution facilities to reduce number and length of food distribution trips across the city	T1.2.4.	Optimize placement of infrastructure facilities that generate significant traffic
F4.5.3.	Regulate Food Distribution itineraries so as to mitigate traffic congestion	T1.4.2.	Optimize itineraries within other infrastructure systems
F4.6.1.	Reduce unnecessary food packaging	SW1.1.4.	Reduce Solid Waste Generation by other infrastructure systems
F4.6.2.	Promote use of recyclable materials for food packaging	SW1.1.4.	Reduce Solid Waste Generation by other infrastructure systems
F4.7.1.	Promote water-efficient technologies in Food Processing (Provide incentives for Water recycling in Food Processing)	W1.4.1.	Reduce water required for Food Production and Food Processing
F4.9.1.	Promote on-site processing of food waste in Food Processing facilities	SW1.1.3.	Promote on-site processing to reduce incoming Solid Waste
F4.9.2.	Identify potential for using Food Processing and Distribution residues and by-products as feedstock for other infrastructure systems	E2.1.1.	Identify all potential Energy sources
		L4.10.3.	Incorporate Solid Waste and wastewater Treatment residues and by-products in regular Landscape Maintenance if available
F4.9.3.	Organize Waste recovery systems through all Food Processing and Distribution stages	SW4.3.4.	Solid Waste and Food systems to work in synergy for the Collection and Transfer of food waste

NOTES

1. Robert L.Paarlberg, *Food Politics* (New York: Oxford University Press, 2013), 10.
2. "Agriculture and the Green Economy," accessed on October 29, 2015, http://www.farmingfirst.org/green-economy/(data source: water data by IWMI, land use data by Foresight, GHG emissions data by WRI, labor data by CIA and World Bank).
3. FAO, "How to Feed the World in 2050," 3, accessed on October 29, 2015, http://www.fao.org/fileadmin/templates/wsfs/docs/expert_paper/How_to_Feed_the_World_in_2050.pdf.
4. B. Halweil, "Home Grown: The Case for Local Food," Worldwatch Paper 163 (2002), http://www.nlc.org/, retrieved September 20, 2012.
5. FAO, "Food Wastage Footprint: Impacts on Natural Resources," 2013, 6, accessed on October 29, 2015, http://www.fao.org/docrep/018/i3347e/i3347e.pdf.
6. FAO, "Food Agriculture and Cities," 30, accessed on October 29, 2015, http://www.fao.org/fileadmin/templates/FCIT/PDF/FoodAgriCities_Oct2011.pdf.
7. WWF, accessed on October 29, 2015, http://assets.wwf.org.uk/custom/foodstory/
8. FAO, "Food and Nutrition in Numbers," 2014, 20, accessed on October 29, 2015, http://www.fao.org/3/a-i4175e.pdf.
9. Richard Dobbs, Corinne Sawers, Fraser Thompson, James Manyika, Jonathan Woetzel, Peter Child, Sorcha McKenna, and Angela Spatharou, "Overcoming Obesity: An Initial Economic Analysis," McKinsey Global Institute, Discussion Paper, November 2014, p. iii.
10. Halweil, "Home Grown."
11. M. Gold, "What Is Sustainable Agriculture?" United States Department of Agriculture, Alternative Farming Systems Information Center, July 2009, accessed on October 29, 2015, http://www.nal.usda.gov/afsic/pubs/agnic/susag.shtml#define.
12. FAO, "Food Agriculture and Cities,"21.
13. E. Engelhaupt, "Do Food Miles Matter?"*Environmental Science and Technology* 42 (2008), 3482.
14. Kelly Rae Chi, James MacGregor, and Richard King, "Fair Miles: Recharting the Food Miles Map," IIED/Oxfam, 2009, 16.

IMAGE CREDITS

P. 250: Photograph by Michal Kubicek, accessed on October 29, 2015, https://unsplash.com/photos/0AbVUUorULU. **P. 255:** (Top) Photograph by Pavel Voinov, https://unsplash.com/photos/2f_CnEeZsVg, accessed on October 29, 2015. **P. 256:** Adapted from http://www.farmingfirst.org/green-economy/, section 3, accessed on October 29, 2015, data sources: water IWMI, land Foresight, GHG emissions WRI, labor CIA, World Bank. **P. 258:** Photograph by Crucifix Jean-Luc, accessed on October 29, 2015, https://unsplash.com/photos/19tQv51x4-A. **P. 259:** Photograph by Dierregi, accessed on November 30, 2015, http://mrg.bz/Ar1cAA. **P. 260:** Photograph by Wiselywoven, accessed on October 29, 2015, http://www.morguefile.com/archive/display/940499. **P. 261:** Data source WWF, accessed on October 29, 2015, http://assets.wwf.org.uk/custom/foodstory/. **P. 265:** "Food Agriculture and Cities," FAO, October 2011, p. 21, accessed on October 29, 2015, http://www.fao.org/fileadmin/templates/FCIT/PDF/FoodAgriCities_Oct2011.pdf. **P. 266:** (Top) Data source: http://www.bbc.co.uk/education/guides/zf6fr82/revision/1, accessed on October 29, 2015.

PART 3

EXAMPLES OF PLANNING WITH THE GUIDELINES

Part 3 presents the application of the Planning Guidelines in the development of master plans for two different urban projects. The first is the new DCK City in Karachi, Pakistan, with a projected population of 600,000, that is currently under construction. The guidelines are used as tool to seamlessly incorporate infrastructure needs in the plan. The second project concerns the reuse of the former Athens International Airport in Greece. The 1500 acres seafront site, once in the outskirts of the city, is currently engulfed by densely populated suburbs. Specific design decisions and actions are given for the Transportation and Landscape systems, using the Guidelines as checklists.

PROJECT NAME: Master plan for DHA City Karachi (DCK)
LOCATION: Karachi, Pakistan
AREA: 47,000,000 sq m (11,640 acres)
PLANNING: Prof. S. N. Pollalis, chief planner of the partnership Osmani & Company (PVT.) LtD, RMJM, Doxiadis Associates

Examples of Planning with the Guidelines

New City
in Asia

Chapter 11

11.1 | DHA CITY KARACHI (DCK)

DHA City Karachi (DCK) is situated 56 km from the city core area of Karachi in Pakistan. The master plan of the city was completed in 2012 and the physical execution is planned to proceed in phases, with expected completion by 2030.

The city lies on a site of 11,640 acres and is planned to accommodate 600,000 people. Initially, the new city was conceived as a bedroom community to Karachi. However, its distance from the center of Karachi dictated a more integral approach toward a self-sufficient mixed-use city.

The master plan comprises 50,000 commercial and residential plots and includes a variety of uses and amenities such as health and education cities, golf course, sports facilities, parks, lakes, outlet mall, cultural uses, and commercial centers in neighborhoods, serving both DCK's own residents as well as Karachites. The city's land uses were carefully selected aiming to reinforce social and economic growth as the city develops in time.

The site of DCK is quite interesting in terms of terrain and topography, as the minimum elevation point is 348 feet above mean sea level (MSL) and the highest about 961 feet. With varied topography, it features a network of small and large streams and river valleys and is dominated by large areas of sporadic shrubs and grassland along with a few thickets of trees. The preservation of the topography of the natural landscape, including the relief, the streams, the ridge, and the flood areas, and their incorporation into the urban fabric became the fundamental design strategy. The master plan approach for the new city handled physical constraints and site issues while ensuring adaptation to environmental, social, and economic context.

With limited planning and a constant influx of migrants, Karachi, an urban center with a population of 19 million, ranks 135th among 140 international city centers in terms of the quality of its livable environment.[1] In contrast to developed western countries and cities with higher-quality standards, Karachi lacks upgraded infrastructures and utilities, needing improved public health, water supply, and energy.

The vision for DCK, therefore, was for it to become a prominent suburb of Karachi, functioning as a model residential development with the aim of providing its citizens a safe, comfortable, efficient, and beautiful lifestyle, while developing toward social, environmental, and economic sustainability. The incorporation of supra-regional facilities within a network of local activities would form the connective tissue of the development. The creation of lively communities within the city was essential for promoting DCK as a safe urban hub.

The aspirations and needs of both the future residents of DCK and visitors played a key role in the overall concept and planning of the city, as they determined the land use strategy. Unlike in a developed country, where certain parameters are taken for granted among the desired living standards, people of Pakistan in general and of Karachi in particular aspire for the following major parameters: quality infrastructure, safety and peace of mind, proximity, healthy and pleasant environment, efficiency, and sustainability.

Sustainability was one of the main aspirations for the new city, so the project brought together an interdisciplinary and international team of experts to propose innovative solutions for the context of Pakistan based on the Zofnass Guidelines framework. Sustainability consultants were involved right after the initial conceptual master plan proposal was approved and urban planning directions were decided. Although Zofnass Guidelines were not applied from the very beginning of the master plan design, as is advised, the goals set were innovative for Pakistan in terms of infrastructure technologies and management as well as sustainable strategies with respect to infrastructure and planning.

The Zofnass Guidelines were provided to project managers, planners, and engineers as a tool for applying as many sustainable strategies and actions as possible with regard to their respective fields of study. The intention also was to detect and carefully manage interactions and synergic relationships among proposals for diverse infrastructure systems, thus contributing to the city's integrating planning.

The engineers and consultants worked very closely with the project owner throughout the process, and their objectives for a model development aligned from the beginning. Regular workshops took place with the participation of representatives from both the project team and the DCK in which progress and pending issues were presented and discussed.

The justification of sustainable strategies for every infrastructure system was made through extensive analytical studies and the use of state-of-the-art tools like GIS and parametric computational models.

This chapter aims to present a sample of the sustainable approach of DCK planning, which was based on Zofnass planning guidelines.

11.2 | IMPLEMENTATION OF PLANNING GUIDELINES IN DCK

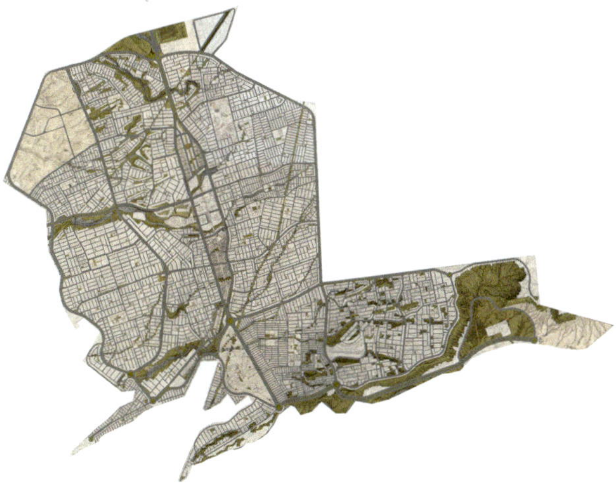

As planning was based on Zofnass Guidelines, integration was a central theme in the work flow and during design decisions. The master plan addressed the city as a system consisting of multiple social, economic, and environmental variables. Land use selection and placement, urban fabric morphology, accessibility options and infrastructure all aimed for high-quality living, economically driven development, state-of-the-art facilities, and environmental sustainability. Adaptation to natural systems formed the basis of the design. The plan was adapted to the landscape, particularly to the existing water systems and flows such as the nallas and water bodies. As the city aims to be mostly self-sustaining, solutions for all basic infrastructure systems were promoted during basic planning and land use allocation. For example, the decision for a pedestrian-friendly city was implemented through the creation of local mixed-use centers in each residential neighborhood, framed by an extensive network of pedestrian routes.

Each infrastructure category was resolved as an integrated system, as shown by the production of integrated technical reports for each category. Particular emphasis was given to locating and establishing synergies between systems. A platform collecting information at various scales was continuously developing in order to provide a database for the control and management of the project in the next phases of detailed planning, construction, and operation. The database contained basic metrics on land uses, areas, zoning, population, FAR, regulations, phasing strategy, etc. related to the master plan. Geographic information systems were used to initially upgrade the collected data to data with identified geographic location in order to facilitate the processes of automated mapping, visualizing, and interpreting the data related to the infrastructure systems at any scale. In combination with the master plan database, metrics on noise pollution, materiality, shading, CO_2 emissions, proximities to city attractors, energy consumption, water demand, waste generation, and landscaping were carefully analyzed and mapped in order to create a tool that allows for a continuous evaluation of infrastructure planning alternatives with regard to the sustainable objectives. Through this multiple-criteria-decision-making tool, an overall surveillance of infrastructure interrelations, synergies, and impacts was secured, contributing to holistic sustainable planning across all infrastructure systems and scales throughout the whole city.

Apart from the state-of-the-art tools used for the planning of the city, it should be noted that all infrastructure systems were decoded, analyzed, and studied through the Zofnass framework and more specifically through the lens of system levels. Infrastructure planning was approached first by decoding and second by evaluating the infrastructure systems and the interconnections, conflicts, and relationships of the entities. Their final design was based on Zofnass objectives. Each infrastructure system was approached by its respective system levels, and Zofnass Objectives and Guidelines were addressed on this basis.

This chapter does not aim to present all the sustainable strategies and the degree of sustainable performance of the new city. It focuses rather on showing how the Zofnass

Planning Guidelines framework was applied in the planning of DCK and to each of its infrastructure system analyses and sustainable proposals. A brief description of each infrastructure system together with its basic characteristics will be first provided in order to present what DCK infrastructure systems include and which are the major issues to be managed. Then the basic elements and points of all infrastructure systems in DCK will be summarized in a table organized by system levels. The table reflects the spectrum of entities, networks, decisions, and strategies that have to be holistically analyzed and managed in order to achieve efficiency for all infrastructure systems, sustain economy, provide prosperity for the new development, and protect the environment. A very important subject of the Zofnass framework is synergic relationships between infrastructures. Two infrastructure systems, Energy and Solid Waste, will be used as examples in the chapter to show diagrammatically a sample of synergies that appear in certain phases of their flowing function. Not all synergies will be shown, as the aim is to realize at which system level synergies appear. For example waste-to-energy facilities, included in Waste Treatment, are an alternative form for Energy Generation. The main part of the chapter comes next, in which indicative guidelines have been selected to show how DCK planners have come up with sustainable actions to fulfill the respective goals.

TRANSPORTATION

DCK road hierarchy

The sustainability strategy for DCK's transportation infrastructure system is based on trip reduction, promotion of nonmotorized transport, enhanced city connectivity, congestion prevention, and minimization of landscape and natural systems alteration. Clusters of basic amenities are equally spread across DCK within short walking distance for all the residents, and in combination with extensive networks of pedestrian and bicycle infrastructure, green and nonmotorized transportation is promoted. The use of private cars and the length of trips are also reduced as the city center (central business district) is estimated to be reachable from any location within approximately 15 minutes. A variety of means of public transportation are provided, including bus and tram routes, to reinforce the city's connectivity and restrain traffic congestion.

WATER

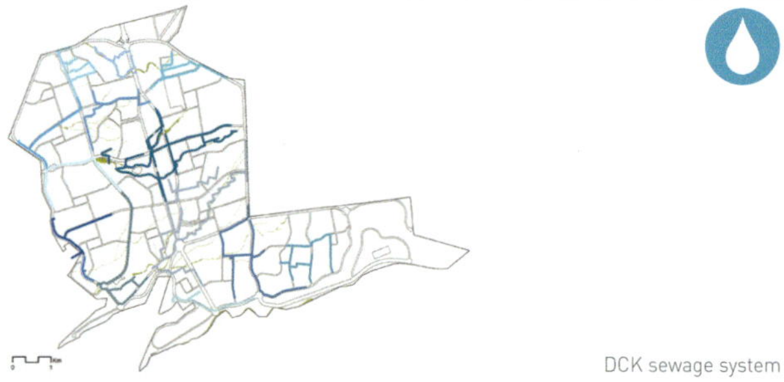

DCK sewage system

Water is a scarce source in the region of DCK; thus a careful plan for sustainable water management was essential. The plan briefly included the following:

a. Water supply from groundwater and recycled water. Whenever needed, supporting water supply will be provided by surface water transmitted from Keenjhar Lake (water from Indus River);
b. Wastewater treatment and rainwater management on site;
c. Bodies of water integrated in master plan as a valuable amenity for flood prevention and runoff treatment;
d. Gray water reuse strategies at house level;
e. Decentralized treatment plants at sector level with reuse for irrigation;
f. Water conservation strategies and methods for economy in use of potable water.

ENERGY

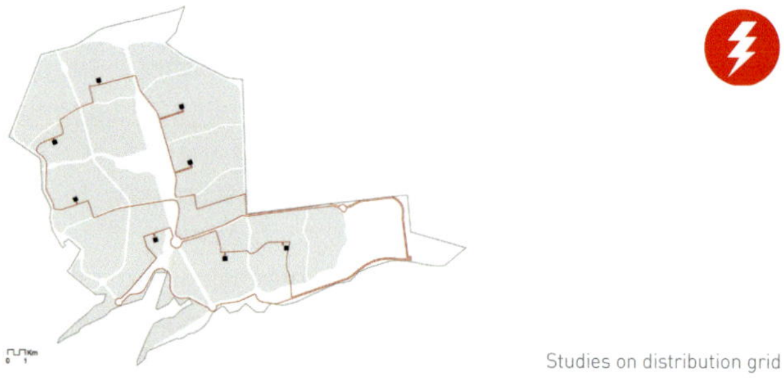

Studies on distribution grid

The main concepts for the city's energy system are:

a. Reduction of energy demand from buildings and infrastructure;
b. Use of renewable energy in public and private sectors;
c. Mitigating heat island effect by providing heat sinks (lakes, green areas etc.) at strategic locations, shading from trees/buildings, efficient planning to benefit from prevailing winds.

The main target for energy production is that at least 30% of the total energy demand will be produced by renewable sources and the remaining 70% by conventional sources. Solar energy at macro and micro scale (solar plants and decentralized home production respectively) as well as wind farms contribute in energy generation. The biomass and city waste plant produce an amount of energy to be added to the city's total energy production system.

SOLID WASTE

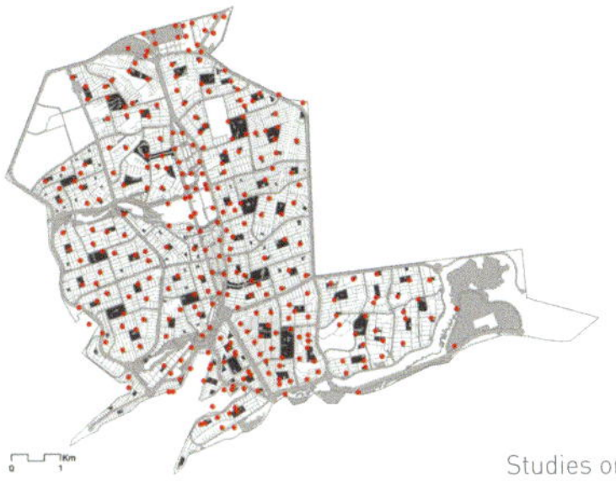

Studies on separate waste collection points

A carefully designed system of waste management will be applied in DCK, which will include strategies for homes and businesses to reduce the amount of waste they produce as well as facilities for recycling, composting, and recovering energy. All waste from DCK will be treated before disposal to a landfill.

Waste will proceed through a combined transfer station and MRF facility for material recovery and volume reduction. For further treatment, the option of anaerobic digestion and incineration with energy recovery have been examined and identified as the most suitable for DCK. As a result, a massive diversion of waste from landfill will be achieved. The percentage of diversion has been calculated at between 59% and 87% depending on the options of waste treatment selected during the final design of the system.

LANDSCAPE

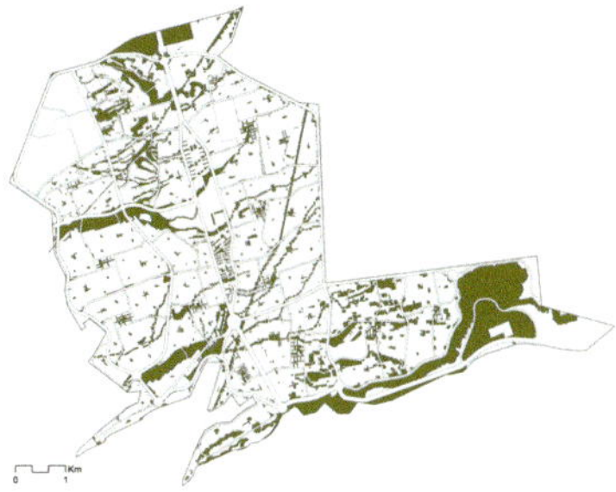

DCK landscape entities

The aim for DCK was to create multifunctional landscape areas offering various activities and services. The landscape entities were conceived as a whole system in which appropriate roles, functions, and services are appointed.

Overall, the extensive analysis of landscape infrastructure in DCK defined the services to be offered in the appropriate areas with reference to the natural and socioeconomic context as well as the macro-context conditions and pressures. The landscape therefore hosts ecological, productive, supporting, and recreational activities that include rainwater/stormwater management, air quality improvement, carbon sequestration, water infiltration, enhancement of local productivity, high-quality recreation, and accessibility to all residents.

INFORMATION

DCK information networks study

Information infrastructure in DCK has a dual significance:

a. It enhances social and economic sustainability through state-of-the-art information and communication networks;
b. It becomes a factor for optimizing sustainability performance in all other infrastructure systems, transforming DCK to a smart city.

With respect to the above roles, a comprehensive database of residents, employees, and frequent visitors is formulated for background checks and security purposes. In government administration, the implementation of information infrastructure (e.g., cloud computing) is expected to reduce operational costs by 30%.

FOOD

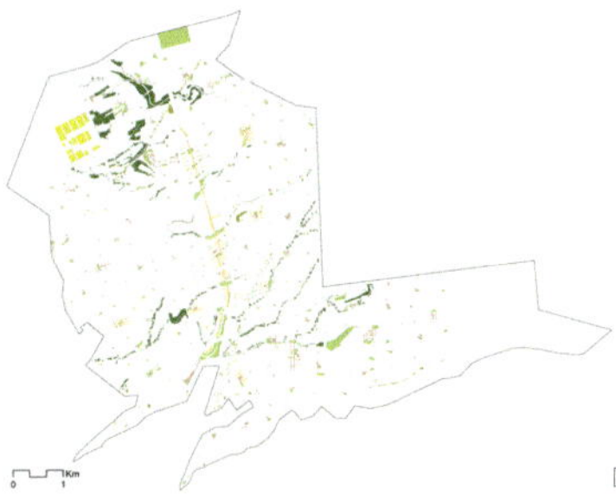

DCK areas with agricultural activity

Despite the fact that the soil in most areas of DCK is infertile and inadequate for agriculture, the food infrastructure system aims for maximum food security through sustainable agricultural development and balanced growth of the food sector. Special areas with farmhouses will be exclusively dedicated to food production. A number of landscape entities will function as food production entities, which will create, preserve, and enhance areas appropriate for agricultural activities and poultry and cattle raising, combined with low-impact manufacturing and supporting commercial uses.

In addition, several parks, playgrounds, piazzas, and pedestrian ways will also contribute to food production. Also, areas that are expected to be developed in later phases of the project will be exploited at an early stage for temporary food production.

INFRASTRUCTURE SYNOPSIS THROUGH SYSTEM LEVELS

LEVEL 1 DEMAND ANALYSIS	**SERVICES DEMAND** Provisioning/regulating/cultural: - Enhancement of local food production - Water filtration and purification - Rainwater management (preservation of natural drainage nallas) - Erosion regulation at ridges - Habitat protection (several indigenous plants and animal species, some endangered) - Local climate regulation - Regulation of atmospheric composition - Noise and dust control - Recreation and ecotourism - Aesthetic values	**TRIP GENERATION** Trip generators and attractors: - Community centers C5 - CBD - Gateway - Subsectors - Higher education - Parks and recreation - Energy park - Agriculture	**WATER CONSUMPTION** - Total water demand: 43,945,402 gpd - Avg. water demand: 0.35 g/sqft - Highest water demand: 17,254,708 gpd - Residential plots (39%) - Highest water demand per sqft observed in: mixed use plots (1.4 gpd per sqft)
LEVEL 2 STRATEGIC DECISIONS	**ACTIVITIES PROGRAM** - Natural: preservation of existing vegetated areas, streams, and ridges - Productive: seasonal agriculture, communal gardens, allotments, plant nurseries - Recreational/cultural: lakes, parks, playgrounds, piazzas, pedestrian ways, golf course, sports fields, theme parks, botanical garden, birds aviary - Supporting: protective zones for existing gas and oil pipelines and other infrastructure systems	**TRANSPORTATION MODES** - Walking: pedestrian paths - Cycling: cycling paths - Bus: 7 bus lines - Tram: 3 tram lines - Private vehicles - Private VS public transport: 70%/30%	**WATER SUPPLY** Two sources of potable water: - Groundwater (on-site wells) - Installation of a 124 km water pipeline, conveying water from a lake 122 km away from Karachi and a new water treatment plant. Nonpotable water supply: - STP's recycled water - Rainwater
LEVEL 3 NETWORKS	**PATTERNS** Definition of all entities' principal function based on their location, program, spatial pattern, and size and spatial configuration. In general: - Productive program entities support the functions of source and sink - Natural program entities support the functions of conduit, habitat, sink, or barrier - Recreational program entities support the functions of filter, sink, and conduit - Supporting program entities support the function of barrier	**TRANSPORTATION NODES** - Checkpoints - Transit stations - Parking areas - Depots - Logistics center	**WATER TREATMENT** - 15 decentralized sewage treatment plants are planned to treat 31 MGD - Projected sewage flow is 80% of average daily water demand - Projected sewage volumes are estimated for residential, commercial, mixed use, and amenities land uses
LEVEL 4 NODES	**MAINTENANCE CYCLE** Adaptation of effective methods for regulation of maintenance needs: - Drip irrigation adapted to plant needs - Use of recycled wastewater for irrigation - Integration of rainwater collection for irrigation - Seasonal trimming for proper plant growth - Recycling of green manure and trimming residues for compost - Revegetation of spaces - Supporting of young plants till established - Hard surfaces and structures maintenance	**TRANSPORTATION NETWORKS** - Cycle network - Pedestrian network - Road hierarchy analysis - Public network routes (bus and tram) - Intersections - Crossings between networks and other corridors - Drainage - Services reservation corridors - Traveler information system and VMS	**WATER NETWORK** - 12 overhead service tanks - 3 overhead service tanks and pumping stations - Supply pipes of 50.5 km total length - Supply network total length: 43 km - Wastewater total network: 72 km

ENERGY CONSUMPTION

- Projected residential energy use: 82.6 KWh/m^2 /year
- Projected nonresidential energy use: 189.2 KWh/m^2/year
- Projected buildings electricity consumption: 3073 GWh/year
- Electricity consumption in residential buildings is estimated about 52% out of total buildings' consumption

ENERGY SUPPLY

Energy sources:
- 5.9% solar energy
- 3.9% wind energy
- 8.3% biomass
- 81.9% residual fuel oil

ENERGY GENERATION

Mixed sources system of electricity generation, with specific areas are dedicated to the following plants:
- Conventional energy generated by a residual fuel oil plant (350 MW capacity)
- Biomass and waste plant (40 MW capacity)
- Wind farm with 17 windmills (50 MW capacity)
- Photovoltaic park (20 MW capacity)
- Solar thermal park (30 MW capacity)

ENERGY DISTRIBUTION

- Bus bar power house switchyard
- 7 grid stations (gas-insulated stations)
- Step-down distribution transformers
- Transmission lines
- Distribution lines
- Meters
- Separate meters for buildings with hybrid solar
- Total load for transmission: 500 MW
- Reduced transmission losses due to distributed generation

WASTE GENERATION

- Residential areas: 487.72 t/day
- Commercial areas: 149.95 t/day
- Amenities areas: 104.35 t/day
- MSW: 741.7 t/day

DIVERSION MODEL

- Waste mix: food and green 50%, paper and cardboard 15%, plastics 12%, glass 5%, aluminum and metal 3%, others: 15%
- Set a goal of 100% waste treatment
- Create new controlled landfill for DHA
- Anaerobic digestion plant
- Transfer station
- Material recovery facility
- Trash bins and mixed recycle bins in residential areas

TREATMENT AND FINAL DISPOSAL

- Controlled deposit with landfill gas flaring
- Location of landfill within a radius of 10–40 km from DCK
- Combined transfer station with material recovery facility
- MRF services: control of origin/weight, sorting of recyclables, compaction, biological treatment, temporary storage limit of the city toward the landfill location

COLLECTION AND TRANSFER

- Transfer station combined with clean and dirty MRF and anaerobic digestion plant
- Optimization of collection routes by minimizing length and by combining routes
- Selected vehicles capacity: 40 receptacles of 1,100 lt
- Total of 23 vehicles for waste collection and 6 vehicles for separate waste collection
- Determining 3709 points of waste collection (trash) and 318 points of separate waste collection (recyclables)

INFORMATION DEMAND

- Demand on enhancing the infrastructure systems of: water, energy, waste, transportation
- Demand on serving as infrastructure itself for economic development
- Demand on serving as infrastructure itself for improving the quality and reach of services to the public
- Demand on serving as infrastructure itself for improving governance

COLLECTION AND COMMUNICATION MODES

- Internet (fixed and mobile broadband)
- Mobile cellphones
- Fixed telephones
- Digital security system
- Infrastructure systems
- Traffic management/transportation

INFORMATION NODES

- Central control center (CCC)/IT museum
- ICT units / central offices (COs)
- Information pavilion spaces (IP)
- Advance traveler information system (ATIS)
- Security control centers

INFORMATION NETWORKS

- ICT optical fiber network
- Network rings
- Network manholes
- Service corridors

DEMAND

- Promotion of local and seasonal products from the on site farmhouses and the adjacent foodshed
- Reduction of food waste generation through the promotion of on site food waste composting

SUPPLY AND PRODUCTION

- Farmhouses exclusively for food production
- Seasonal productive landscape areas relative to streams
- Communal gardens and allotments
- Integration of productive activity within the program of other landscape entities (such as open spaces, parks, playgrounds, public squares, pedestrian paths) will offer part of their areas for food production
- Residential plots of large areas, education city, and amenities open spaces will also contribute to food production

TRADE

- DCK commercial plots are placed in community centers, CBD, and City Gateway complex
- Prioritization in efficiency and accessibility: all markets are found within 10 minutes walk from every residence

PROCESSING AND DISTRIBUTION

- Provision for processing and storage facilities at the logistics center of the City Gateway
- Strategic location of logistics center serves to avoid heavy goods vehicles (HGVs) circulating within the city
- Control check to all HGVs at the city gate
- Distribution from logistics center to city centers and commercial areas will be performed by light goods vehicles (LGVs)

WASTE SYSTEM SYNERGIES AND RELEVANT GUIDELINES

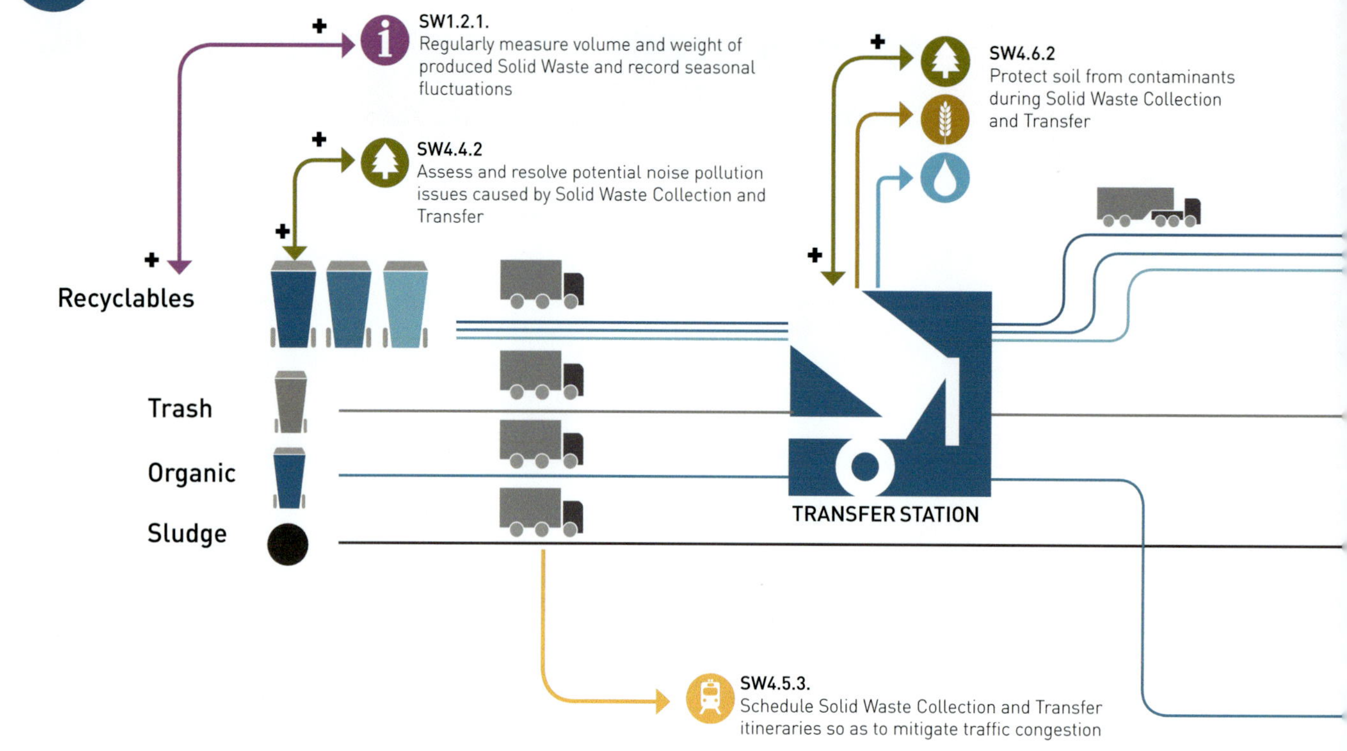

ENERGY SYSTEM SYNERGIES AND RELEVANT GUIDELINES

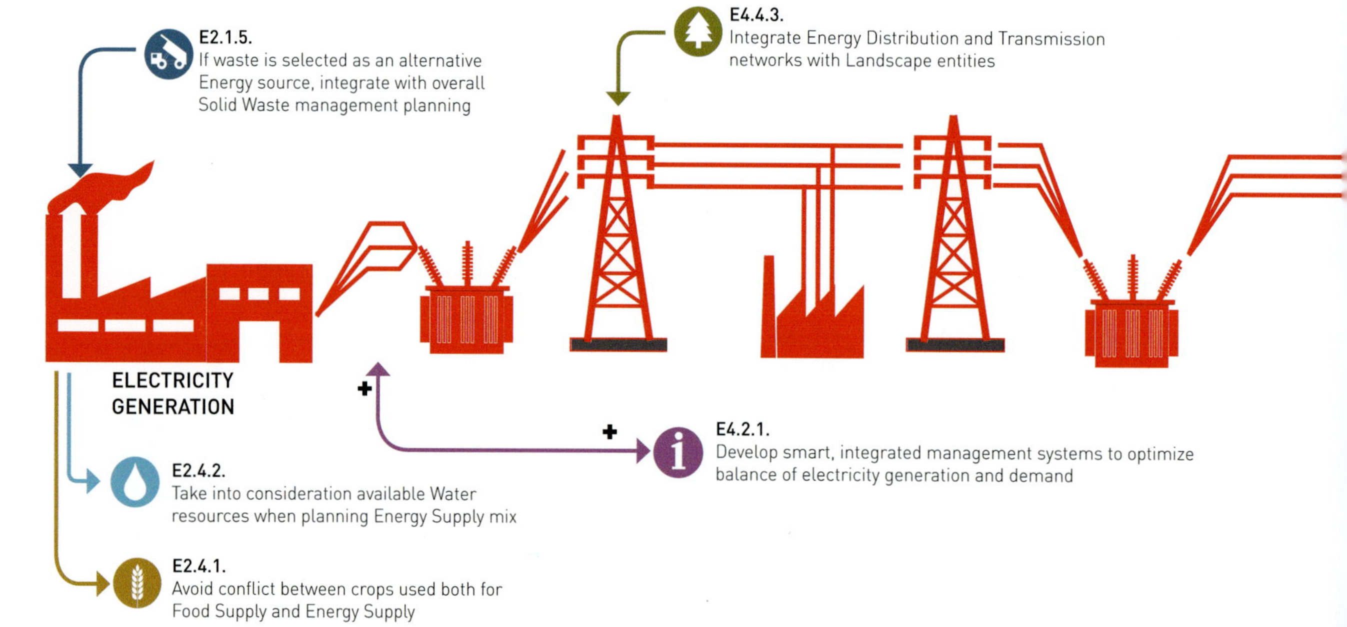

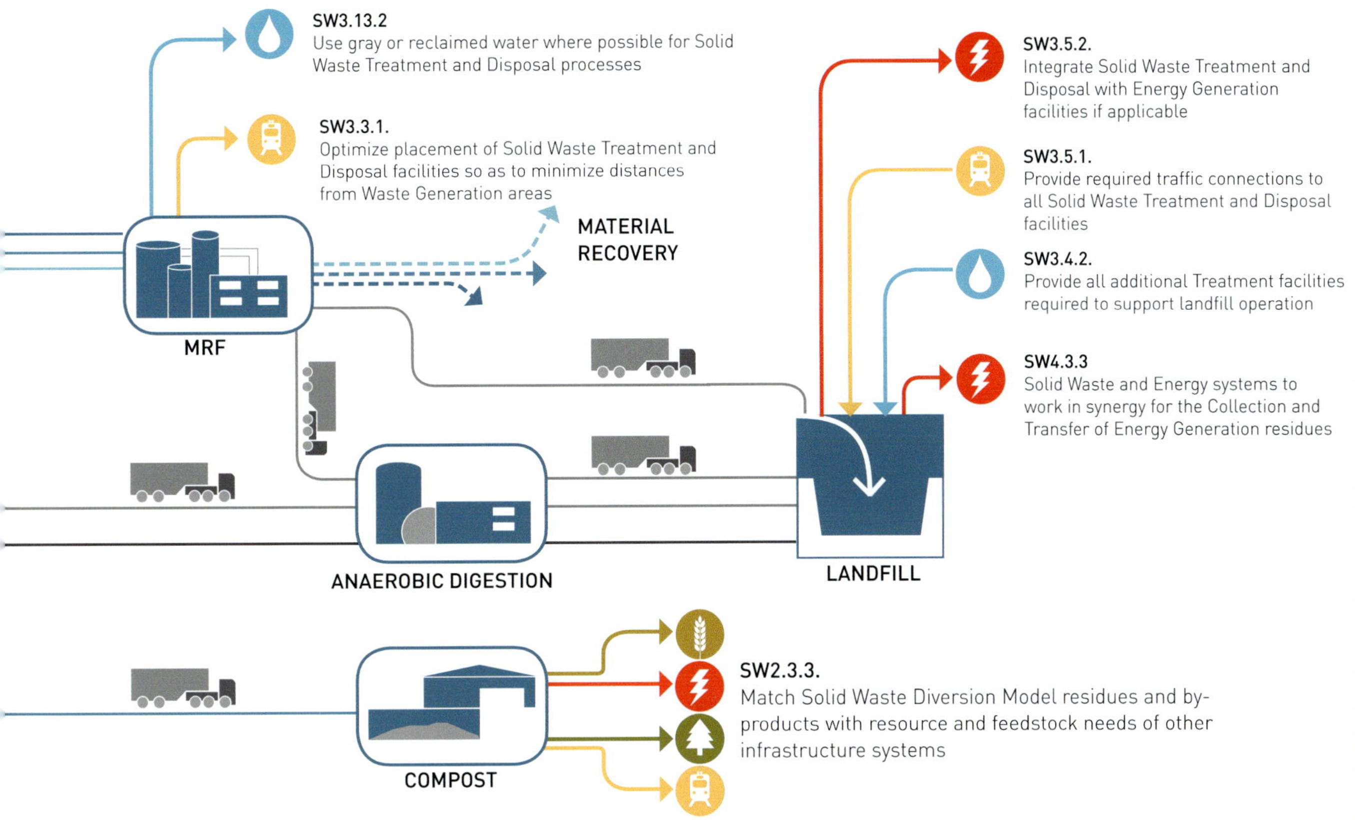
SW3.13.2
Use gray or reclaimed water where possible for Solid Waste Treatment and Disposal processes
SW3.3.1.
Optimize placement of Solid Waste Treatment and Disposal facilities so as to minimize distances from Waste Generation areas
MATERIAL RECOVERY
MRF
ANAEROBIC DIGESTION
LANDFILL
SW3.5.2.
Integrate Solid Waste Treatment and Disposal with Energy Generation facilities if applicable
SW3.5.1.
Provide required traffic connections to all Solid Waste Treatment and Disposal facilities
SW3.4.2.
Provide all additional Treatment facilities required to support landfill operation
SW4.3.3
Solid Waste and Energy systems to work in synergy for the Collection and Transfer of Energy Generation residues
COMPOST
SW2.3.3.
Match Solid Waste Diversion Model residues and by-products with resource and feedstock needs of other infrastructure systems

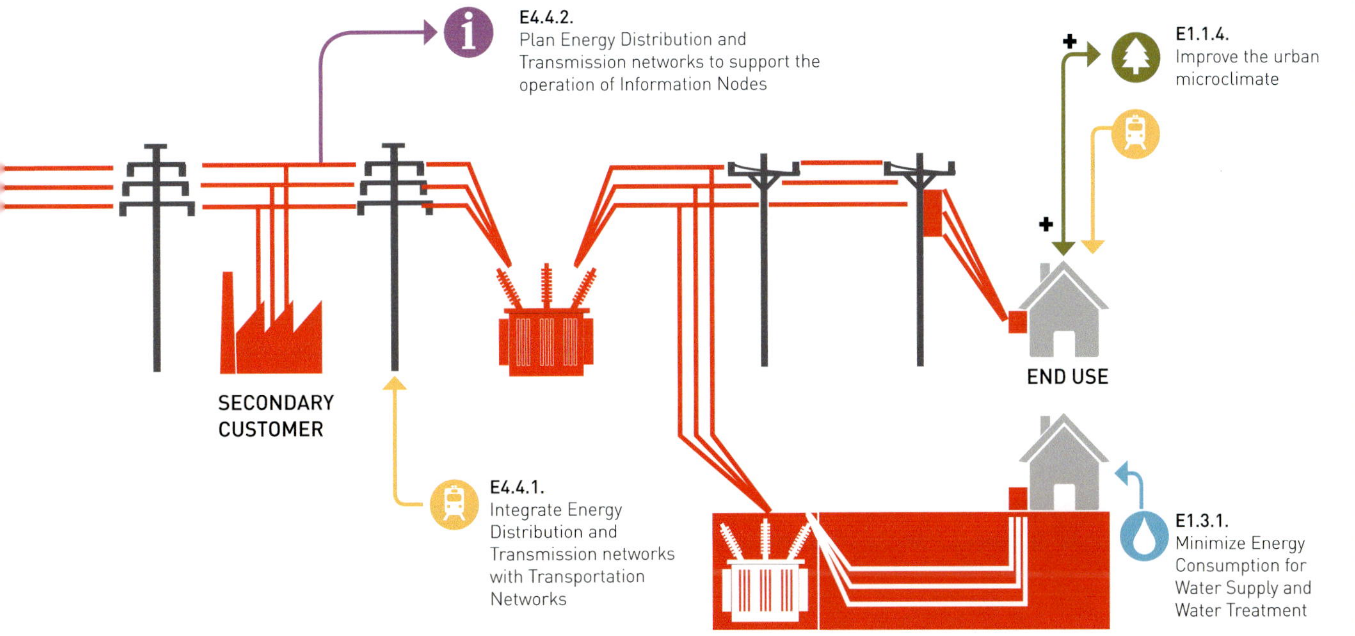
E4.4.2.
Plan Energy Distribution and Transmission networks to support the operation of Information Nodes
E1.1.4.
Improve the urban microclimate
SECONDARY CUSTOMER
END USE
E4.4.1.
Integrate Energy Distribution and Transmission networks with Transportation Networks
E1.3.1.
Minimize Energy Consumption for Water Supply and Water Treatment

SUSTAINABLE ACTIONS ACCORDING TO PLANNING GUIDELINES

OBJECTIVE

L3.9.
Preserve and enhance biodiversity in Landscape corridors and patches

GUIDELINE

L3.8.1
Preserve the natural topography

SYNERGIES

▲ TO
+ WITH
▼ FROM

SUSTAINABLE ACTION

Earthwork was minimized and extensive cut-and-fill was avoided. Construction on ridges was carefully managed as 64% of the ridges were integrated with the landscape.

The main features of the DCK site's natural topography are steep ridges and slopes. Extensive topographical analysis was done in order to provide the basis for deciding on the development of the city's networks and constructions with the least intervention in these landscape features. The majority of the ridges were conserved and integrated with the landscape system, allowing large areas to contribute to the services and functions of the landscape infrastructure system.

OBJECTIVE

L3.3.
Assign and establish the appropriate ecological functions to Landscape system entities according to identified services, location, etc.

GUIDELINE

L3.3.1.
Support/establish the functions of conduit, sink, and barrier through topography/microtopography of Landscape entities

SYNERGIES

▲ TO
+ WITH
▼ FROM

SUSTAINABLE ACTION

Each entity's principal function is mapped and justified with respect to the context.

Another characteristic of the natural landscape in the DCK site is the existing natural drains, called nallas. These natural corridors are "linear landscape elements that provide for movement between habitat patches" and play "a key role in connectivity."[2] Together with all landscape entities, they were appropriately recorded as conduits which would be supported by hydrogeological works. With the aim of preserving the watershed, hydrogeological simulations were included in the infrastructure studies in order to define the most suitable section for the nalla areas.

OBJECTIVE

T4.8.
Maintain the integrity of natural-world elements and systems

GUIDELINE

T4.8.5.
Develop permeable paving and vegetation configurations for Transportation Networks to provide natural drainage

SYNERGIES

▲ TO
+ WITH
▼ FROM

W2.3.6
L3.4.2

SUSTAINABLE ACTION

Permeable paving materials such as porous asphalt and concrete will be used in pedestrian and bicycle routes, cul-de-sacs, and parking areas.

52% of road network will be paved with permeable materials

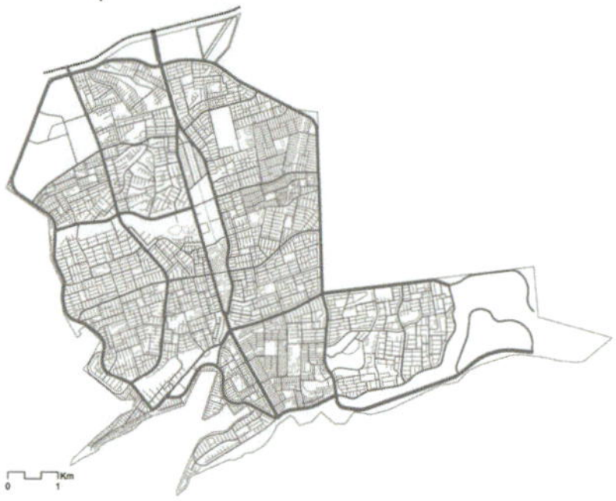

DCK transportation network

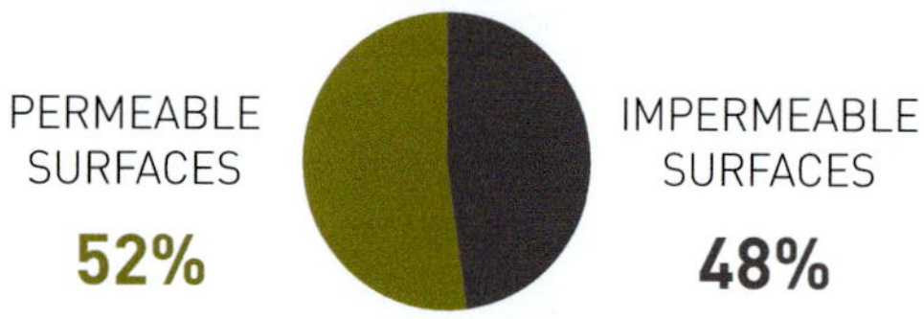

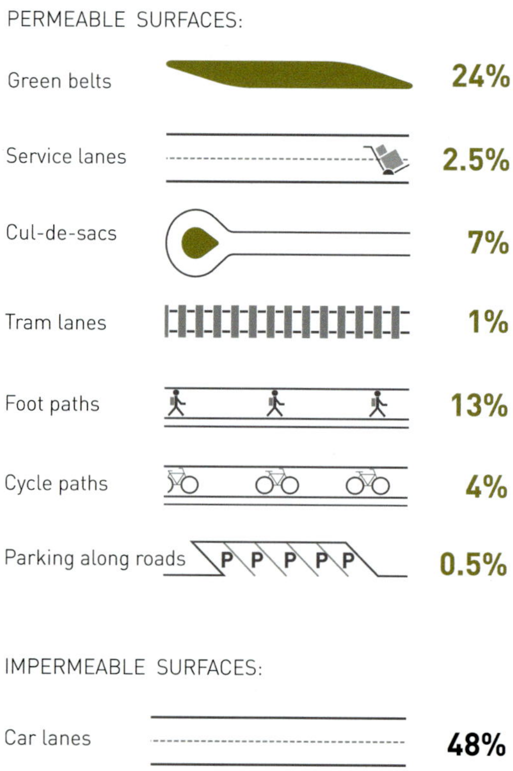

Transportation network entities constitute a variety of uses other than car lanes. In DCK it is proposed that all areas except for the car lanes be paved with permeable materials in order to protect the natural cycle of water in the city. It is estimated that the proposed permeable areas, including greenbelts, service lanes, cul-de-sacs, tram lanes, footpaths, cycle paths, and parking areas, possess more than 50% of the total area of transportation infrastructure networks. The planning of transportation infrastructure is in strong synergy both with landscape and water as it contributes to these systems' sustainable performance.

OBJECTIVE

T4.7.
Promote community well-being

GUIDELINE

T4.7.3.
Assess and resolve noise pollution issues caused by Transportation Networks

SYNERGIES

▲ TO
+ WITH
▼ FROM

+ L1.7.2

SUSTAINABLE ACTION

The percentage of residents living on streets with speed limit greater than 50 km/h is minimized.

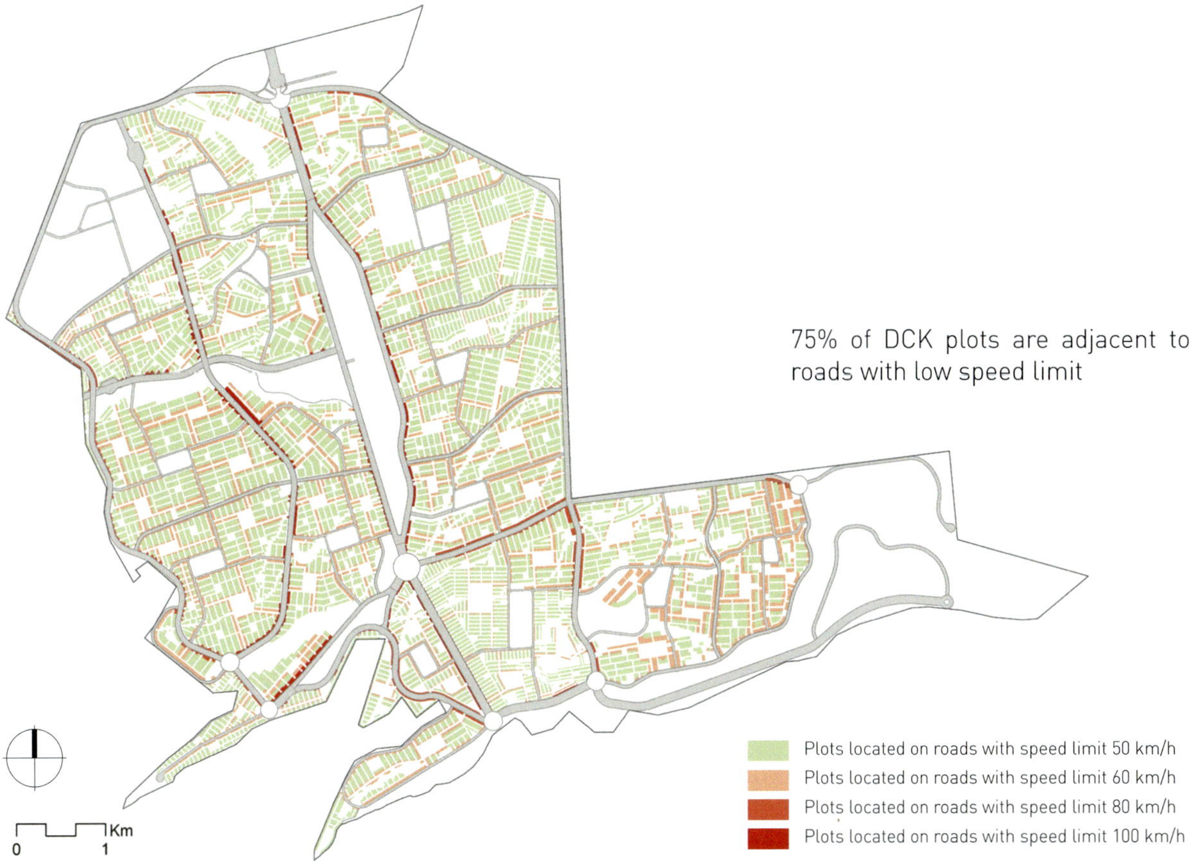

Extensive transportation networks and high-speed roads can cause discomfort to residents due to the increased noise pollution. Although the city's transportation infrastructure includes several roads of high circulation, an effort was made to secure the residents' well-being by locating most of the houses on roads with the lowest speed limits. Specifically, 75% of the city's plots are situated on roads with low speed limits where noise levels are rather low and noise pollution is not an issue.

OBJECTIVE

T4.1.
Develop Transportation Networks for optimized Transportation system performance

GUIDELINE

T4.1.2.
Provide adequate coverage and length of routes for public transport modes implemented

SYNERGIES

▲ TO
✚ WITH
▼ FROM

SUSTAINABLE ACTION

Bus and tram stops are located within walking distance of the majority of the houses.

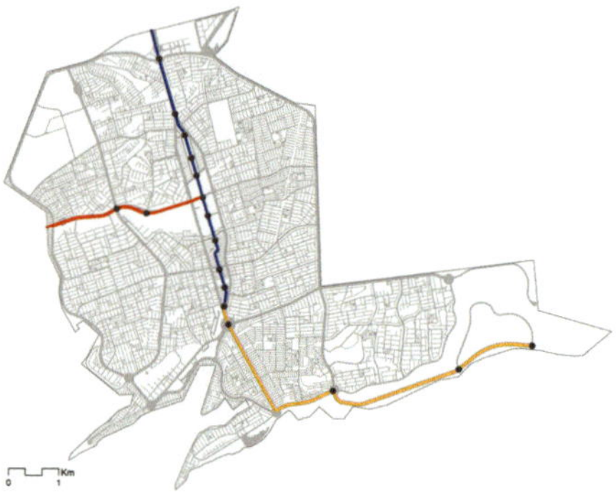

Tram routes

Bus routes

Tram

Shuttle bus

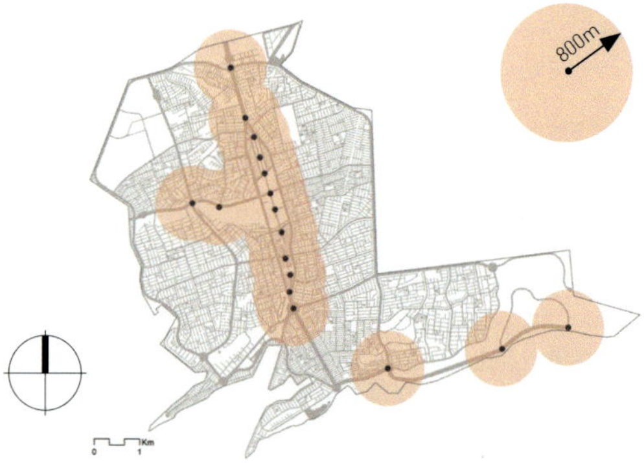

Areas highlighted are within a radius of 800 meters from tram stops

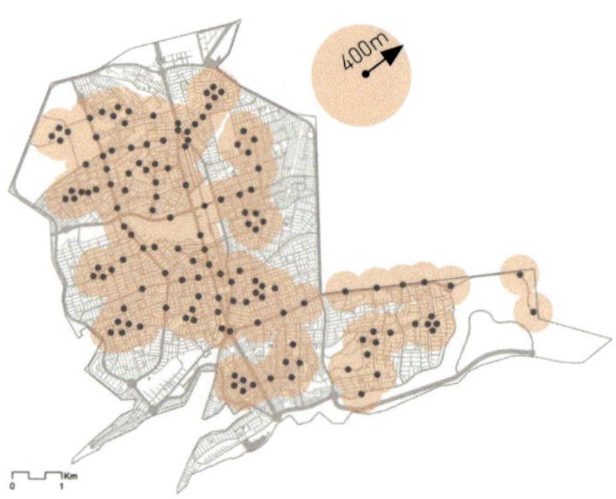

Areas highlighted are within a radius of 400 meters from bus stops

DCK's transportation network aims to provide efficient services to all residents by covering the whole city with public transportation. Bus routes extend to all neighborhoods, and they all connect to the central tram line which connects the main city attractors like the central business district, outlet mall, amusements parks, and the health and education city. The majority of the population (72%) can walk to either a tram stop or a bus stop within 5 to 10 minutes. Appropriate analysis took place in order to properly arrange bus and tram stops around the city.

OBJECTIVE	GUIDELINE	SYNERGIES
T4.1. Develop Transportation Networks for optimized Transportation system performance	T4.1.1. Provide adequate coverage and length of Networks for all Transportation modes implemented	▲ TO + WITH ▼ FROM

SUSTAINABLE ACTION

Extensive pedestrian and bicycle routes have been planned.

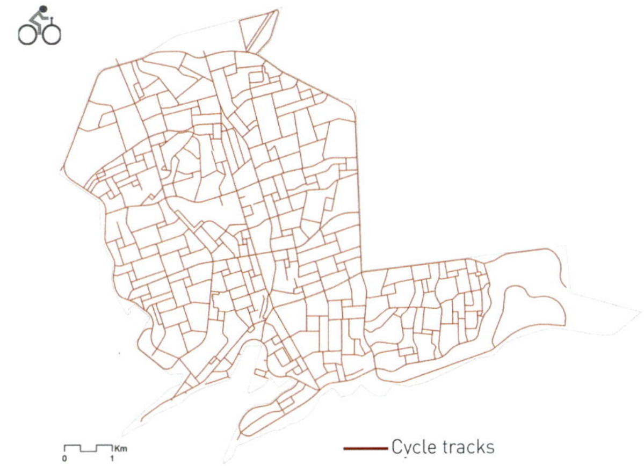

Cycle tracks

For 98.5% of DCK's total area, a bicycle track is found within a maximum of 400 meters

Public spaces, squares, pedestrian routes in centers

Pedestrian network

The pedestrian-friendly city concept was implemented through the creation of multiple mixed-use centers in neighborhoods, framed by an extensive network of pedestrian routes. This strategy contributes to GHG emissions reduction, mitigation of traffic congestion, and promotion of healthy living. As for cycling, although it is usually neglected when road corridors are being planned, bicycle riding in DCK is being projected as one of the main transport modes. The bicycle network is highly accessible, allowing for 98.5% of DCK areas to lie within a distance of 400 meters from a bicycle track.

OBJECTIVE

W2.1.
Secure city Water Supply in the long term

GUIDELINE

W2.1.4.
Promote use of Water resources available within project boundaries

SYNERGIES

▲ TO
✚ WITH
▼ FROM

✚ L1.5.2

SUSTAINABLE ACTION

On-site resources are exploited by using groundwater.

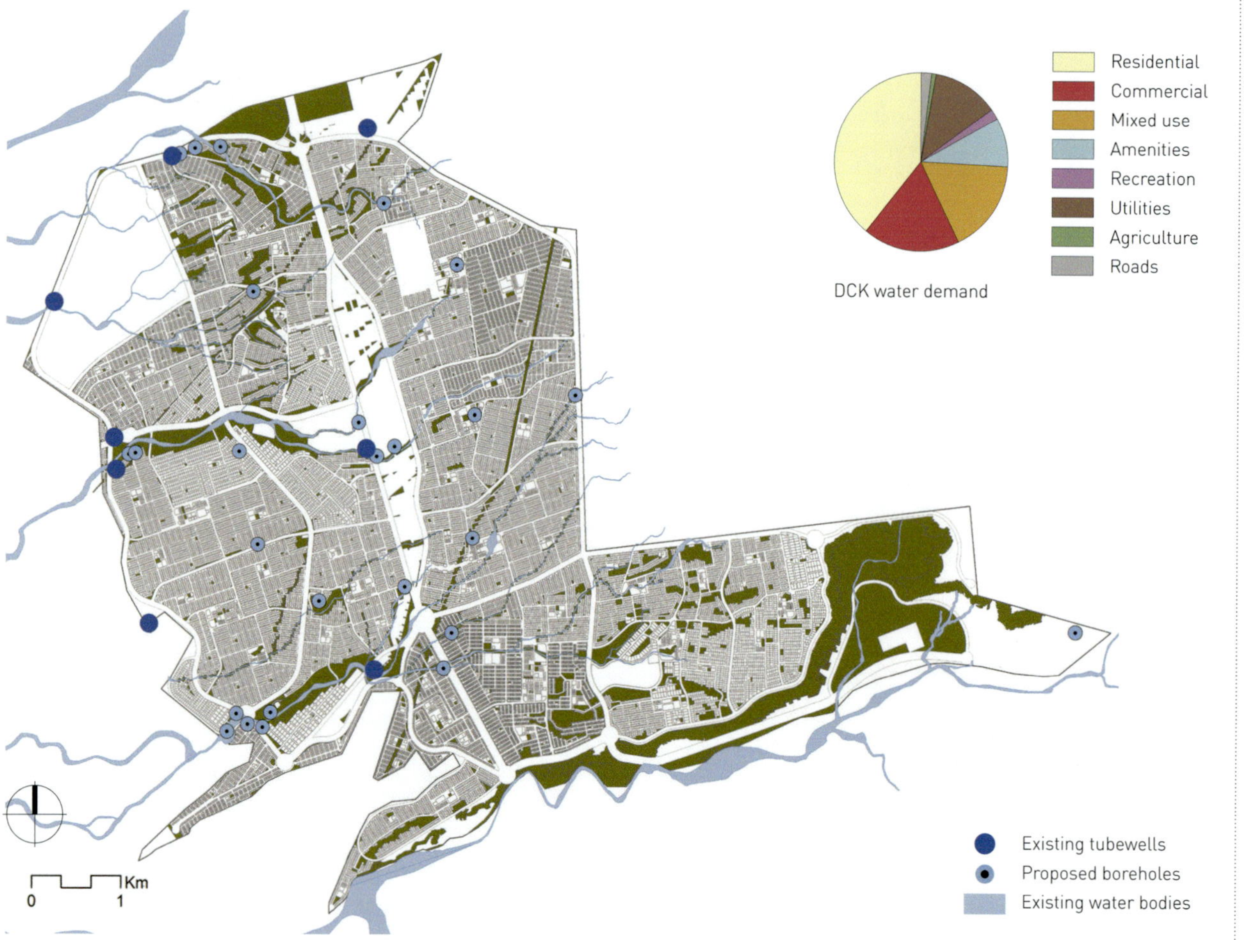

Water supply is a major issue for Pakistan's cities due to the country's water scarcity. Karachi is planning to supply itself with water transmitted from Keenjhar Lake. Although DCK could also benefit from this source, its planners decided to cover the city's water needs by using groundwater from the site, in order to be more autonomous and secure the long-term water supply. Several tube wells will be used and work continues with proposed boreholes to support the supply system. Appropriate analysis provided by landscape infrastructure will contribute in pursuing this action.

OBJECTIVE

W2.1.
Secure city Energy Supply in the long term

GUIDELINE

E2.1.2.
Promote use of multiple and diverse Energy sources to cover city needs in the long term

SYNERGIES

▲ TO
+ WITH
▼ FROM

SUSTAINABLE ACTION

A varied mix of energy sources are exploited, introducing special land uses such as energy park, biomass energy production area, and wind farms.

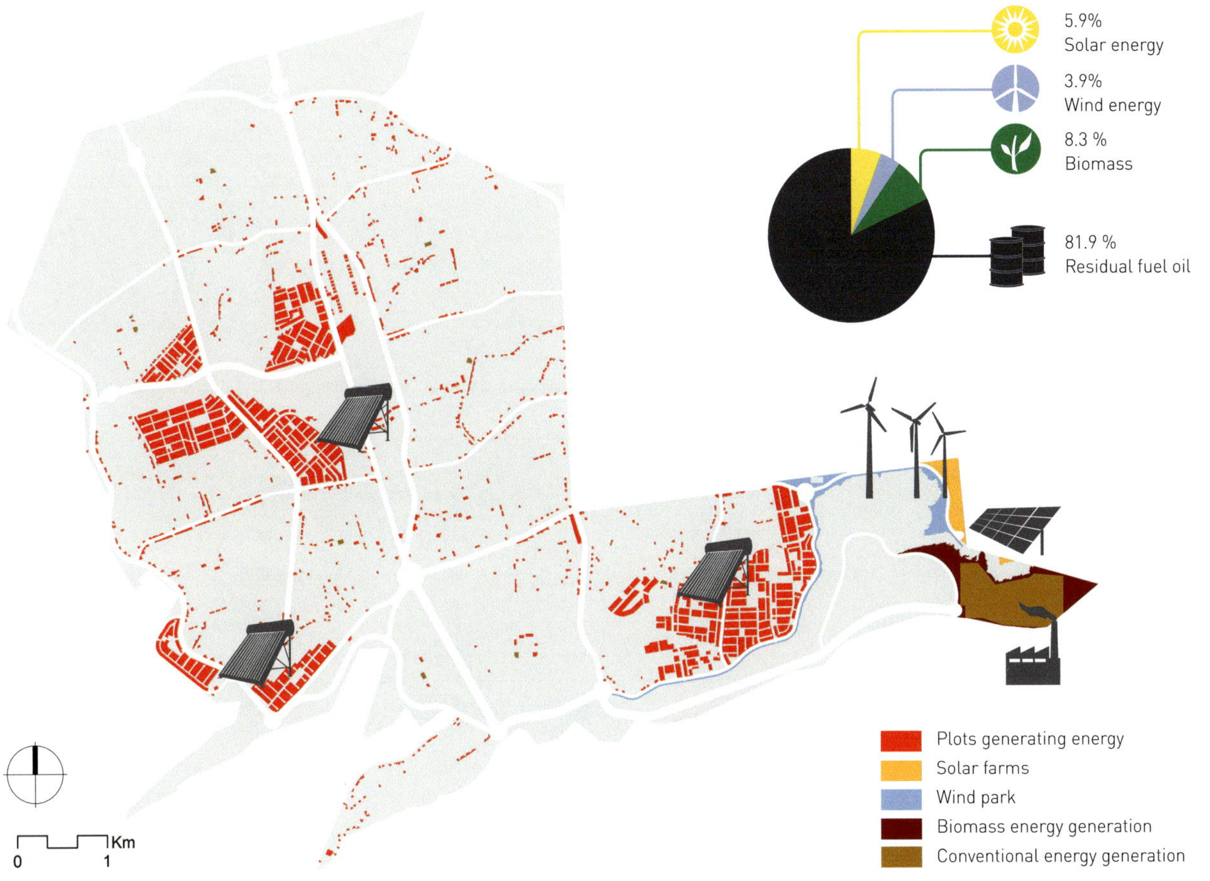

The DCK energy strategy is a holistic plan for sustainable energy generation at both macro (solar and wind parks) and micro (small house hybrid systems) levels. This energy strategy is in line with the latest energy approach, both internationally and also in Pakistan. Aiming to secure the city's energy supply in the long term, DCK promotes the use of multiple and diverse energy sources. The energy supply for electricity generation combines diverse energy sources. Apart from residual fuel oil, other sources of supply include solar energy, wind energy, and biomass energy production.

OBJECTIVE

SW4.2. Improve Solid Waste Collection and Transfer efficiency

GUIDELINE

SW4.2.3.
Provide adequate type, location, and capacity of Solid Waste Collection points based on land use, population data, and population distribution

SYNERGIES

▲ TO
✚ WITH
▼ FROM

SUSTAINABLE ACTION

All treatment and disposal entities (waste bins, recycle points, etc.) are recorded with all appropriate data to provide a database for proper management.

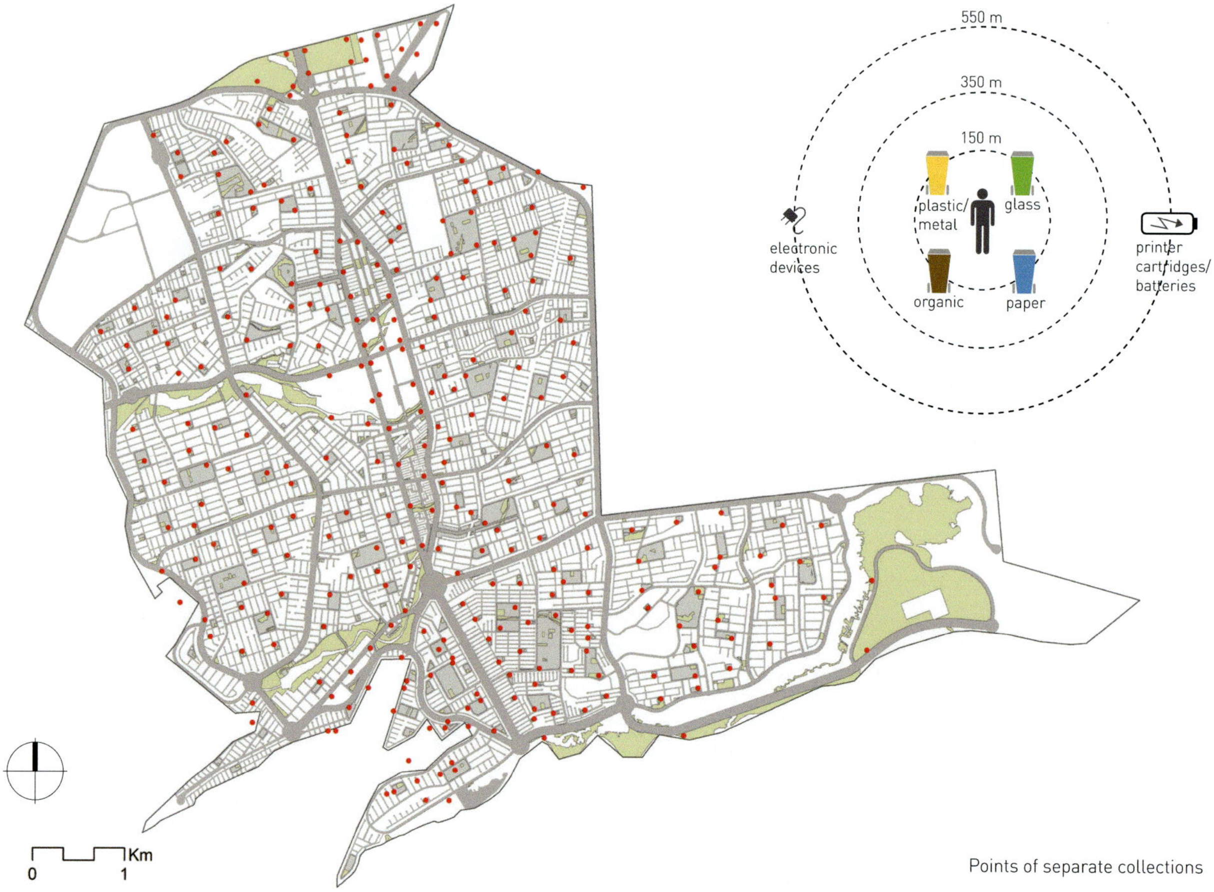

Points of separate collections

An extensive analysis of projected waste production and waste mix per neighborhood has been carried out based on population statistics and projections. The goal is to promote efficient waste collection and transfer by determining appropriate locations and characteristics of waste receptacles. In order to facilitate recycling for residents, recycling is organized through neighborhood centers, resulting in receptacles being located within 150 meters of each residential plot.

OBJECTIVE

F2.4.
Promote urban agriculture

GUIDELINE

F2.4.2.
Provide space for communal gardens or allotments

SYNERGIES

▲ TO
+ WITH
▼ FROM

L2.2.1

SUSTAINABLE ACTION

Public and private open spaces are exploited for various land uses in order to maximize productive and agricultural land within the site.

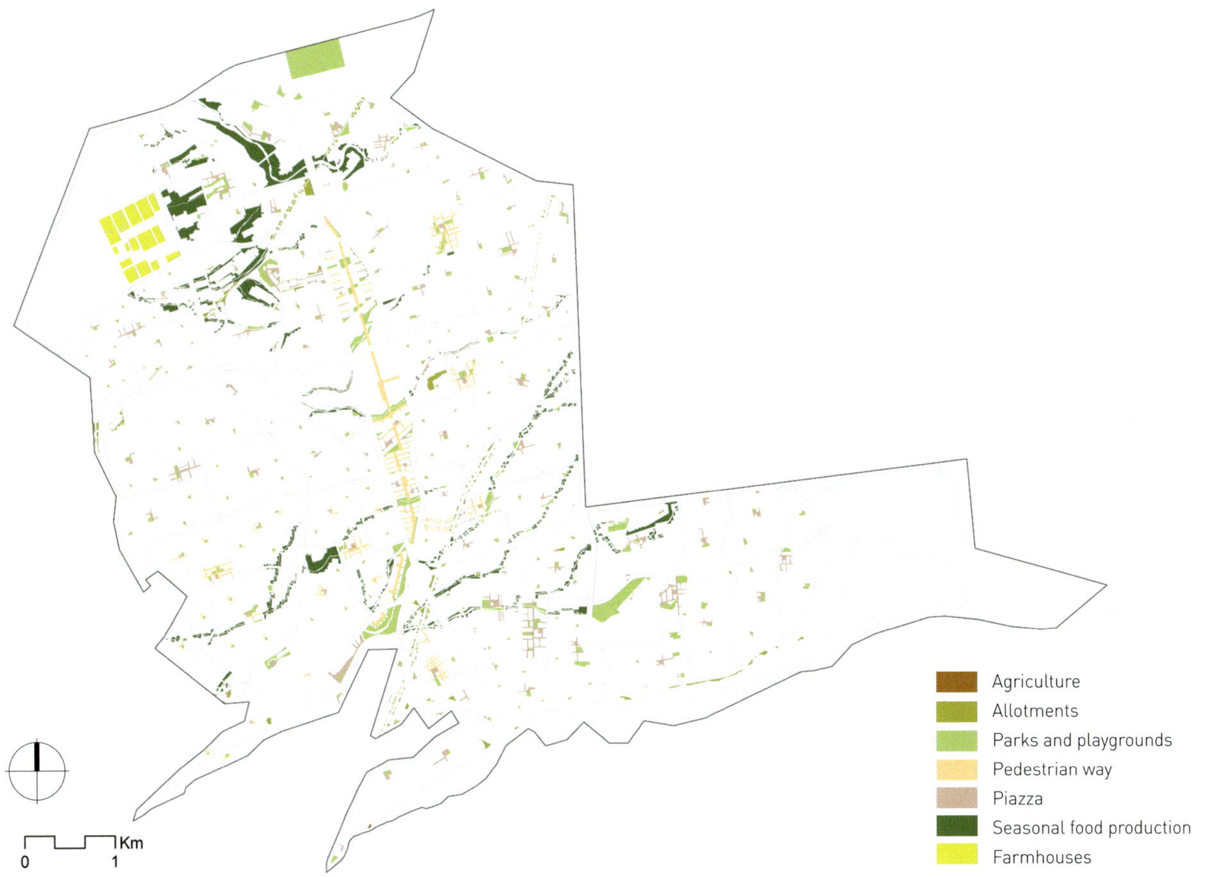

Apart from selected plots that will be dedicated to farmhouses, another strategy for DCK's food production is strongly linked with landscape infrastructure. All landscape entities are classified according to their potential for productivity. Selected public areas, such as landscape related to canals, communal gardens, unused allotments, public piazzas, and parks, will be partly or totally used for food production. Also, large private plots will dedicate part of their open space (about 15%) to productive activities. It is estimated that about 550 acres will be used for urban agriculture.

OBJECTIVE

I3.2.
Place and distribute Nodes for optimized Information system performance

GUIDELINE

I3.2.1.
Place Information Nodes facilities in correlation to data Collection and Communication needs

SYNERGIES

▲ TO
+ WITH
▼ FROM

SUSTAINABLE ACTION

Information pavilion spaces are strategically located in community centers throughout the city (where they serve as modern libraries, information centers and outlets for IT services, exhibition spaces, etc.).

The information infrastructure of DCK includes primary control centers that will cater to all of DCK and will control and monitor the city gates, CCTVs in the city, observation towers, the peripheral boundary, mobile security vehicles, emergency kiosks, police patrolling, etc. It should have direct links to all the individual buildings' central security control. Three primary control centers located in the Gateway and City Center will be linked to all secondary control centers. In order to comply with long-term city development, DCK's information infrastructure will be structured in phases.

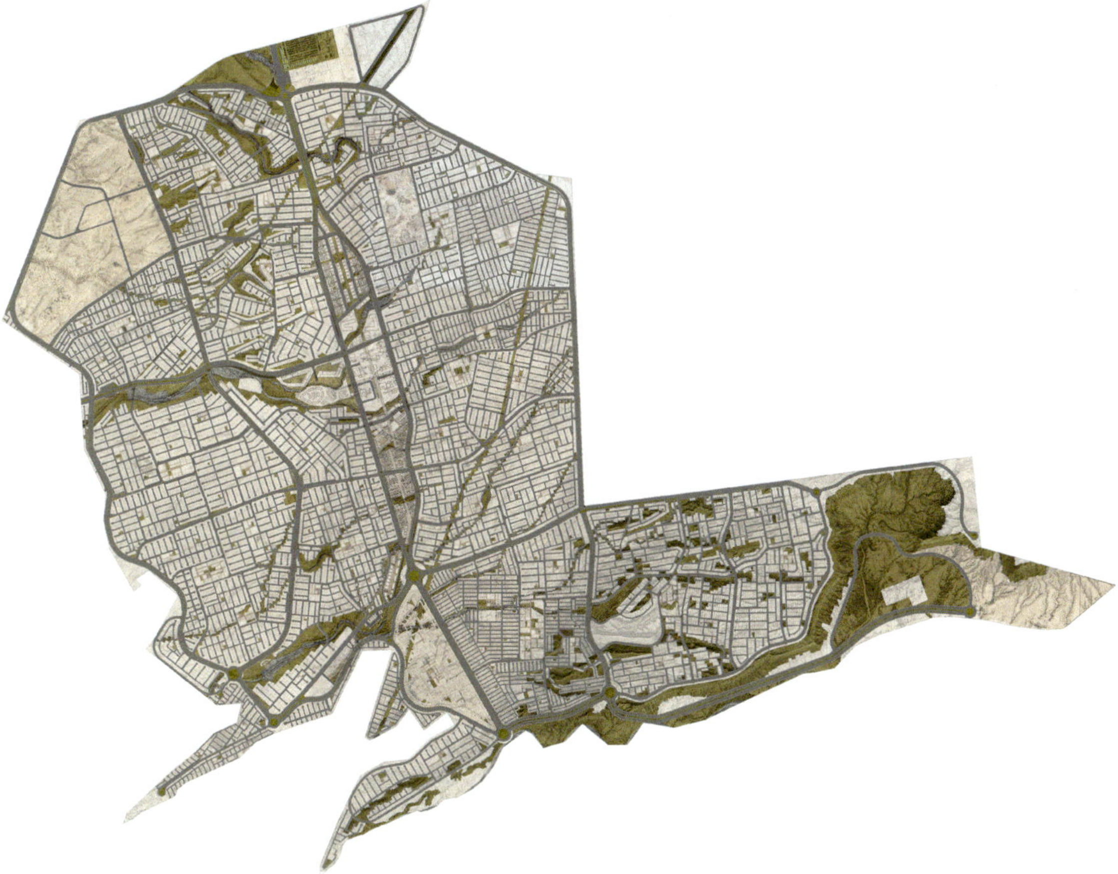

DCK final master plan

DCK site image showing the topography

DCK transportation works in progress

NOTES

1. Economist Intelligent Unit, "A Summary of the Livability Ranking and Overview," August 2012, www.eiu.com.
2. Anthony B. Anderson and Clinton N. Jenkins, *Applying Nature's Design: Corridors as a Strategy for Biodiversity Conservation* (New York: Columbia University Press, 2006), 3.

IMAGE CREDITS

Pp. 288-289: (Bottom) Adapted from "Basic Structure of the Electric System," United States Department of Energy, accessed on February 22, 2012, http://www.ferc.gov/industries/electric/indus-act/reliability/blackout/ch1-3.pdf. **P. 301:** Google Earth, imagery date January 21, 2014.

PROJECT NAME: Hellinikon Urban Development Model
LOCATION: Athens, Greece
AREA: 6,000,000 sq m (1,482.63 acres)
PLANNING: Prof. S. N. Pollalis and Associates

Examples of Planning with the Guidelines

City Expansion in Europe

Chapter 12

12.1 | HELLINIKON URBAN DEVELOPMENT MODEL

The redevelopment of the former Hellinikon airport[1] of Athens, Greece, provides a unique opportunity for a sustainable urban development of a large site situated within a metropolitan area. The 1,500-acre (600-hectare) site, surrounded by a dense urban fabric, includes a 3.5 km long pristine coastal zone. The site is located 18 miles (25 km) from the new International Airport of Athens, 5 miles (8 km) from the city's historic center, and 7 miles (10 km) from the port of Piraeus, the gateway to the islands and an important commercial hub in the Mediterranean. Hymettus mountain forms its northwest boundary. The large scale and the central location of the former airport make Hellinikon a most challenging real estate development.

The plan presented in this chapter has not been implemented. It is a proposal developed by a state agency that submitted it as an option for revitalizing Athens in the midst of Greece's unprecedented economic recession when this project started in 2011. This proposal balances international competitiveness and financial growth with public benefit while addressing integration with the surrounding urban environment.[2]

The Hellinikon site is mostly flat, with two small hills and environmentally degraded streams crossing it. It has hosted various activities in the past 75 years. The country's first major civilian airport,[3] inaugurated in 1938, altered the site's topography. A few decades later a US military base was located at the southeast part of Hellinikon.[4] Increased air traffic led to the construction of a second terminal on the east side of the site, designed by the famous architect Eero Saarinen.[5] In the 1970s, the airport had reached the limit of its capacity, and research[6] started to plan for its relocation and for the future use of the existing site. The airport was eventually relocated in 2001, and an international competition[7] was held for the empty site, amidst a public debate on what would be a best use of the site.

The construction of Olympic venues for the 2004 Olympic games within the site further fragmented it. Buildings, athletic facilities, and parking areas occupied large parts of the property. Furthermore, with a lack of greenery in the region, urban sprawl has climbed up the slopes of Hymettus.

The discussion on what to do with the empty site took a turn toward creating a huge green area, subsequently supported by a law that called for a metropolitan park for almost the entire area.

In 2011, the redevelopment of the area came again to the fore, amidst the economic crisis and the realization that a 1,500-acre park was not feasible for a city the size of Athens, a view expressed publicly by Professor Spiro Pollalis, based on a study at Harvard University.[8] In March 2011, the Greek state established Hellinikon SA,[9] a company responsible for the development of a metropolitan pole and the management and maintenance of the property's premises, and created the opportunity for the development of a unique site. In June 2011, Professor Pollalis was appointed the company's first chairman and CEO.

This chapter, a case study for a sustainable city expansion, is based on the master plan and the technical report,[10] developed in 2011–2012[11] under the guidance of Professor Pollalis. Following an integrated sustainable approach based on the Harvard University Zofnass Program principles, the master plan was accompanied by a phasing plan and the corresponding business plan. Following Ekistics principles, experts from different disciplines worked together with the common goal of creating a plan for the site to benefit Athens in the long term.

Project Vision[12]

The main goal was to create a new vibrant and safe urban environment, with a metropolitan park and state-of-the-art infrastructure systems within the existing dense urban fabric of southeast Athens. It was to be a city within a city, responding to the needs of Athenians as well as the needs of the adjacent communities, with the public's endorsement.

The second goal addressed Athens as a destination, not only for recreational tourism but also for educational, professional, health, and even culinary tourism. Athens is within 4 hours' flight from 3 continents,[13] has a mild climate, historic monuments, and a magical coastline, with a plethora of alternative landscapes and experiences.

The third goal was to provide investment opportunities, business innovation, and high-added-value activities and to ensure the development's financial and social viability for the next two decades and beyond.[14]

Planning and Legislation Framework

The site was not subject to a specific building law, which was required to proceed with the development. An unorthodox process was followed. After the planning was completed and deemed satisfactory following semipublic opinion polls, the proposed land uses were incorporated in the Greek State Law 4062/2012[15] under the heading "Development of the Athens former international airport Hellinikon." The law defined the land uses and their distribution within the site and the limits on height,

PROJECT VISION

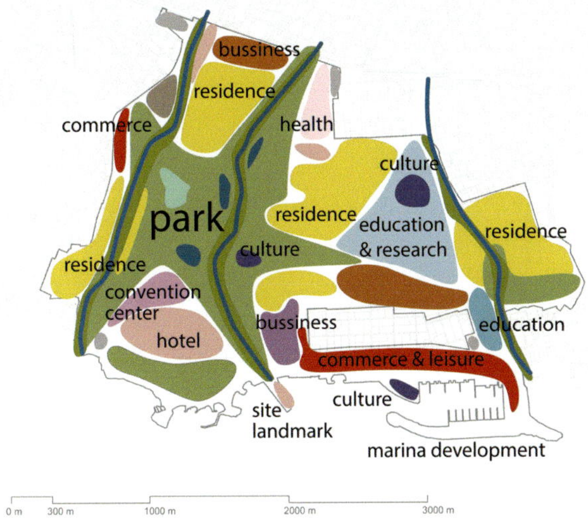

SUSTAINABILITY CRITERIA

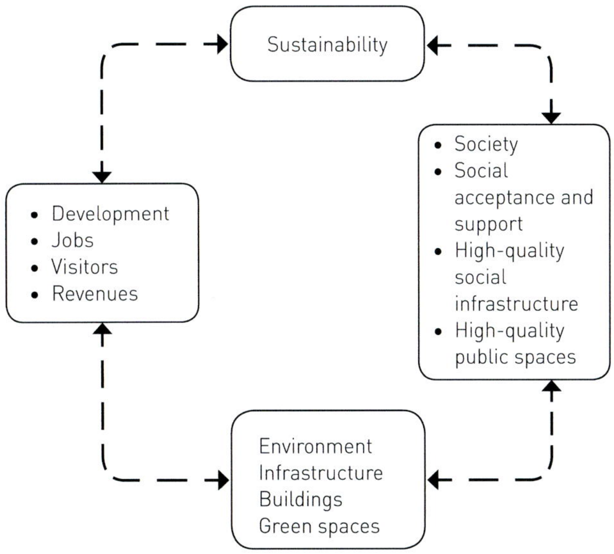

HELLINIKON MASTER PLAN

HELLINIKON LAND USE PLAN

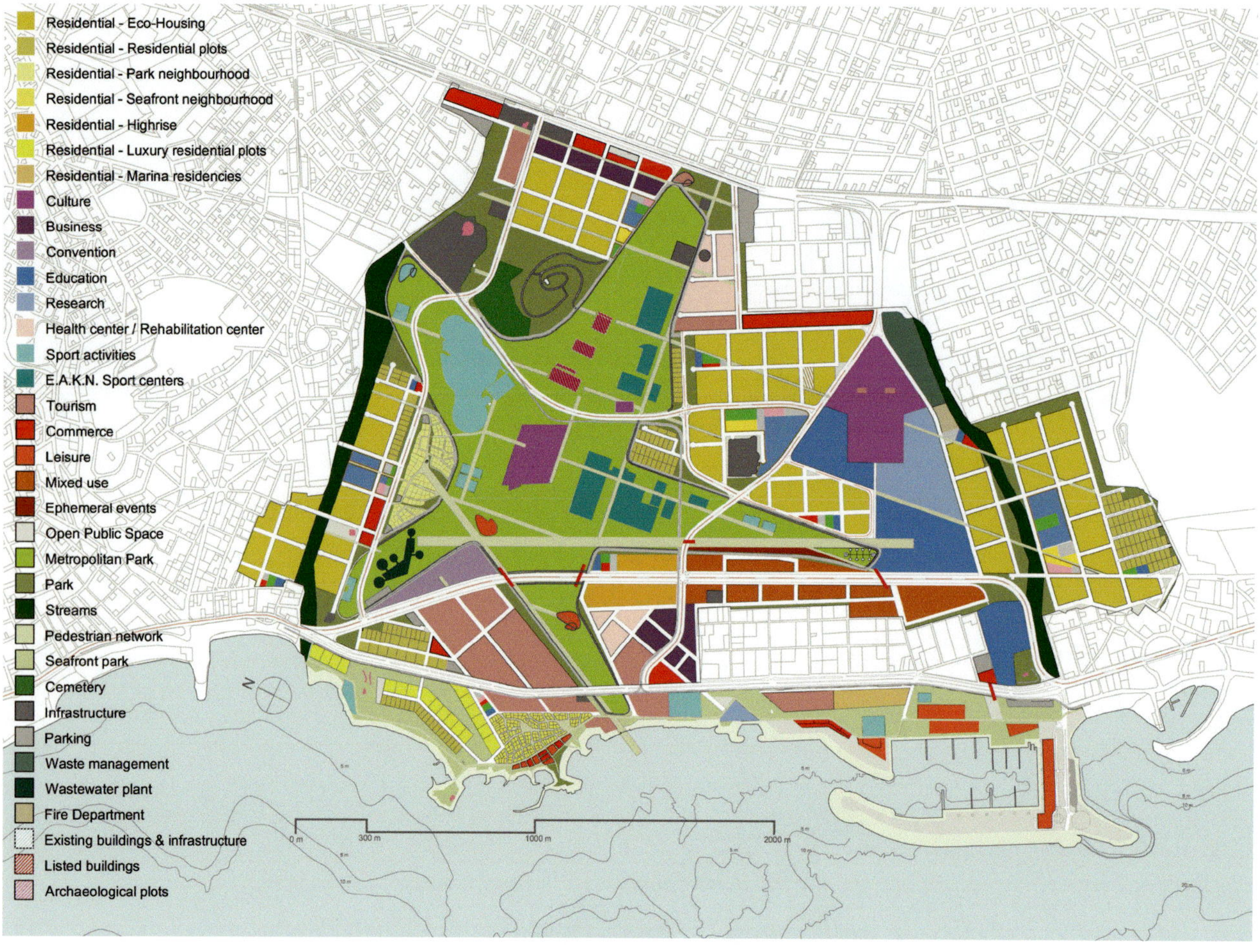

footprint, density, etc., consistent with the plan; and experts characterize it as balanced and rather conservative in terms of the allowed built area, adjacencies, and especially the treatment of the waterfront with low-rise buildings.

According to both the plan and the law, the total area of the metropolitan park is to be a third of the site, or 200 hectares. A spectrum of land uses are to be located in the rest of the site, including education, business, tourism, retail, leisure, waste transfer stations, waste recycling and treatment, health services, and residential areas. For the overall area, the coverage ratio should not exceed 35%, and the floor-to-area ratio should be less than 0.60, including the existing infrastructure premises and the listed buildings to be preserved.[16] While the entire area is up for for-profit development, the state revenue of the floor-to-area ratio above 0.50 (i.e., 0.10) is devoted by law to the creation of small green areas in the Athens metropolitan area.

Sustainability

The site's development will take long. Both a strong vision of the final outcome and flexibility are prerequisites, as needs may change in the course of implementation. The adaptability to new demands can be safeguarded by applying sustainability criteria, which are inherent in the project's planning principles, to all scales and phases of the project.[17]

The interconnection and balance among environmental, social, and economic aspects of sustainability are central to the Hellinikon planning strategies and a challenge for Athens. The protection and preservation of natural functions within the site, the enhancement of greenery volumes, the introduction of state-of-the-art infrastructure systems and high-standards buildings, the creation of quality spaces and a dense network of open public areas for the well-being of the community, the public's involvement and support, as well as the creation of jobs and new innovative ways to attract short- and long-term visitors and inhabitants, have been considered in order to ensure a viable implementation process.

12.2 | PLANNING STRATEGIES AND INFRASTRUCTURE SYSTEMS FOR HELLINIKON

The planning decisions of the project across its different spatial scales, ranging from the city level to the configuration of neighborhoods and the building guidelines, incorporated from the very beginning the five focus areas defined in the Zofnass Program: Leadership, Quality of Life, Natural World, Resource Allocation, and Climate and Risk.

In brief, the main planning strategies[18] have been:

1. A self-sufficient city integrated with its surrounding context,
2. Minimization of car use and connectivity with the adjacent municipalities,
3. Making use of the existing infrastructure and the waterfront,
4. Creation of a metropolitan park and a publicly accessible waterfront,
5. Land use proximities and synergies,
6. Sustainable infrastructures and renewable energy sources,
7. Compact neighborhoods with local centers, connected to the surrounding areas, and
8. "Eco-housing," promoting a prototype of urban blocks.

This chapter focuses on the infrastructure planning guidelines as applied in the master plan. Landscape and Transportation have been selected for analysis. Both are strongly connected to the demand for the optimum development of the area so as to benefit the Athens region:

- Introducing appropriate open spaces within the site, enhancing and revitalizing the natural functions and cycles of the existing surrounding green mosaic;
- Converting the existing abandoned urban void to a multiply-accessible area with a variety of open public spaces with connections to the wider metropolitan region, the surrounding areas, and the waterfront through a range of transportation modes and organized networks.

The planning objectives and guidelines have been put in a checklist form. Each checklist provides a comparison of actions implemented in the master plan with the objectives in the four system levels of each infrastructure. Furthermore, the infrastructure assessment provides an alternative view of the project's scale, context, and phase of implementation and highlights the project's infrastructure system potentials and priorities.

PLANNING STRATEGIES AND RELATED ZOFNASS CATEGORIES

1. A SELF-SUFFICIENT CITY INTEGRATED WITH ITS SURROUNDING CONTEXT

The site's main characteristics are its scale, its close proximity to the historic city center of Athens, and its present condition: a fenced and fragmented area, not functioning in continuity with the surrounding municipalities.

In this context, the design's primary objective was to propose a series of activities and create the proper infrastructure systems to provide a seamless urban continuum, not only beneficial to the development's future residents and visitors, but also a vital part of the city and the Athens region. The proposal also suggests the extension of the surrounding transportation network within the site, with the appropriate adjustments to anchor the site within the existing urban grid.

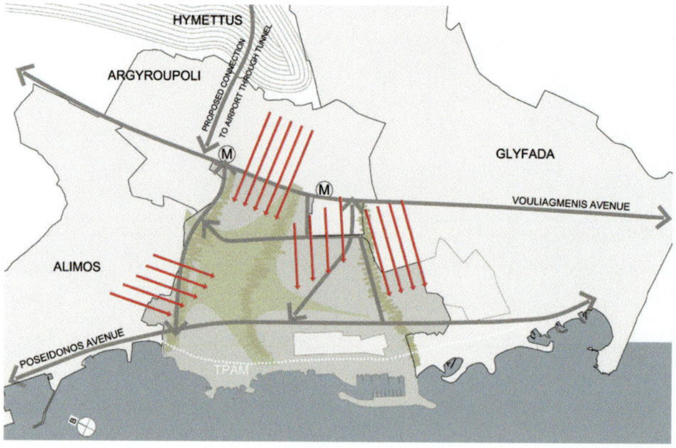

2. MINIMIZATION OF CAR USE AND CONNECTIVITY WITH THE ADJACENT MUNICIPALITIES

The provision for easy and quick access while mitigating car use was central to the design. A dense network of cycling and pedestrian paths is developed across the area, which is dotted with multiple points of interest and activities. The suggested path configuration is a rectangular grid of 1000 ft x 1000 ft (300 m x 300 m), a 4-minute walking distance. Superimposing this layer on the site's topography and distributing land uses, an alternative and safe crossing of the site has been created, promoting a healthier way of life.

Furthermore, the existing public transportation networks (tram, metro, and bus lines) along and near the boundaries of the site are extended throughout the site, enhanced with the distribution of parking areas and transit stations along them, aiming to reduce energy consumption and greenhouse gas emissions.

3. BENEFITS FROM THE EXISTING AVENUE RELOCATION AND FROM RESTORATION OF THE COASTLINE'S CONTINUITY

Two main roadways, parallel to the coastline, define the site. Vouliagmenis Avenue forms its east boundary, while Poseidonos Avenue, on the west, crosses the site and divides the coastal area from the former airport area, acting as a barrier between the two.

An important design decision was to realign Poseidonos Avenue toward the center of the site and locate it on the footprint of an existing former airport runway, while maintaining the existing avenue and converting it into a low-impact supplementary road. The advantages are multiple: the area in close proximity to the sea is doubled and the existing impermeable surface of the runway is reused. Potential scenarios proposed by others to underground the existing axis so as to provide direct access to the beach would require long ramps, would be more expensive by having to reconstruct and alter the various infrastructure systems and networks that lie beneath the road, and would also require dependable drainage as the roadway would be below sea level.

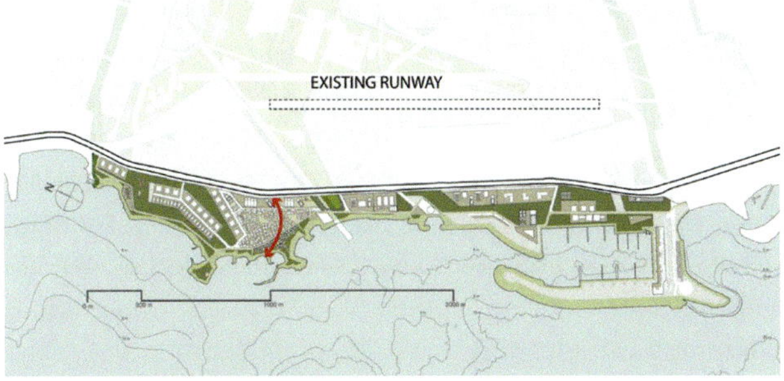

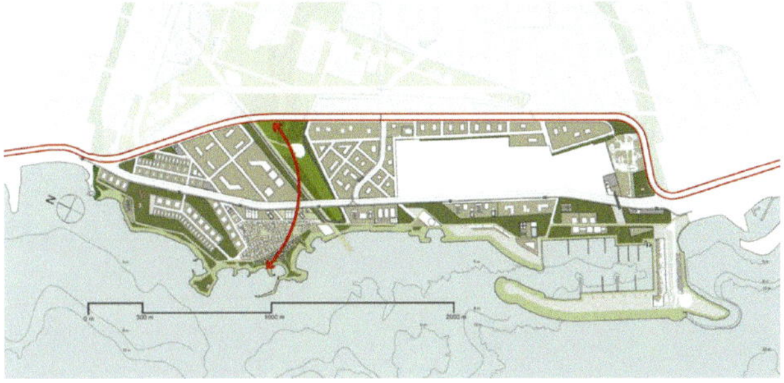

4. CREATION OF A METROPOLITAN PARK AND PROVISION FOR AN ACCESSIBLE WATERFRONT

Apart from being a legal prerequisite and most Athenians' key demand, the metropolitan park's large green area functions as the connecting hub between smaller public and private green spaces, dispersed within the site and along the surrounding land mosaic. The park's extension from Vouliagmenis Avenue on the eastern boundary of the site to the coastal front on the west and from Trahonon stream on the northern boundary to the center of the site's area improves the connection between the mountain and the seafront.

The park's starfish shape increases its perimeter and the number of properties and land uses overlooking its greenery. Its dynamic form creates a strong interaction with the existing and proposed urban area, and at the same time the built adjacent areas that penetrate the park promote accessibility and connectivity, while the inhabitants' and visitors' sense of safety is enhanced.

5. LAND USE SYNERGIES

The development was envisioned from the very beginning as a metropolitan pole and as a network of uses of national caliber and international reference. The dynamic of the new activities emerges from their strategic arrangement on the site, their appropriate interconnection, their synergy with those of the immediate and regional urban network, and the added value of being adjacent to the metropolitan park, rather than from the influence of each land use alone. A world-class museum in the heart of the park, a cultural center hosted in the former terminal by Eero Saarinen, a convention center, sports and leisure facilities, health and rehabilitation research centers, an aviation museum, the Agios Kosmas Marina, high-quality hotels and office spaces, entrepreneurship incubators: these are expected to attract Athenians and foreign visitors and ensure a vivid and sustainable environment for the local communities and the site's inhabitants as well.

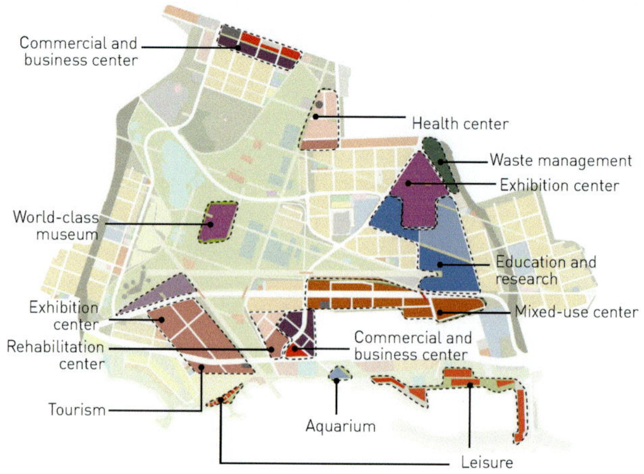

6. SUSTAINABLE INFRASTRUCTURES AND RENEWABLE ENERGY SOURCES

The proposed state-of-the-art infrastructure facilities, such as a solid waste transfer station, a material recovery facility, a compost facility, a wastewater treatment facility, and water storage tanks, do not merely complement the urban model but are key concepts of the master plan. Placing them on site aims to minimize the proposal's environmental impact and load on the existing infrastructure of Athens and the surrounding areas. Instead of being diverted to Athens' central management facilities, the urban development's sewage and organic waste will provide reclaimed water and compost for the park's maintenance. Therefore, further to its recreational and ecological function, the park becomes a key element for successful and balanced resource management and an example for greater Athens. Additionally, the promotion of water and material recycling enhances societal awareness and improves everyday life.

7. COMPACT NEIGHBORHOODS WITH LOCAL CENTERS AND CONNECTIVITY WITH SURROUNDING AREAS

The concept of neighborhood compactness and self-sufficiency in terms of everyday needs and activities has been inspired by the subsectors of Ekistics theory[19] of the Greek urban planner Constantinos Doxiadis.[20] The estimated 40,000[21] future inhabitants will be housed in residential areas designed by following the pattern of the surrounding neighborhoods so as to create a sense of orientation and integration within the existing urban fabric. Additionally, two types of mixed-use local centers of different scale and density of activities, according to the demographics and the estimated population, are distributed within the various residential areas. The design and distribution of local centers strengthens the viability of the development by mitigating car use, therefore lowering energy consumption, noise and air pollution, while improving traffic flow and quality of life.

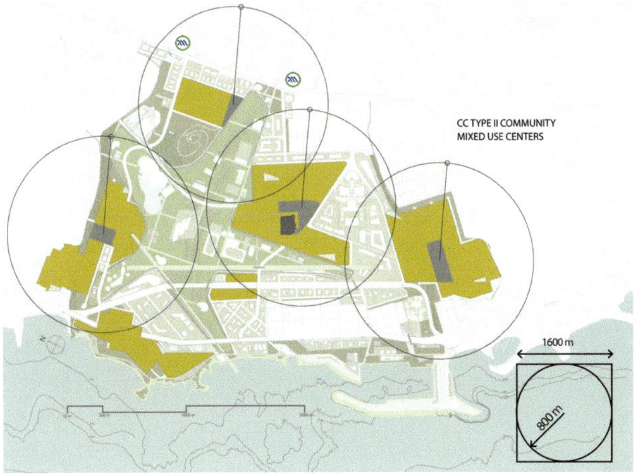

8. "ECO-HOUSING," THE DEVELOPMENT'S MODEL URBAN BLOCKS

A new type of urban block is introduced, with a prototype dimension of 130 m x 130 m. However, the eco-housing plot adapts to the site's topography, to the street network, and to the 300 m x 300 m pedestrian grid, which extends throughout the site, providing an interesting residential tissue. The buildings are placed at the block's perimeter with their footprint covering only 50% of each plot, therefore creating a large courtyard at the center. The courtyard provides a functional open space with greenery and a range of facilities for its residents. The positions of buildings and courtyard in relation to the block's adjacent street width facilitate the natural lighting and ventilation of the buildings. The block's built and open space act as a self-standing sustainable system with solar panels, native plants, green roofs, enhanced shell insulation, water-permeable surfaces, and water elements.

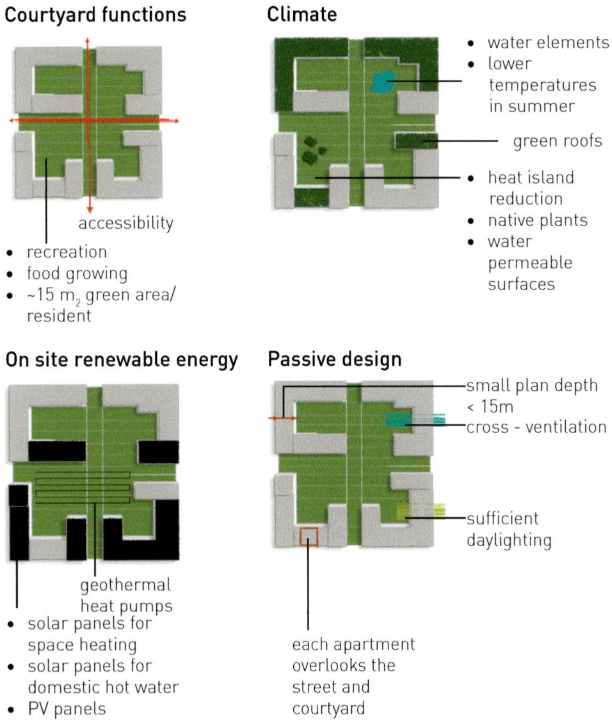

12.3 | GUIDELINES AND ACTIONS FOR SUSTAINABLE PLANNING

LANDSCAPE INFRASTRUCTURE

The starting point for landscape infrastructure is the identification of services that the landscape will provide. The site's immediate relation with the sea and its proximity to the large natural environment of Hymettus mountain highlight the potential of the area and at the same time the issues that emerge both from the nearby urban tissue and the site's particularities and identity.

The metropolitan park and the open and green spaces have been envisioned to assume a multifunctional role, becoming an integral part of the surrounding ecosystem. The challenges were the following:

- The creation of a 2,000,000 sqm public park that will enhance the wider region of Athens, which lacks greenery, and the site's proposed urban development.
- The restoration of existing natural entities, such as the degraded streams,[22] the coastline, and the abandoned green spaces of the site, and thus the enhancement of the microclimate.
- The reclamation of the altered and leveled site's topography, and the inversion of the site's current 65% rate of hardscape and impermeable surfaces.[23]
- The provision of a system of smaller open spaces, private or public, within the area so as to promote continuity with the adjacent environment.

TRANSPORTATION INFRASTRUCTURE

The Hellinikon site is a void in the existing urban tissue. Therefore, in addition to creating a metropolitan park, the site's connectivity with the rest of the city was central to the overall planning.

An urban development is defined by the system of roads, arteries, axes, and paths that allow for interconnections of activities and services. The proposed new networks exploit the existing infrastructure, roads, and transportation and focus on:

- The enhancement of urban compactness with the incorporation of mixed uses, as a key concept to ensure the long-term growth of the site and the minimization of car use, and integration with the immediate and wider surrounding environment;
- The creation of a network of dispersed local centers that host various local activities and reduce unnecessary trips and mobility while covering the everyday needs of the residents;
- The promotion of nonmotorized modes and enhancement of pedestrian and bicycle routes with the design of a new grid exclusively for walking and cycling, which expands across the site;
- The improvement of the efficiency of the existing transportation system with the implementation of a dense network of nodes and their interconnection to the existing ones and with the promotion of different, proposed and available means of transportation;
- The adaptation of the transportation infrastructure to the existing context and appropriate reuse of the existing paved surfaces.

LANDSCAPE INFRASTRUCTURE—LANDSCAPE DEMAND CHECKLIST

	GUIDELINES	ACTIONS
	L1 LANDSCAPE DEMAND	
	OBJECTIVE L1.1. Understand urban planning opportunities, vulnerabilities, and limitations through the analysis and documentation of local context and processes	
L1.1.1.	Identify large-scale ecological flows and geographic patterns expanding beyond the city's boundaries	Mapping of the main surrounding natural components that compose the local identity, microclimate, and landscape, such as Hymettus mountain in proximity to the site's east limit, the Saronic Gulf coastline configuring the area's west boundary, the hills and Trahonon stream on the north, and the old Hellinikon stream and Glyfada Golf Course forming the south limit of the former airport area.
L1.1.2.	Collate local and regional environmental data	Identification of the project's weaknesses and strengths through the analysis of the activities hosted by the site and the surrounding dense urban fabric, the alterations of the site's topography, aquifer permeability, topsoil and air quality, climate characteristics, and current flora and fauna.
L1.1.4.	Collate local and regional social/economic data	Extensive analysis of the surrounding municipalities' population data, built density, existing land uses, social amenities, and current real estate market that define the project's investment risks and level of competitiveness.
L1.1.5.	Collate local and regional data on cultural aspects	Identification of the site's landmarks, such as the listed former east terminal designed by Eero Saarinen and the former US Air Force hangars, the archaeological findings, and the old Agios Kosmas Orthodox church.
	OBJECTIVE L1.2. Define potential and required Landscape services preserving the ecological balance of natural communities, habitats, and environmentally critical lands	
L1.2.1.	Define needs for protection and restoration of habitats from loss, fragmentation, and degradation through Landscape infrastructure	Extensive analysis of the site highlighted the need to maintain Hassani hill's current topography and vegetation, restore the two streams crossing the site, reclaim the coastline, and protect the existing waterfront vegetation.
L1.2.2.	Define needs for protection and restoration of ecological flows and wildlife migration routes through Landscape infrastructure	Restoration of the two existing streams and enhancement of the natural connection between Hymettus mountain and the sea, reclamation of the coastline, and protection of the site's existing flora and fauna.
L1.2.4.	Define needs for restoration of contaminated areas through Landscape infrastructure	Conducting of preliminary studies for potential soil and groundwater contamination, due to the site's various past activities.
	OBJECTIVE L1.3. Define potential and required Landscape services regulating city areas' response to natural phenomena	
L1.3.1.	Define needs for rainwater/stormwater management through Landscape infrastructure	Identification of the catchment areas connected to the streams.
	OBJECTIVE L1.4. Define potential and required Landscape services regulating and improving conditions within the urban fabric	
L1.4.1.	Define needs for climate adaptation and microclimate improvement through Landscape infrastructure	Introduction of extensive green areas in every neighborhood throughout the site (of which 65% is currently covered by water-impermeable surfaces) so as to mitigate the heat island effect.

	GUIDELINES	ACTIONS
L1.4.2.	Define needs for air quality improvement through Landscape infrastructure	Creation of a metropolitan-scale park and provision for local-scale greenery dispersed across the development's site to help improve the city's air quality.
	OBJECTIVE L1.5. Define potential and required Landscape provisioning services	
L1.5.1.	Identify opportunities for the provisioning of food, fiber, and feedstock provided by the natural context	Designation of specific areas within the metropolitan park for the use of urban agriculture.
L1.5.2.	Define strategies to enhance (benefit from) the productivity of landscape through Landscape infrastructure	On-site compost production from organic waste to be used as fertilizer.
	OBJECTIVE L1.6. Define potential and required Landscape cultural services	
L1.6.1.	Identify recreational/social services opportunities provided by the natural context	Creation of significant opportunities for leisure and recreation, benefiting from the nearby sea.
L1.6.2.	Define strategies to offer recreational/ social services required by demographics, population distribution, and culture through Landscape infrastructure	Provision of access to a large-scale metropolitan park and an open public shoreline, while at the same time providing additional sport and leisure facilities for the surrounding municipalities.
	OBJECTIVE L1.7. Define potential and required Landscape services regulating the function and performance of other infrastructure systems	
L1.7.2.	Define needs for mitigating and compensating for other infrastructure systems' impact on community life through Landscape infrastructure	Optimum configuration of landscape elements to mitigate visual discontinuity and noise pollution caused by the wastewater treatment facility, the tram depot, and the waste transfer station. The restored streams, the preserved Hassani hill, the adequate metropolitan park shape, and the use of arboreous vegetation will serve as barriers between the facilities and the new and existing residential areas.
L1.7.3.	Define needs for protection of infrastructure systems against projected climate change and extreme events through Landscape infrastructure	Introduction of extensive open green areas and water-permeable surfaces, restoration of streams, creation of a metropolitan park, and minimization of the building footprint and road network surface, providing potential protection from flooding.

Landscape Demand System Level Summary

The assessment of the Landscape Demand system level showed that 16 out of a total of 24 guidelines had been taken into consideration in the master plan design process.

The extensive analysis of the Hellinikon site, which highlighted the features integral to the local identity of the area, provided the framework for the design strategies required to preserve and restore ecological balance, thus contributing to the inhabitants' quality of life and to the development's competitiveness. The range of potential and required services identified highlighted the significant multifunctional role of Landscape infrastructure.

The analysis included:

- The identification of the current topography and the natural elements such as the existing hills and vegetation, the streams' current situation, as well as the shoreline's current characteristics,
- A survey of potential soil and groundwater contamination, considering the various activities hosted by the site in the past, so as to specify restoration requirements,
- The identification of the catchment areas connected to streams, with the aim of determining the rainwater/ stormwater management strategy,
- Highlighting the landscape's cultural services that may be provided through the appropriate waterfront design and the metropolitan park configuration concept,
- The requirement to maintain the landscape's entities and ensure their self-sufficiency, through the production and use of compost on site,
- The need for the provision of urban agriculture activities within the metropolitan park.

However, guidelines referring to the need for protection against extreme events, for water infiltration/purification, and to address soil erosion etc. have been considered as strategies related to the next phase of the project.

SUSTAINABLE ACTION

Understand the site's scale within the surrounding natural and man-made context and identify the project's weaknesses and strengths.

RELATED GUIDELINES

L1.1.1, L1.1.2, L1.1.4

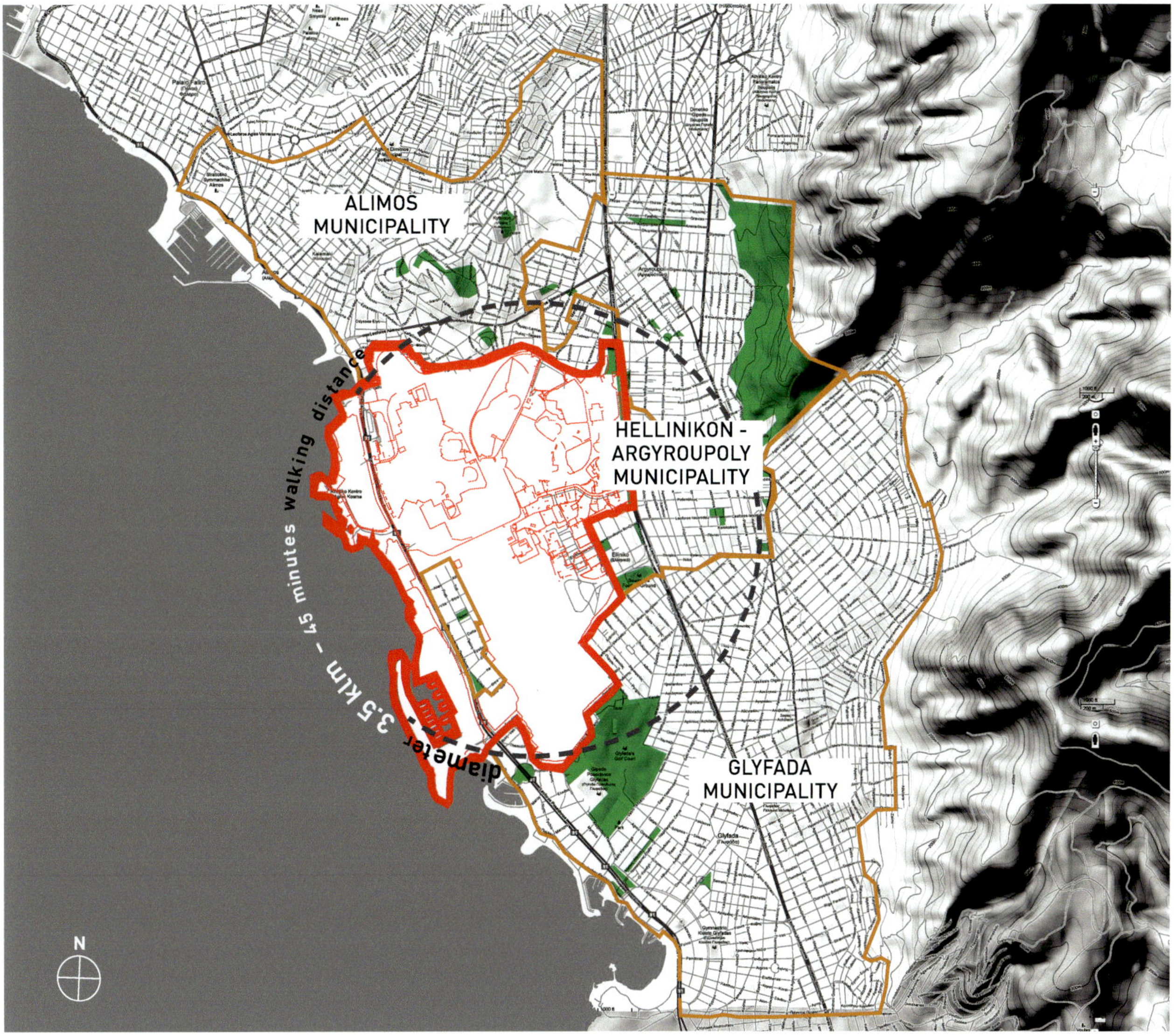

SITE'S CHARACTERISTICS

- Surrounding municipalities' green areas
- Surrounding municipalities' boundaries
- Site's boundary
- Fences and barriers in site's current arrangement

SUSTAINABLE ACTION

Incorporate spaces responding to the landscape's required regulating, provisioning, and cultural services.

RELATED GUIDELINES

L1.5.2, L1.6.1, L1.6.2, L1.7.2

SITE'S LANDSCAPE SERVICES

Residential areas

Provisioning services: Reuse produced organic waste on site to improve the park's soil

Cultural services: Arrange sport facilities along the coastline and the park

Regulating services: Create protective buffers

LANDSCAPE INFRASTRUCTURE—LANDSCAPE PROGRAM CHECKLIST

GUIDELINES		ACTIONS
L2	**LANDSCAPE PROGRAM**	
	OBJECTIVE L2.1. Develop an integrated Landscape infrastructure strategy	
L2.1.1.	Hierarchize Demand for Landscape services based on city goals for sustainable development	Provision of a metropolitan park and various recreation facilities along the coastline, responding to Athens residents' demands. Restoration of the existing streams, meeting the need for flood protection and physically connecting Hymettus mountain with the sea. Further improvement of the site's microclimate and rainwater and stormwater management through the provision of vegetated areas.
L2.1.2.	Trace correlation of alternative Landscape infrastructure strategies with city competitiveness and economic sectors	Promotion of a new identity and provision of various outdoor facilities and accessible public coastline. Support for activities according to the financial pillars, such as tourism, health, education, recreation, and culture.
	OBJECTIVE L2.2. Select the optimum activities to form the Program of Landscape infrastructure	
L2.2.1.	Translate identified Landscape services to specific activities	Ensuring a sustainable and safe metropolitan park by including various recreational facilities, cultural landmarks, sports venues, communal gardens, and playgrounds that will create a sense of safety, and in parallel adding infrastructures that will support the park's self-sufficiency.
L2.2.2.	Integrate long-term city growth and development goals with Landscape activities Program	Increase of public green areas and establishment of an adequate connection and direct access to the coastline, as part of Athens' goals for international competitiveness.
L2.2.3.	Secure long-term financial viability of Landscape activities Program	Proposal for various supraregional and local activities and infrastructures along the coastline, inside the park, and across the local parks that will promote the landscape system's self-sufficiency and provide safe and lively public spaces, thus securing their economic viability.
	OBJECTIVE L2.3. Make Landscape entities as multifunctional as possible	
L2.3.1.	Define possible combinations of compatible Landscape activities to create social, financial, and ecological catalysts	Promotion of combined different landscape activities across the site such as sports facilities, playgrounds, cultural spaces, food growing, and small-scale retail.
L2.3.3.	Combine Landscape activities to enhance daily patterns for safe, lively environments	Dispersal of various activities that will operate throughout the day, across all landscape areas and mostly in the new metropolitan park.
	OBJECTIVE L2.4. Define the optimum distribution of Landscape activities across the city	
L2.4.1.	Adapt the scale and distribution of Landscape activities to the size of communities and areas they intend to serve	Distribution of smaller parks with low-impact activities for everyday use in each neighborhood, and integration of larger entities such as the park, the streams, and the coastline with significant activities that will attract visitors from all over the city.
	OBJECTIVE L2.5. Ensure compatibility of Landscape activities with locations	
L2.5.1.	Ensure the protection of areas with special natural and cultural features through the regulation of Landscape activities within them and their adjacent areas	Restriction of high-nuisance activities to the Hassani hill area; the restored streams and the coastal zone and buffer zones will also be defined around them.

	GUIDELINES	ACTIONS
L2.5.2.	Adjust the intensity of Landscape activities according to scale and context	Provision for large-scale sports and cultural facilities in the park to respond to metropolitan-scale needs, and provision for lower-intensity activities along the coastline and for playgrounds and smaller courts in the neighborhood parks. Preservation of the restored streams' natural landscape and prevention of any activity along them.
L2.5.3.	Integrate Landscape activities with existing natural properties and processes where possible	Enhancement of landscape activities such as stormwater management and biodiversity preservation through the metropolitan park's location in the northern part of the site, where the most unpaved area is found, through the preservation of Hassani hill's current topography that will act as a natural barrier between the various infrastructure facilities, and through the streams' restoration and their conversion into natural corridors between mountain and sea.

SUSTAINABLE ACTION

Promote multifuncionality and safety by introducing new landscape system components intertwined with various landscape activities.

RELATED GUIDELINES

L2.1.1, L2.1.2, L2.3.1, L2.3.3

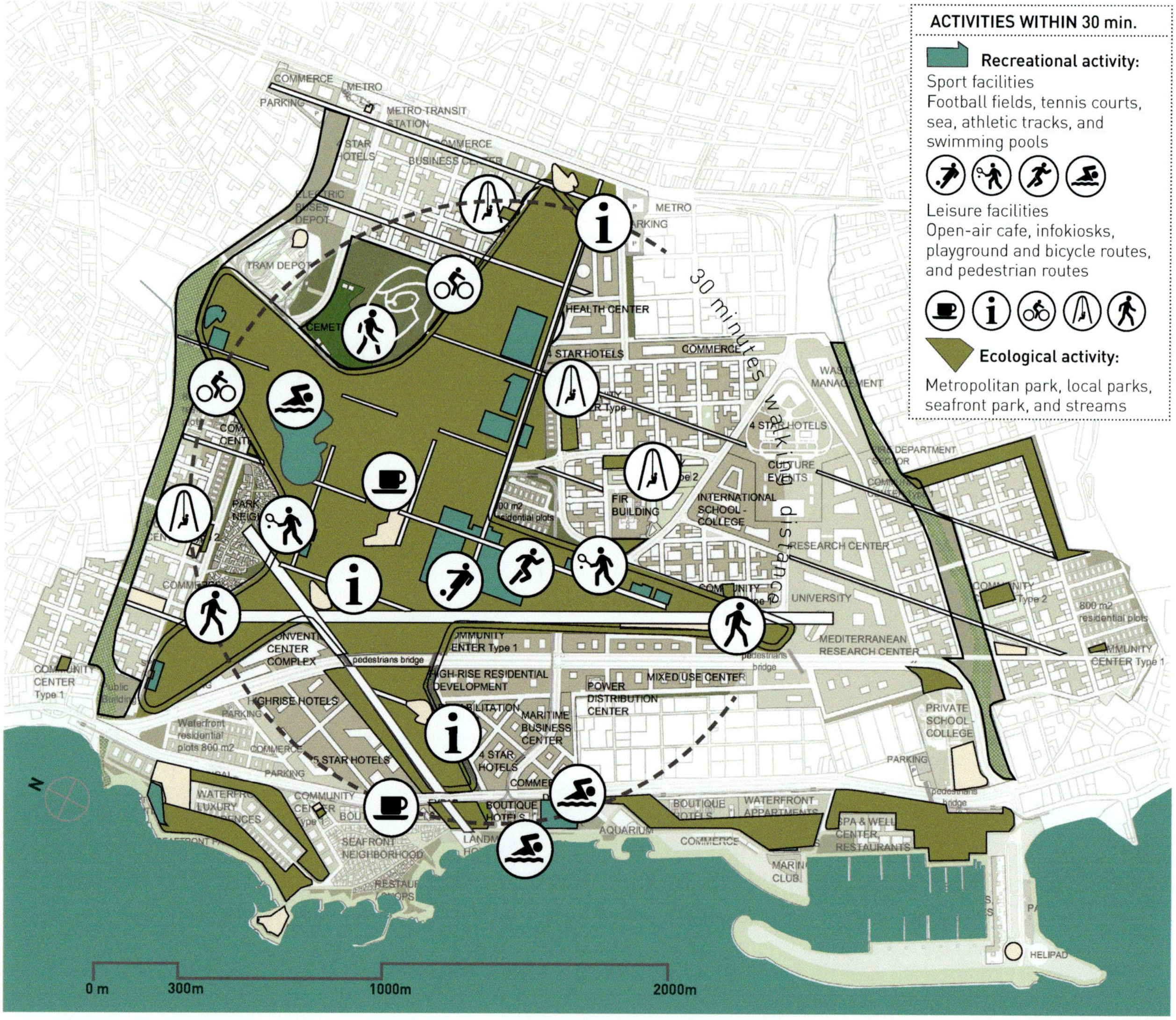

Landscape Program System Level Summary

On the Program level, which promotes the multifunctional role of landscape infrastructure, 11 of a total of 13 guidelines have been considered in the development's design phase.

According to the identification of the required services described in the Demand system level, the design strategies should focus on promoting the site's new identity by introducing activities related to the project's financial pillars and goals. Indicatively, the design provides outdoor leisure, sports, and recreational facilities within the metropolitan park or the local open spaces. Furthermore it addresses the interconnectivity of the landscape entities, first by providing various transportation means and a dense pedestrian network that allows for their easy access and second through their appropriate configuration and distribution across the proposed urban tissue. The latter strategy will also enhance the sense of safety of visitors and residents.

Finally, the objective to ensure compatibility of landscape activities to locations will be achieved with the provision of adequate scales of facilities according to the landscape pattern that will host them. Thus the design situates large-scale sports and cultural facilities within the park to respond to the metropolitan-scale needs, while lower-intensity activities are located along the coastline and playgrounds and smaller courts in the neighborhood parks.

LANDSCAPE INFRASTRUCTURE—LANDSCAPE PATTERN CHECKLIST

	GUIDELINES	ACTIONS
	L3 LANDSCAPE PATTERN	
	OBJECTIVE L3.1. Identify main Landscape system entities to support identified services and selected activities	
L3.1.1.	Identify and place main patches in connection to land use planning, demographics, and natural context	Distribution of the local parks along the new neighborhoods and placement of the metropolitan park's main volume in the site's northern part, where the maximum of unpaved area and Hassani hill are found.
L3.1.2.	Identify and place main corridors in connection to land use planning, demographics, and natural context	Restoration of the current streams' traces and reclamation of the fragmented and partly reserved coastline.
L3.1.3.	Integrate corridors and patches to function as an overall network of open spaces	Using the flexible, dynamic starfish shape of the metropolitan park to promote the maximum interconnection among the other landscape entities, such as the restored streams, the preserved Hassani hill, and the reclaimed coastline.
	OBJECTIVE L3.2. Delimit main Landscape system entities	
L3.2.1.	Define shape and area of main patches for better integration with the natural processes and the urban fabric	Using the park's starfish form and its optimum location, where most of the unpaved areas are concentrated, to provide safety, connectivity between the existing and proposed urban areas, and integration of the current natural topography and the suggested patches and corridors of the development.
L3.2.3.	Avoid fragmentation of large patches	Only a few axes of the transportation network cross the metropolitan park and the preserved Hassani hill.
	OBJECTIVE L3.3. Assign and establish the appropriate ecological functions to Landscape system entities according to identified services, location, etc.	
L3.3.1.	Support/establish the functions of conduit, sink, and barrier through topography/ microtopography of Landscape entities	Enhancement of the conduits by recreating the site's two streams; provision of visual barriers among the other infrastructures by means of the restored streams and the preserved Hassani hill volume. At the same time, preservation of the existing artificial lake enhances the network of sinks in the area.

	GUIDELINES	ACTIONS
L3.3.2.	Support/establish the function of filter through the selection of surface cover materials in Landscape entities	Extensive use of green areas to maximize the permeable surfaces and reverse the existing hard-surface topography.
L3.3.3.	Support/establish the functions of sink, habitat, filter, and barrier with adequate vegetation density, patterns, vertical vegetation layering, and plant selection in Landscape entities	Provision for three different context-adapted types of vegetation: the coastline vegetation, the greenery along the streams, and a specific variety of plants in the metropolitan park.
L3.3.4.	Support/establish the function of source for the creation of new resources through Landscape entities	Design of adequate spaces for the creation of communal gardens or urban agriculture areas within the metropolitan park.
L3.3.5.	Add multiple secondary ecological functions to Landscape entities	Preservation of the existing Hassani hill natural barrier and its enhancement with arboreous vegetation to block the view to the tram depot from the residential areas. The metropolitan park functions both as the site's main patch and as a barrier between the surrounding urban environment and the wastewater treatment and compost facilities that are located within the park.
	OBJECTIVE L3.4. Expand the system of Landscape corridors and patches through a supporting matrix	
L3.4.1.	Exploit opportunities for additional Landscape supportive patches and corridors in the private space	Increase of private green space by introducing a new type of urban block with large inner courtyards and promoting the use of green roofs, and by distributing extensive green spaces within large properties such as the university and research center.
L3.4.2.	Exploit opportunities for additional Landscape supportive patches and corridors given by Transportation entities	Introduction of tree-lined streets and other landscape elements adapted to street type.
	OBJECTIVE L3.5. Integrate Landscape with other infrastructure systems' entities and facilities	
L3.5.2.	Investigate needs of Landscape attractors' integration with Transportation Nodes	Provision of a new tram line that extends to the nearest metro station and bus lines that cross the metropolitan park, thus improving the park's accessibility and connectivity with the shoreline, the surrounding areas, and Athens' city center.
	OBJECTIVE L3.6. Enhance quality of place	
L3.6.1.	Use the configuration of Landscape entities to strengthen each location's sense of place	Provision for three different context-adapted types of vegetation: the coastline vegetation, the greenery along the streams, and a specific variety of plants in the metropolitan park.
L3.6.2.	Ensure openness and easy access to Landscape entities for all members of the community	The pedestrian and bicycle perimeter road forming the starfish shape of the park facilitate its accessibility. The promenade along the coastal front contributes to its interconnection with the surrounding areas.
L3.6.3.	Cater to the needs of different community groups	Enhancement of adults' and children's everyday activities through various sports and leisure facilities within small- and large-scale landscape entities, providing a relaxing and high-quality environment.
	OBJECTIVE L3.8. Preserve and enhance biodiversity in Landscape corridors and patches	
L3.8.1.	Preserve the natural topography	Maintenance of Hassani hill's current topography and the configuration of the coastline, and preservation and restoration of the existing streams.
L3.8.3.	Enhance biodiversity, mimicking natural vegetation patterns in Landscape entities	Provision for three different context-adapted types of vegetation: the coastline vegetation, the greenery along the streams, and a specific variety of plants in the metropolitan park.

GUIDELINES		ACTIONS
	OBJECTIVE L3.10. Take advantage of locally available resources for development of Landscape patches and corridors	
L3.10.2.	Use materials that are readily and locally sourced for the construction of Landscape patches and corridors	Reclamation on site of building materials from the demolition of existing structures, for use in landscaping entities, relief restoration, and road construction.

Landscape Pattern System Level Summary

Of a total of 29 guidelines forming the Pattern system level checklist, 19 have been taken into consideration in the project's preliminary design phase. Most of them refer to:

- The identification of the main landscape system components, embracing the optimum placement and shape of the metropolitan park, the appropriate distribution of local parks among the new neighborhoods, the restoration of the streams, and the reclamation of the fragmented coastline;
- The preservation of the existing topography of hills and streams, providing visual barriers among the other infrastructures; the preservation of the artificial lake that enhances the area's network of sinks; and the design of adequate spaces for the creation of communal gardens or urban local agriculture spaces inside the metropolitan park, with the aim of establishing ecological functions of landscape system components and preserving and enhancing biodiversity in landscape corridors and patches;
- Design strategies to create a supporting matrix for better landscape infrastructure performance, such as increasing private green space by introducing a new type of urban block, locating additional private facilities with large open green areas near the main landscape patches, and introducing tree-lined streets and different vegetation according to street type;
- The introduction of a pedestrian and bicycle perimeter road which forms the starfish shape of the park, and inclusion of various sports and leisure facilities along small- and large-scale landscape entities, to enhance adults' and children's everyday activities, facilitate accessibility to green areas, promote a sense of safety, and enhance quality of place.

SUSTAINABLE ACTION

Optimize design of new patches and corridors and integrate in the existing green areas.

RELATED GUIDELINES

L3.1.3, L3.2.1, L3.2.3

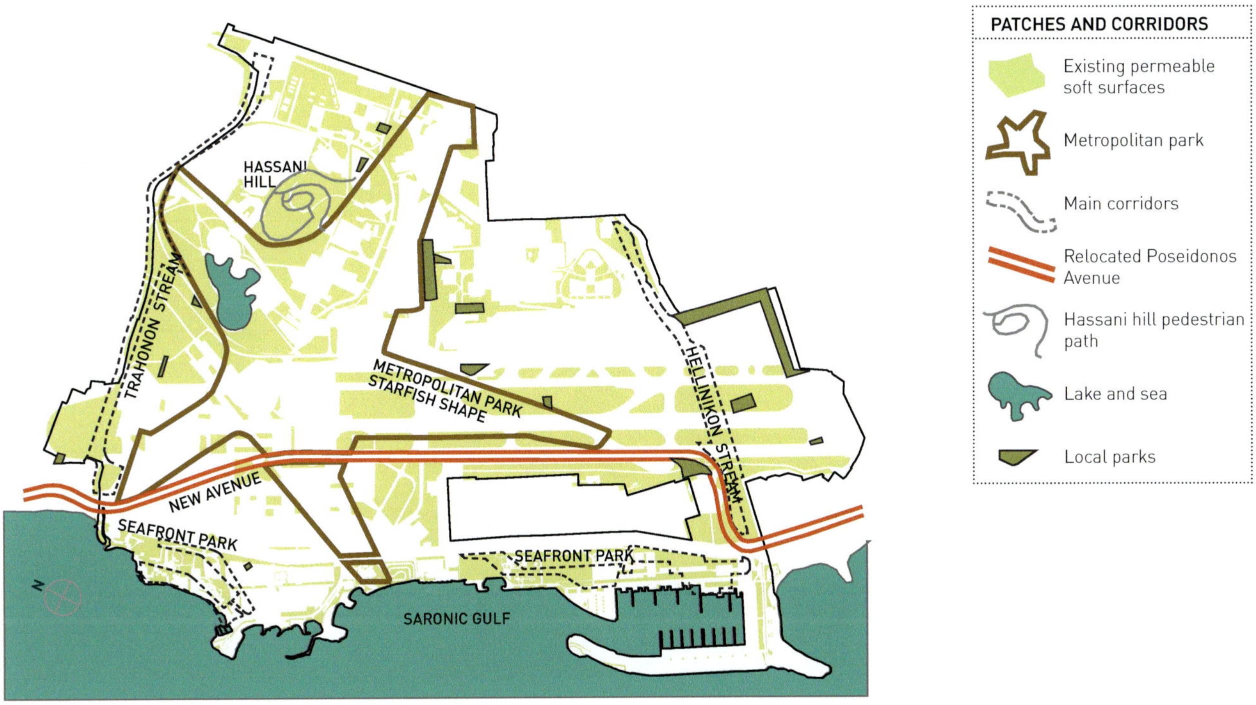

SUSTAINABLE ACTION

Develop design strategies to maximize green spaces on private lots.

RELATED GUIDELINES

L3.3.2, L3.4.1

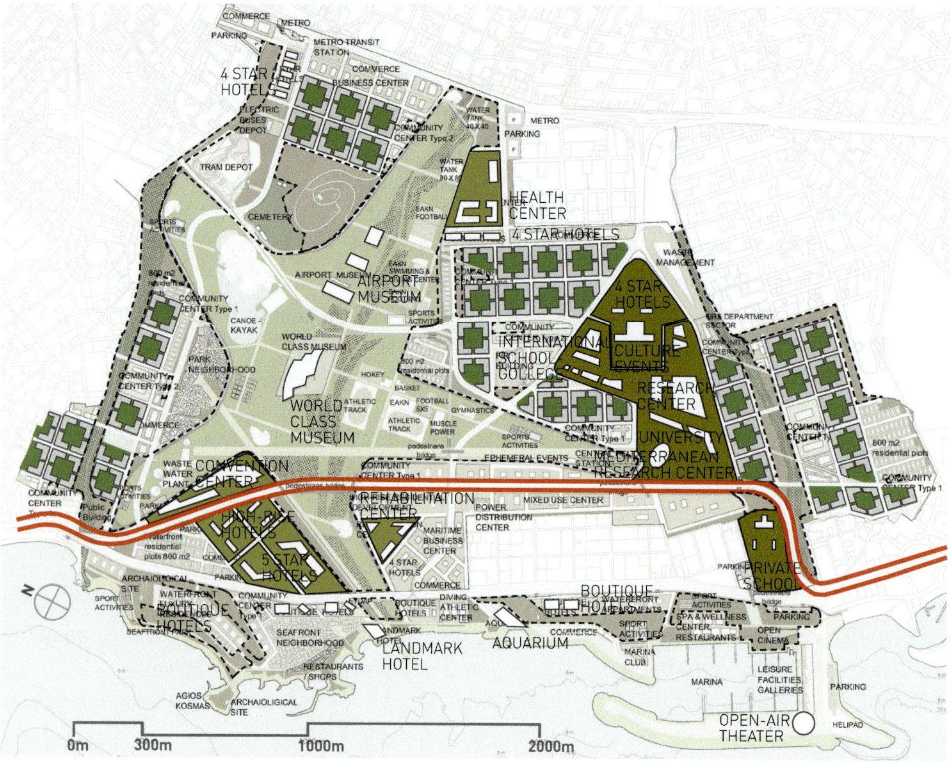

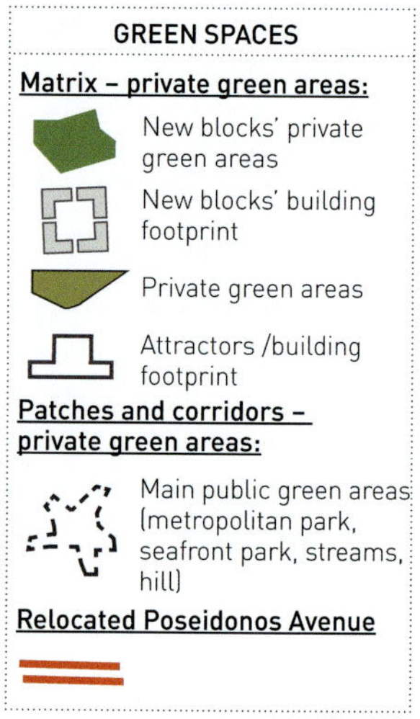

SUSTAINABLE ACTION

Avoid fragmentation of large patches implementing walking and cycling axes for access and placing other infrastructures' networks and nodes at the perimeter.

RELATED GUIDELINES

L3.2.3

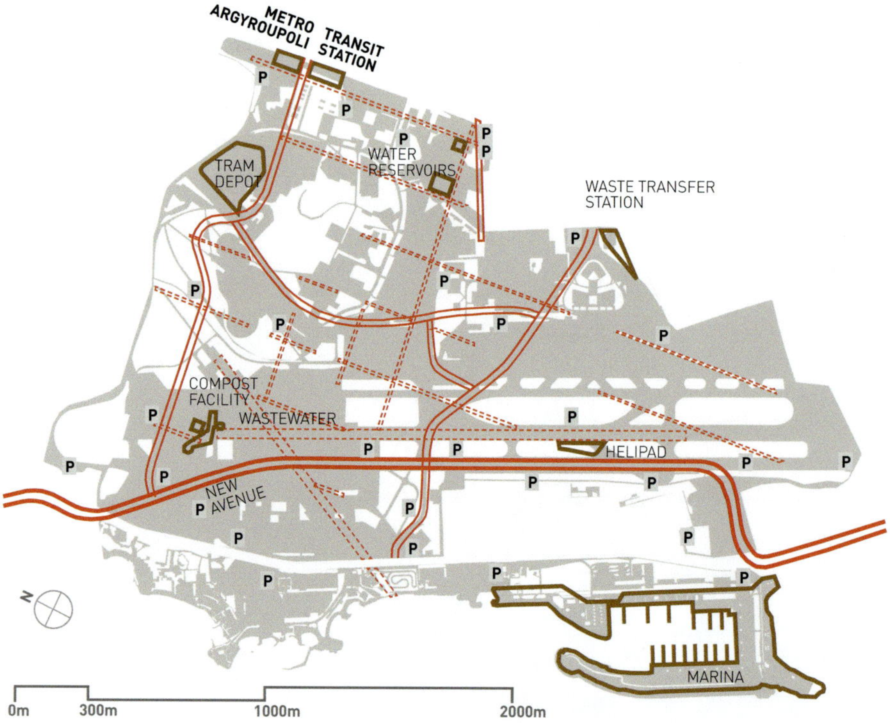

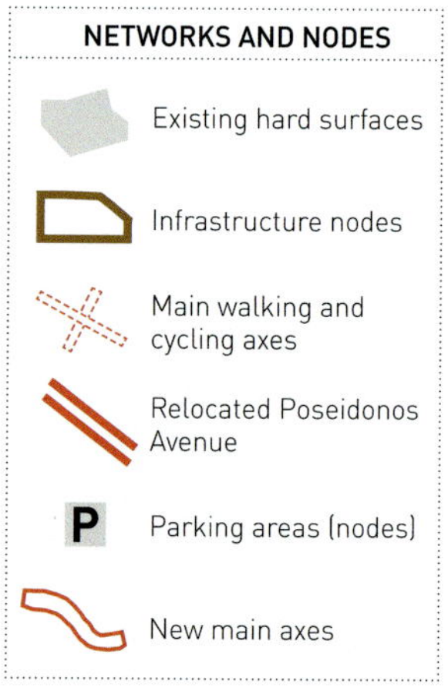

GUIDELINES		ACTIONS
L4	**LANDSCAPE MAINTENANCE**	
	OBJECTIVE L4.1. Minimize intensity of required Landscape Maintenance services, aim for self-sufficiency	
L4.1.1.	Promote use of plants and materials adequate to local climate, soil, and exposure	Increase of the local, Mediterranean plant variety both for corridors and patches.
	OBJECTIVE L4.4. Enhance natural processes	
L4.4.2.	Support nutrients cycle during Landscape Maintenance	Compost production from organic waste and reuse of it to improve the park's soil.
	OBJECTIVE L4.8. Optimize Water use in Landscape Maintenance	
L4.8.2.	Use native plants of low water requirements	Provision for three different context-adapted, native types of vegetation: the coastline vegetation, the greenery along the streams, and a specific variety of plants in the metropolitan park.
L4.8.4.	Harvest rainwater for Landscape Maintenance where possible	Rainwater harvesting in reservoirs created along the streams.
L4.8.5.	Use gray or reclaimed water for cleaning and irrigation in Landscape entities	Location of the wastewater treatment facility inside the metropolitan park, for the collection of reclaimed water and its reuse for the park's irrigation.
	OBJECTIVE L4.10. Define potential and required Landscape services regulating and improving conditions within the urban fabric	
L4.10.2.	Use residues of Landscape maintenance practices as supply resources for other infrastructure systems	Organic waste treatment from landscape maintenance, such as reuse of pruning and seasonal trimming residues and recycling of green manure to produce high-quality compost.
L4.10.3.	Incorporate Solid Waste and wastewater Treatment residues and by-products in regular Landscape Maintenance if available	Use of the reclaimed water produced by the wastewater treatment facility for irrigation, and of the compost produced through the management of organic waste as fertilizer. Both promote the development's self-sufficiency.

Landscape Maintenance System Level Summary

In the Landscape Maintenance system level, 19 of a total of 26 guidelines have been considered in the Hellinikon project, while other guidelines referring to detailed technical issues, such as the introduction of monitoring systems, the management methods, and the assessing techniques, are related to the implementation and operation phases of the development rather than the master plan design phase.

Key objectives for the promotion of landscape self-sufficiency and optimum water management, for solid waste reduction and resource recovery, and for the enhancement of natural processes have been implemented through the following specific design decisions:

- Introduce native Mediterranean plant varieties both for corridors and patches and organize them according to the landscape feature entities, meaning providing coastline vegetation, alternative greenery along the streams, and a specific variety of plants in the metropolitan park;
- Promote compost production from organic waste and reuse it to improve the park's soil;
- Propose the use of water reservoirs along the streams;
- Locate the wastewater treatment facility inside the metropolitan park, for the collection of the reclaimed water produced by the various facilities of the development and its reuse for irrigation;
- Facilitate on-site organic waste collection, pruning and seasonal trimming residues, and recycling of green manure, which will be used to produce high-quality compost.

SUSTAINABLE ACTION

Select configurations suitable for each area in order to promote site's self-sufficiency.

RELATED GUIDELINES

L4.1.1, L4.8.2

	TYPE OF SPATIAL PATTERNS	CATEGORIES OF ACTIVITIES	FUNCTIONAL PATTERNS	CONFIGURATION
PATCHES	NEW METROPOLITAN PARK 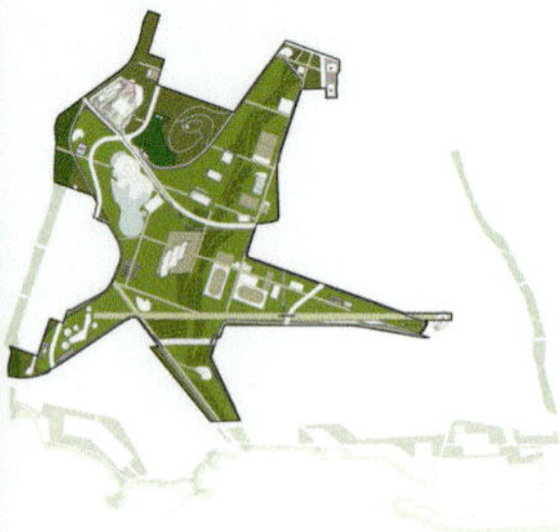	1. ecological 	 habitat, sink	Maintenance of existing hill and lake topography and extensive use of native plants and arboreous vegetation
		2. recreational 	 filter	Promotion of permeable materials in open-air leisure and sport facilities' surfaces
		3. productive 	 source	Use of communal garderns and urban agriculture parcels
		4. supporting 	 barrier	Conflict mitigation among residential areas and tram depot and wastewater plant
CORRIDORS	RESTORATION OF EXISTING STREAMS 	1. ecological 	 habitat, conduit	Introduction of linear, dense and native vegetation and water reservoirs/catchment areas for stormwater management
		2. recreational 	 filter, conduit	Provision for pedestrians' pathways and leisure areas along streams' banks and for use of permeable materials
		4. supporting 	 barrier	Adequate barrier design between residential areas and waste transfer station
	NEW SEAFRONT PARK AND COASTLINE RECLAMATION 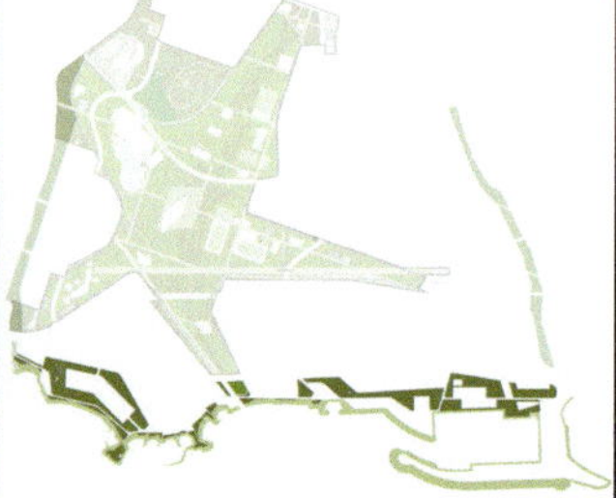	1. ecological 	 conduit	Coastal front upgrade by enhancing existing vegetation along the seafront pedestrians' path
		2. recreational 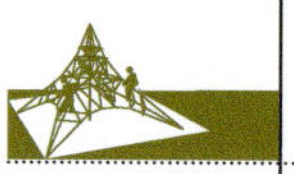	 conduit	Reclaimed, public coastal front and creation of open-air leisure and sports facilities along the pedestrians' path

TRANSPORTATION INFRASTRUCTURE—TRIP GENERATION CHECKLIST

GUIDELINES		ACTIONS
T1	**TRIP GENERATION**	
	OBJECTIVE T1.2. Reduce length of Trips	
T1.2.1.	Promote mixed-use development	Introduction of supraregional facilities to serve metropolitan requirements. Placement of local activities within the neighborhood centers to cover existing and future residents' everyday needs at walking distance.
T1.2.2.	Include basic amenities within each residential area	Creation of new local centers hosting various activities and services according to the emerging demographic data of each neighborhood.
T1.2.3.	Promote compactness in urban fabric	The proposal for a specific urban block size for residential buildings, which will be dispersed throughout the site, and the introduction of high-rise mixed-use buildings located along the main avenue.
T1.2.4.	Optimize placement of infrastructure facilities that generate significant traffic	Location of the waste transfer station and material recovery facilities near the urban model's new main avenues.
	OBJECTIVE T1.3. Monitor Trip Generation trends	
T1.3.1.	Identify main city Trip attractors	The 2,000,000 sqm metropolitan park, the Convention Center, the world-class museum, the Incubator center, and the marina's leisure facilities will become the development's main trip attractors.

Transportation Trip Generation Summary

For the identification of Trip Generation parameters in the Hellinikon urban model, it was important to understand the diversification of the interregional land uses that will attract visitors, along with the capacity of each land use and the estimated population number. Together these set the design strategies for trip reduction.

Minimizing unnecessary and everyday trips requires the appropriate distribution of interregional facilities along with the creation of a dense network of local centers with amenities that would cover everyday needs.

However, the 5 (of a total of 10) guidelines implemented in the Trip Generation system level indicate that those guidelines were not addressed in synergy with the Information infrastructure during the design phase of the master plan.

SUSTAINABLE ACTION

Identify land use properties, capacities, estimated population data, and their impact on everyday trips.

RELATED GUIDELINES

T1.3.1

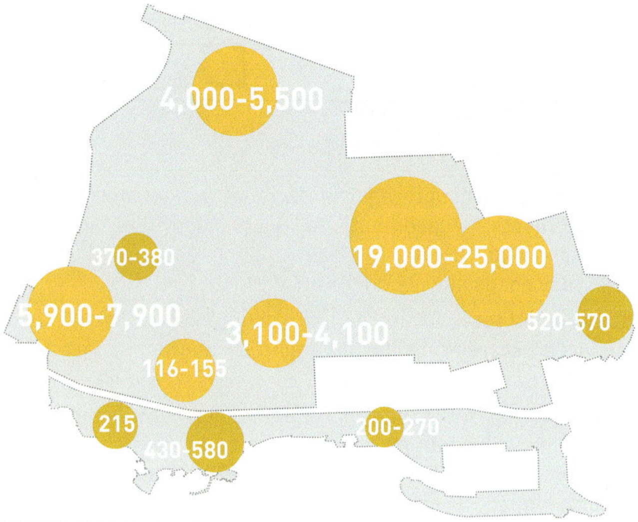

TRIP PRODUCTION (ORIGIN)

LAND USE GENERATOR

Home-Based Other
- Residential apartments
- Residential houses/plots

ESTIMATED PASSENGERS

35,000–42,000 residents

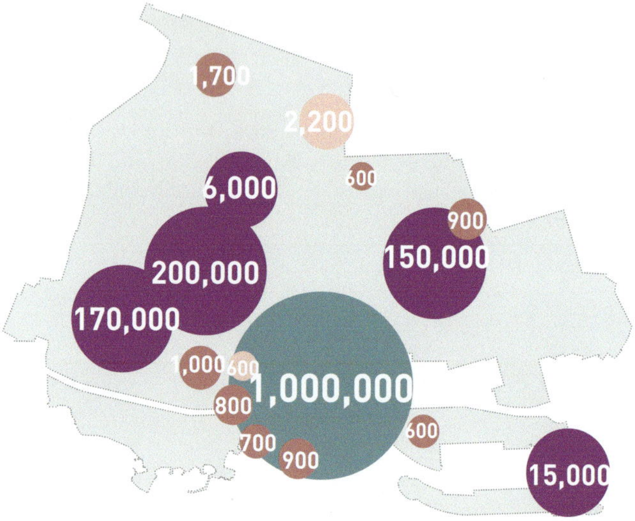

TRIP ATTRACTION (DESTINATION)

LAND USE GENERATOR

Non-Home Based
- Museums
- Aquarium
- Hotels
- Hospitals

ESTIMATED PASSENGERS

1,500,000 visitors
7,200 hotel beds
2,800 hospital beds

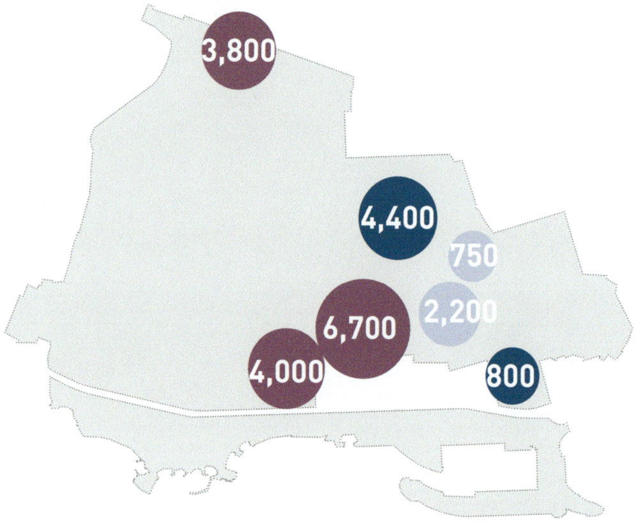

TRIP ATTRACTION AND PRODUCTION

LAND USE GENERATOR

Home-Based Work
- Business centers
- University/research center
- Private schools

ESTIMATED PASSENGERS

15,000 employers and employees
2,200 university students
750 researchers
5,200 private school students

	GUIDELINES	ACTIONS
	T2 TRANSPORTATION MODES	
	OBJECTIVE T2.1. Secure long-term city connectivity at local and supraregional level	
T2.1.1.	Ensure coverage of passenger and freight needs through multiple and alternative Transportation Modes	Interconnection of existing and proposed transportation modes, such as the extension of the current tram line to the metro station. Interconnection of local centers through the creation of an extensive pedestrian and bicycle network and multiple bus lines. Full exploitation of the existing marina and location of a helipad facility near a main avenue.
T2.1.2.	Trace correlation of different Transportation Modes with city competitiveness and economic sectors	Promotion of the development's connectivity to the islands through the marina and helipad nodes, and enhancement of its role as tourist destination by connecting its proposed cultural facilities to Athens' historic center via mass transit. Reinforcement of the urban model's job-generating role by locating alternative attractors, such as business centers, healthcare facilities, and educational services, along the main avenues.
	OBJECTIVE T2.3. Reduce traffic congestion	
T2.3.1.	Provide and promote reliable public and mass transit	Development of new bus lines, conversion of the existing auxiliary tram line into a fully functional passenger line, and their interconnection with the site's two existing metro stations.
	OBJECTIVE T2.4. Reduce greenhouse and air pollutant emissions	
T2.4.2.	Promote nonmotorized Transportation Modes	Design of a pedestrian and bicycle grid of 300 m x 300 m, which will extend throughout the site connecting the neighborhood centers with the supraregional facilities.

Transportation Modes System Level Summary

In the Modes system level, 4 of a total of 14 guidelines were applicable to the design of the project. The assessment demonstrates two noteworthy issues related to this phase of the project:

First of all, the objectives related to alternative-fuel vehicles, their regular maintenance, etc., prove the strong interdependence of transportation infrastructure with energy strategies, and could not have been considered applicable at the master planning phase, as they are connected to the next phases of the project.

Secondly, the close association of the transportation system with monitoring services provides a range of objectives, deriving from the Information infrastructure, which again could not have been considered as strategies related to the master planning phase.

However, the distribution of landmarks and trip attractors along nodes and the provision for their interconnection, the upgrading of existing networks and nodes, and the introduction of new ones were key concepts in the design, responding to issues of long-term city connectivity at local and supraregional levels, traffic congestion, and reduction of greenhouse and air pollutant emissions.

SUSTAINABLE ACTION

Promote interconnectivity among alternative public and low-emission transportation modes.

RELATED GUIDELINES

T2.1.1, T2.3.1, T2.4.2

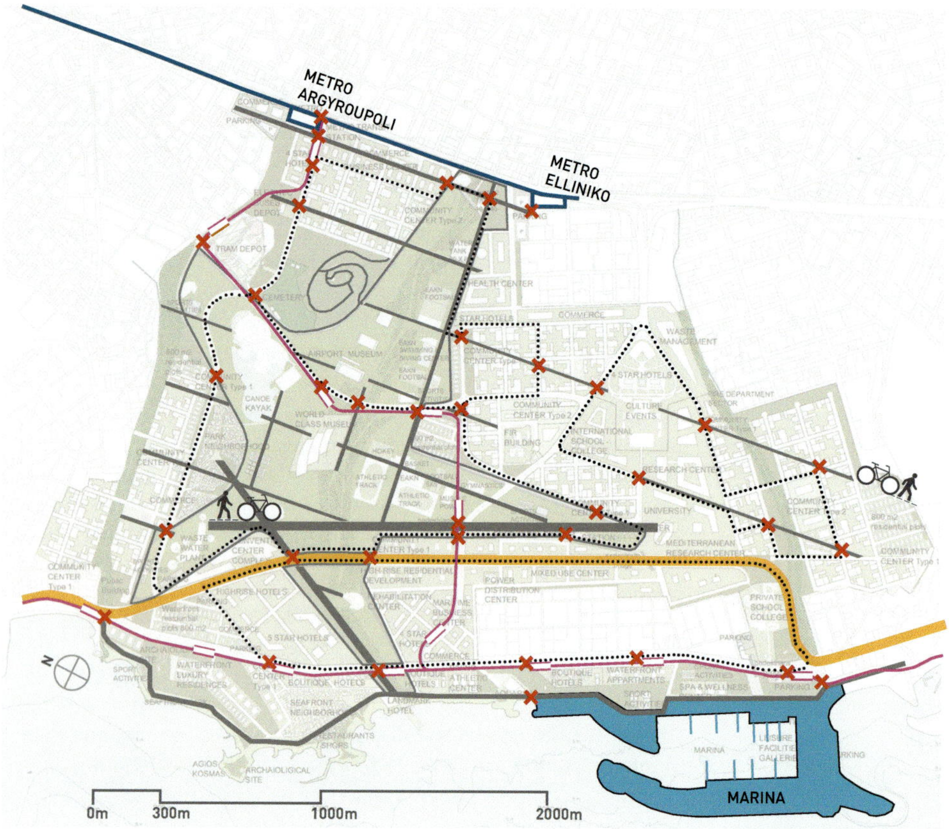

NETWORKS AND NODES

- Modes' crossings
- Bus lines
- Metro stations
- Tram line stations
- Main walking and cycling axes
- Marina node and piers
- Relocated Poseidonos Avenue

TRANSPORTATION INFRASTRUCTURE—TRANSPORTATION NODES CHECKLIST

GUIDELINES		ACTIONS
T3 TRANSPORTATION NODES		
OBJECTIVE T3.1. Place and distribute Transportation Nodes for optimized Transportation system performance		
T3.1.1.	Place Transportation Nodes in correlation to Trip attractors	Interconnection of the main trip attractors with current and new transportation nodes. The health and business centers are placed in proximity to metro and tram line stations, the park's cultural facilities and museums as well as the aquarium are connected through bus stops and tram line stations, while hotels and leisure facilities are approached via mass transit nodes and the supraregional node of the marina.
T3.1.2.	Ensure equal connection opportunities for all areas of the city	The conversion of the auxiliary existing tramway into a fully functional passenger line with multiple stations and its extension to the site's nearest existing metro station, as well as new bus lines with stops placed throughout the development.

	GUIDELINES	ACTIONS
T3.1.3.	Combine and integrate Nodes of different Transportation modes for efficient intermodal transfer / Create hubs	Promotion of interconnection areas among the existing metro stations, the new metro transit station, and the proposed bus and tramway stops, and addition of adequate parking spaces.
	OBJECTIVE T3.3. Integrate Transportation Nodes with other infrastructure systems	
T3.3.3.	Provide required traffic connections to large Landscape infrastructure entities	Making the coastline and the metropolitan park accessible to the city's surrounding area by converting the tramway into a fully functional passenger line and extending it to the nearest metro stations.
	OBJECTIVE T3.4. Enhance quality of place	
T3.4.1.	Resolve fragmentation of urban fabric and pedestrian routes caused by Transportation Nodes	Reclamation of the existing marina and provision for public access along the coastline axis, with the suggestion of new bus lines that reach the coast.
T3.4.2.	Visually integrate Transportation Nodes to the local context	Maintenance of Hassani hill's local character, proposal for an optimum location and form for the new metropolitan park, and use of arboreous vegetation as visual barriers between the tram depot and the surrounding residential areas.
	OBJECTIVE T3.5. Enhance public space	
T3.5.1.	Design Transportation Nodes and equipment to extend and enhance public space	Configuration of the marina transportation node as an open public space and its integration with adjacent leisure facilities and sports activities.
	OBJECTIVE T3.6. Promote community well-being	
T3.6.1.	Assess and resolve noise pollution issues caused by Transportation Nodes	Creation of a natural barrier and protection of the neighboring residential areas from noise pollution by maintaining Hassani hill's current topography, by designing the optimum shape for the new park, and by introducing arboreous vegetation around the existing tram depot.

Transportation Nodes System Level Summary

The Hellinikon project is the redevelopment of an abandoned large space within an existing urban context. In the past its area has hosted various facilities and infrastructures, such as the tram depot and marina on the waterfront. At the same time the east boundary of the site is close to two metro stations. These facilities are nodes of the wider network of Athens and its wider region, so the area did not lack transportation nodes; still, their interconnection is a prerequisite to facilitate traffic flows to and from the site.

At the Nodes system level, only 8 of a total of 33 guidelines have been implemented at the Hellinikon project. Those not considered, due to their association with the project's next phases, refer to:

- Existing nodes' resilience, which has not been an issue addressed in the course of the design;
- The reuse of on-site materials and resources for the maintenance and construction of a node, decisions that will be defined in the next phases of the project;
- The identification of transportation processes that may be monitored in order to facilitate and regulate traffic flows. This would result from synergy with the Information infrastructure, which has not been taken into consideration in the Hellinikon urban model.

SUSTAINABLE ACTION

Develop interconnections among the multiple nodes and the site's main attractors

RELATED GUIDELINES

T3.1.1, T3.1.2, T3.1.3, T3.3.3, T3.5.1

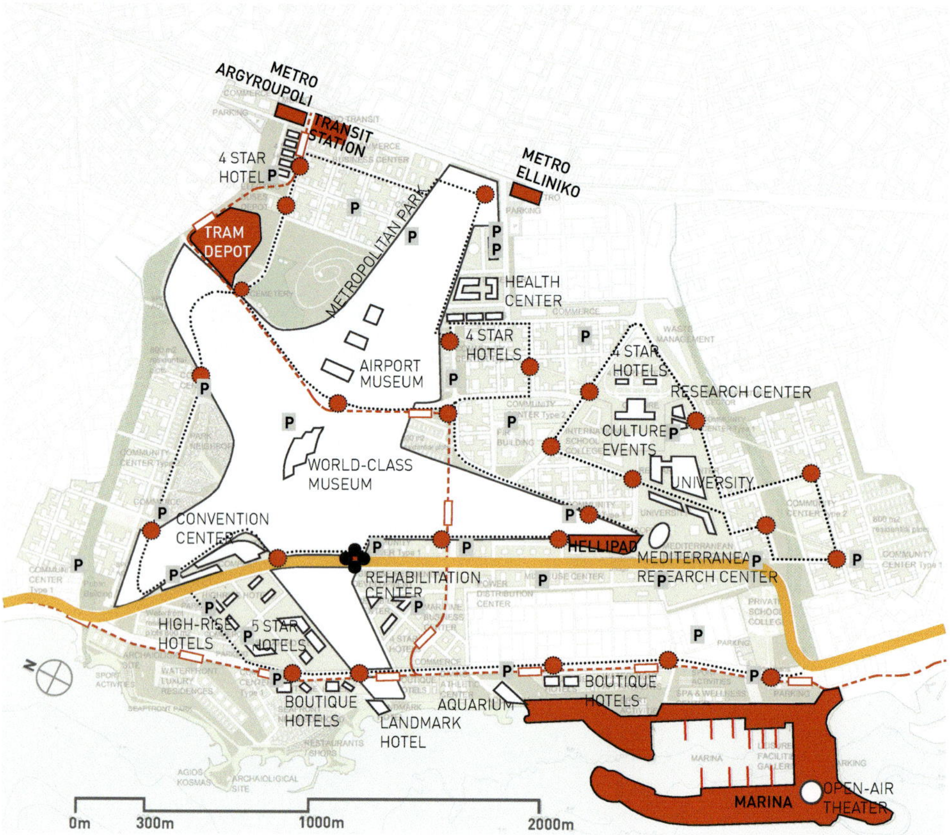

NODES AND ATTRACTORS

- Bus lines and stations
- Tram line stations
- P Parking stations
- **Nodes' large areas:** Metro stations, Tram depot, Marina, Helipad, Transit station
- Attractors / buildings
- Relocated Poseidonos Avenue

TRANSPORTATION INFRASTRUCTURE—TRANSPORTATION NETWORKS CHECKLIST

GUIDELINES		ACTIONS
T4	**TRANSPORTATION NETWORKS**	
OBJECTIVE T4.1.	Develop Transportation Networks for optimized Transportation system performance	
T4.1.1.	Provide adequate coverage and length of Networks for all Transportation modes implemented	Proposal for new bus lines, for a 300m x 300m pedestrian and bicycle grid that will extend across the site, and for a sufficient road network to achieve adequate interconnection among proposed residential neighborhoods, existing surrounding areas, and supraregional facilities.
T4.1.2.	Provide adequate coverage and length of routes for public transport modes implemented	Interconnection of the new network of four bus lines and the converted tram line with the existing metro network.
T4.1.3.	Combine and integrate Networks of different Transportation modes	Incorporation of pedestrian and bicycle lanes in every road type and integration of the 300m x 300m pedestrian/bicycle grid with the new urban block typology.

	GUIDELINES	ACTIONS
T4.1.5.	Locate safe intersections and interconnections at adequate points / Optimize density of intersections	Design of intersection/interconnection areas, stations, and nodes among the public transport network, the bicycle and pedestrian grid, and the new road network. Provision of three pedestrian bridges to cross the new avenue.
	OBJECTIVE T4.2. Facilitate traffic flows.	
T4.2.1.	Define a clear hierarchy in Transportation Networks to serve different levels of flow intensity	Proposal for various different road types according to the total site's estimated demographic data and expected traffic flows.
	OBJECTIVE T4.3. Integrate Transportation Networks with other infrastructure systems	
T4.3.1.	Configure Transportation Networks to provide for needs of Water Networks and facilitate their maintenance	Maintenance of the current level of the main avenue, to avoid affecting the existing water network that lies underneath as would happen if the avenue were routed underground.
T4.3.3.	Configure Transportation Networks to provide for needs of Energy Distribution and Transmission networks and facilitate their maintenance	Maintenance of the existing high voltage cables, which cross the site from west to east. Maintenance of the main avenue's current level without affecting the existing underground energy network.
T4.3.6.	Investigate needs of Solid Waste Collection and Transfer facilities' connection to Transportation Networks	Location of the site's proposed waste transfer station facility along the development's new main road and in close proximity to the existing avenue.
	OBJECTIVE T4.4. Enhance public space	
T4.4.1.	Design Transportation Networks and equipment to extend and enhance public space	Introduction of public leisure and sports facilities at junctions of the 300m x 300m pedestrian grid and along the coastline's proposed pedestrian promenade.
	OBJECTIVE T4.5. Enhance quality of place	
T4.5.1.	Resolve fragmentation of urban fabric and pedestrian routes caused by Transportation Networks	Provision for three pedestrian bridges crossing the new avenue to ensure pedestrians' safe intersections and facilitate their daily routes.
	OBJECTIVE T4.6. Create a pedestrian-friendly environment	
T4.6.1.	Concentrate activities along pedestrian routes that enhance daily patterns for safe, lively environments	Connection between the local neighborhood centers with the design of a dense, 6-minute walking distance, 300m x 300m pedestrian network that extends across the site with the distribution of various cultural, leisure, and retail activities along the grid's junctions.
T4.6.2.	Size urban blocks adequately for a dense pedestrian grid	Mimimizing of the road network area to 16% of the site by the design of the 300m x 300m pedestrian grid to the benefit of additional open public space.
T4.6.3.	Make provisions for pedestrians' protection from harsh climatic conditions	Proposal for the distribution of various plant species and trees in each road type and along the pedestrian network.
	OBJECTIVE T4.7. Facilitate traffic flows.	
T4.7.3.	Assess and resolve noise pollution issues caused by Transportation Networks	Introduction of open green spaces and arboreous vegetation along the new main avenue so as to minimize noise pollution during peak traffic hours.

	GUIDELINES	ACTIONS
	OBJECTIVE T4.8. Maintain the integrity of natural-world elements and systems	
T4.8.1.	Avoid the construction of Transportation Networks on undeveloped land	Reuse of an existing runway to host the development's new main avenue and diversion of the current high traffic flow along the coast.
T4.8.3.	Minimize natural Landscape alteration for the construction of Transportation Networks	Maintenance of Hassani hill's current topography and the canoe-kayak venue's hill and lake land relief.
T4.8.4.	Protect natural habitats and wildlife corridors and compensate for their fragmentation or disruption by Transportation Networks	Restoration of the existing streams, integration of the proposed promenades along their axis with the site's pedestrian network, and minimization of their crossings by other types of transportation networks.
T4.8.7.	Develop vegetation configurations for Transportation Networks to provide habitats	Incorporation of lanes with various plant species or with arboreous vegetation where adequate, along the site's different road types.
T4.8.8.	Adapt plant selection for Transportation Networks to local context	Protect native plants by avoiding the introduction of invasive species along lanes of the different road types.
	OBJECTIVE T4.9. Prepare for short-term hazards and projected future threats	
T4.9.1.	Assess and manage future threats from natural and man-made phenomena affecting Transportation Networks	Promotion of strategies toward the maximization of open green areas and permeable surfaces, such as restoration of the existing streams, creation of a 2,000,000 sqm park, minimization of building footprint, and the reduction of the road network surface to 16% of the site.
	OBJECTIVE T4.12. Optimize Water consumption in Transportation Networks	
T4.12.1.	Use gray or reclaimed water for cleaning and irrigation in Transportation Networks	Use of reclaimed water from the on-site wastewater treatment facility.

Transportation Nodes System Level Summary

In the network system level, comparing the 21 guidelines considered at this phase to the 32 that have not been applied reveals the significance of the transportation network during the design process. A well-designed road network not only provides connections with adjacent and wider areas of a city, but also serves the needs of interconnection and distribution within cities as they grow.

Responding to the main challenges regarding the creation of adequate networks to optimize the Transportation system's performance and its integration with other infrastructure systems, the enhancement of public space and quality of place and the provision of a pedestrian-friendly environment have been addressed through various design strategies and decisions. The key concepts were:

- The proposal for a 300m x 300m pedestrian and bicycle grid that will extend across the site and facilitate daily travel patterns;
- The design of a sufficient road network hosting new bus lines that would promote interconnectivity among the various means of transport;
- The relocation of the traffic flow of the existing main avenue;
- The restoration of the existing streams and their paths that would maximize alternative routes and experiences and minimize crossings by other types of transportation networks.

Finally, as mentioned in the Nodes system level, the objectives for monitoring transportation network processes, enhancing the network's resilience in crisis events, and reusing on-site materials for network construction were not included in this planning phase.

SUSTAINABLE ACTION

Develop clear road hierarchy in relation to the development's main attractors.

RELATED GUIDELINES

T4.2.1

ROADS AND ATTRACTORS

33m.	Relocated Poseidonos Avenue
30m.	Existing Poseidonos Avenue
27.2m.	Main axes
17.6m.	Local axes, Type 1
14.6m.	Local axes, Type 2
7m.	Local axes, Type 3
10.6m.	Service Road
15m.	Pedestrian path, Type 1
10m.	Pedestrian path, Type 2
11m.	Park perimeter path
8m.	Pedestrian path, Type 1
38m.	Reconstructed runways
	Metro stations
	Attractors / buildings

SUSTAINABLE ACTION

Connect local centers with low-emission transportation networks.

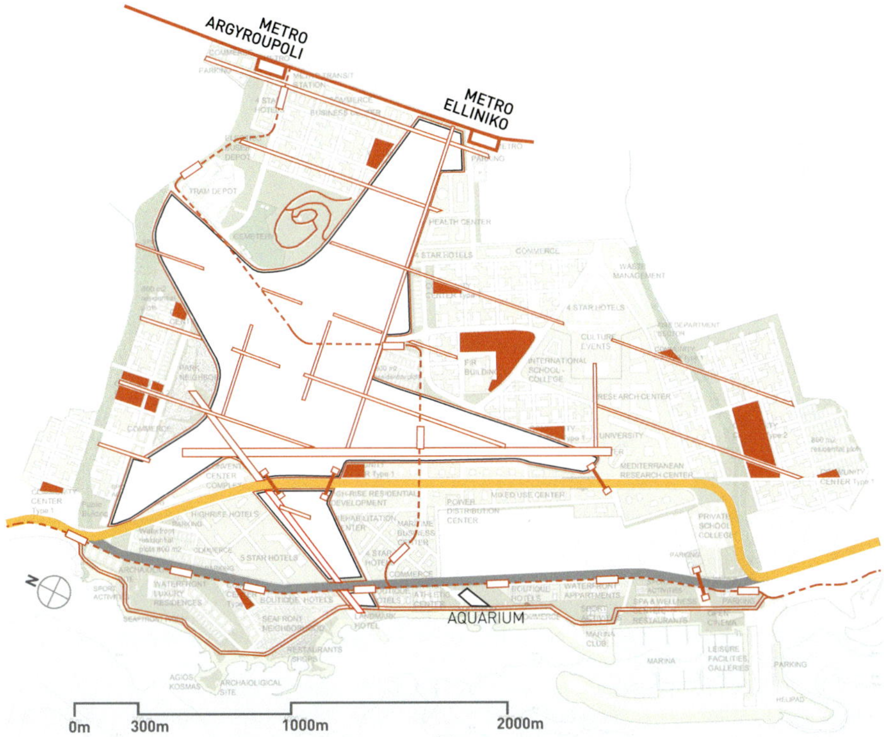

RELATED GUIDELINES

T4.5, T4.3.1, T4.5.1, T3.6.1

NETWORKS AND ATTRACTORS

- Metro stations
- Tram line stations
- Main walking and recycling axes
- Pedestrian bridges
- Local community centers
- Relocated Poseidonos Avenue
- Existing Poseidonos Avenue

CONCLUSION

The Hellinikon project design strategies, based from the very start on the Zofnass Program five focus areas, aimed to achieve a balance among environmental, social, and economic aspects and were not merely stylistic gestures.

In this chapter, the system-level checklists serve as the methodological tool for decoding the design decisions in relation to the project's infrastructures as well as for assessing two selected infrastructures.

The analysis of the results of the checklists highlights that the landscape's multifunctional role and the transportation system's goal of equal access were key elements for the master plan design, as they responded to the project's main challenges and demands. Additionally, comparison of those guidelines that have been taken into consideration with those that have not illustrates that the appropriate infrastructure elements are strongly connected to:

1. The project's nature. In the Hellinikon project, a city expansion case, the proposed infrastructures' features are subject to the capacity of the wider region's existing infrastructure systems and to the type of systems available in the site's surrounding municipalities.
2. The project's scale. In the case of the Hellinikon urban development, the proposal's design process began with land use distribution, the master plan design, and was completed with neighborhood-scale configuration.
3. The project's planning phase. Many guidelines were not considered in the course of the planning process, as they may be better adapted to the next phases, during the urban development's construction or operation.
4. The project's existing context. Although site constraints restricted the range of appropriate infrastructure elements, the constraints offered the potential for integration and synergy among the proposed infrastructures and other existing infrastructure networks, improving the performance of the systems in the wider region.

In conclusion, the main scope of this chapter is to demonstrate the highly significant role of infrastructures for a viable urban environment and to highlight infrastructure configuration as an integral part of the planning process. The use of the checklists decoding tool, introduced in the system level of the Zofnass Program's research for planning sustainable cities, served as the methodological tool for the infrastructures analysis and assessment of the Hellinikon project. The challenges for the Hellinikon project are multiple, particularly with the country witnessing an unprecedented recession since 2010. Yet these same challenges outline the need to create a lively, safe, and sustainable urban environment with state-of-the-art infrastructures which will reactivate Athens' potentials, secure long-term benefits, and promote the development of the former airport site as a most innovative case of city expansion.

NOTES

1. Spiro Pollalis, "About Hellinikon," accessed on February 15, 2015, http://pollalis-hellinikon.com/wp/?page_id=11.
2. Spiro Pollalis et al., "The Urban Development of the Former Athens Airport," paper presented at AESOP-ACSP Joint Congress, Dublin, 15–19 July, 2013.
3. Spiro Pollalis et al, " 'Hellinikon SA' Urban Development Model Technical Report, Part A," p. A.20 and A.21, accessed on February 15, 2015, http://fds.lib.harvard.edu/fds/deliver/46593785/013728962.pdf.
4. "Hellinikon Air Base," accessed on February 15, 2015, http://en.wikipedia.org/wiki/Hellenikon_Air_Base.
5. "Images of the Former Athens International Airport East Terminal," accessed on February 15, 2015, http://images.library.yale.edu/madid/showThumb.aspx?&q=athens%20airport&q1=&q2=&qc1=&qc2=&qf1=&qf2=&qn=&qo=&qm=15&qs=1&sid=&qx=#1778013>

and "List of works by Eero Saarinen," accessed on February 15, 2015, http://en.wikipedia.org/wiki/List_of_works_by_Eero_Saarinen.
6. Urban Environment Laboratory of the National Technical University of Athens, "Hellinikon Metropolitan Park of Athens," research, accessed on May 15, 2015, http://courses.arch.ntua.gr/elliniko.htm.
7. Serero Architectes, "Hellinikon Park" (2003), accessed on October 20, 2014, http://serero.com/projet/hellenikon-park/.
8. Rowe Peter and Hashim Sarkis, "Isopolis," Graduate School of Design, 1997.
9. Hellinikon SA, "The Company," accessed on March 01 , 2015, http://www.hellinikon.com/en/ .
10. Pollalis et al., " 'Hellinikon SA' Urban Development Model Technical Report."
11. Pollalis, "About Hellinikon."
12. Pollalis et al., " 'Hellinikon SA' Urban Development Model Technical Report," Part C, p. C.7.
13. Ibid., Part A, p. A.9.
14. Ibid., Part E, pp. E.32-E.49.
15. Greek State Law 4062/2012 (FEK A'70/30.03.2012), for Greek version site address: accessed on March 02, 2015, http://www.ypeka.gr/Default.aspx?tabid=506, and accessed on January 09, 2015, http://www.hradf.com/sites/default/files/attachments/Law_4062_2012_HELLINIKON.pdf.
16. Pollalis et al., " 'Hellinikon SA' Urban Development Model Technical Report," part A, pp. A.49 and A.50.
17. Ibid., part C, pp. C.10 and C.11.
18. The data for the design strategies presented in this document, have been extracted from a presentation submitted to the Greek newspaper TO BHMA, Prof. Spiro Pollalis, "Innovative Design Strategies," accessed on February 15, 2015, http://pollalis-hellinikon.com/wp/wp-content/uploads/2013/09/2012-12-22_%CE%95%CF%86%CE%B7%CE%BC%CE%B5%CF%81%CE%AF%CE%B4%CE%B1-%CE%92%CE%97%CE%9C%CE%91-%CE%A4%CE%BC%CE%AE%CE%BC%CE%B1-%CE%95%CF%80%CE%B9%CF%83%CF%84%CE%AE%CE%BC%CE%B7%CF%82.pdf, and Kafantaris Tasos, "Hellinikon: A vision for a new city," accessed on February 23, 2015, http://www.tovima.gr/science/article/?aid=492710.
19. As mentioned in Pollalis et al., "The Urban Development of the Former Athens Airport."
20. Alexandros Kyrtsis, *Constantinos A. Doxiadis, Essays, Plans, Settlements* (Athens: Ikaros, 2006)
21. Pollalis et al., " 'Hellinikon SA' Urban Development Model Technical Report," part D, pp. D.53, D.60, and D.61.
22. Image of the 1895 "Vari" map illustrating the streams crossing the Hellinikon and Agios Kosmas site, conducted by J. A. Kaupert, accessed on February 02, 2015, http://digi.ub.uni-heidelberg.de/diglit/curtius1895a/0012?sid=289744cb05a6e04e0d8d451dcd7eacd8.
23. Pollalis et al., " 'Hellinikon SA' Urban Development Model Technical Report," part A, p. A.47.

IMAGE CREDITS

P. 302: Accessed on October 28, 2015, https://hellinikonproject.wordpress.com/2013/09/09/%CF%84%CE%BF%CF%80%CE%BF%CE%B8%CE%B5%CF%83%CE%B9%CE%B1-%CE%B5%CE%BB%CE%BB%CE%B7%CE%BD%CE%B9%CE%BA%CE%BF-location-hellinikon. **P. 305:** (Top left) Accessed on October 28, 2015, http://fds.lib.harvard.edu/fds/deliver/46593785/013728962.pdf. **P. 305:** (Bottom) Accessed on October 28, 2015, http://pollalis-hellinikon.com/wp/wp-content/uploads/2013/12/Masterplan.jpg. **P. 306:** Accessed on October 28, 2015, http://hellinikonproject.files.wordpress.com/2013/02/landuse-plan.jpg. **P. 307:** Accessed on February, 15, 2015, http://fds.lib.harvard.edu/fds/deliver/46593785/013728962.pdf. **P. 308:** (Top right) Accessed on October 28, 2015, https://hellinikonproject.wordpress.com/2013/10/06/%CF%83%CF%84%CF%81%CE%B1%CF%84%CE%B7%CE%B3%CE%B9%CE%BA%CE%B5%CF%83-%CF%83%CF%87%CE%B5%CE%B4%CE%B9%CE%B1%CF%83%CE%BC%CE%BF%CF%85-designstrategies. **P. 308:** (Bottom right) Accessed on October 28, 2015, http://fds.lib.harvard.edu/fds/deliver/46593785/013728962.pdf. **P. 309:** (Top right) Accessed on October 28, 2015, http://fds.lib.harvard.edu/fds/deliver/46593785/013728962.pdf. **P. 309:** (Bottom right) Accessed on October 28, 2015, http://fds.lib.harvard.edu/fds/deliver/46593785/013728962.pdf. **P. 310:** (Top right) Accessed on October 28, 2015, http://fds.lib.harvard.edu/fds/deliver/46593785/013728962.pdf. **P. 310:** (Bottom right) Accessed on October 28, 2015, http://fds.lib.harvard.edu/fds/deliver/46593785/013728962.pdf. **P. 311:** (Top right) Accessed on October 28, 2015, http://fds.lib.harvard.edu/fds/deliver/46593785/013728962.pdf. **P. 311:** (Bottom right) Accessed on October 28, 2015, http://fds.lib.harvard.edu/fds/deliver/46593785/013728962.pdf. **P. 315:** Accessed on October 28, 2015, adapted from http://fds.lib.harvard.edu/fds/deliver/46593785/013728962.pdf. **P. 316:** (Background image) Accessed on October 28, 2015, http://pollalis-hellinikon.com/wp/wp-content/uploads/2013/12/Masterplan.jpg. **P. 318:** (Background image) Accessed on October 28, 2015, http://pollalis-hellinikon.com/wp/wp-content/uploads/2013/12/Masterplan.jpg. **P. 321:** Accessed on October 28, 2015, http://fds.lib.harvard.edu/fds/deliver/46593785/013728962.pdf. **P. 322:** (Top, background image) Accessed on October 28, 2015, http://pollalis-hellinikon.com/wp/wp-content/uploads/2013/12/Masterplan.jpg. **P. 322:** (Bottom, background image) Accessed on October 28, 2015, http://fds.lib.harvard.edu/fds/deliver/46593785/013728962.pdf. **P. 324:** (Background images) Accessed on October 28, 2015, http://fds.lib.harvard.edu/fds/deliver/46593785/013728962.pdf. **P. 326:** Accessed on October 28, 2015, adapted from http://fds.lib.harvard.edu/fds/deliver/46593785/013728962.pdf. **P. 328:** (Background image) Accessed on October 28, 2015, http://pollalis-hellinikon.com/wp/wp-content/uploads/2013/12/Masterplan.jpg. **P. 330:** (Background image) Accessed on October 28, 2015, http://pollalis-hellinikon.com/wp/wp-content/uploads/2013/12/Masterplan.jpg. **P. 341:** (Top, background image) Accessed on October 28, 2015, http://fds.lib.harvard.edu/fds/deliver/46593785/013728962.pdf. **P. 341:** (Bottom, background image) Accessed on October 28, 2015, http://pollalis-hellinikon.com/wp/wp-content/uploads/2013/12/Masterplan.jpg.

BIBLIOGRAPHY AND SOURCES

Alanne, Kari, and Arto Saari. "Distributed Energy Generation and Sustainable Development." Renewable and Sustainable Energy Reviews 10, no. 6 (2006): 539–558.

Amekudzi, Adjo A., Michael D. Meyer, and Catherine Laverne Ross. Transportation Planning for Sustainability Guidebook. Washington, DC: U.S. Federal Highway Administration, 2011.

American Water Works Association. "Resources and Tools for Water Professionals." Accessed April 05, 2014. http://www.awwa.org/resources-tools.aspx.

Anderson, Anthony B., and Clinton N. Jenkins. Applying Nature's Design: Corridors as a Strategy for Biodiversity Conservation. New York: Columbia University Press, 2006.

Arup. "Smart Cities: Transforming the 21st Century City via the Creative Use of Technology." 2011. http://www.arup.com/Publications/Smart_Cities.aspx.

Arup. "The Smart Solution for Cities." 2d Urban Life report. 2011. http://www.arup.com/Homepage_C40_UrbanLife.aspx.

Australian Institute of Landscape Architects (AILA). "Green Infrastructure Report." 2012. Accessed December 10, 2014. http://www.greeninfrastructure.org.au/.

Baby, S. "Assessing Morphological Landscape Carrying Capacity for Coastal Areas in Kuwait." Accessed December 05, 2014. http://www.niscair.res.in/jinfo/IJMS/IJMS-Forthcoming-Articles/IJMS-PR-Aug%202014/MS%202162%20Edited.pdf.

Beyond the Line. "Ecosystem Services." Accessed December 05, 2014. http://www.beyond-theline.org/KeyIdeas/EcosystemServices.aspx.

Biodiversity Information System for Europe. "Glossary of Terms." Accessed June 01, 2015. http://biodiversity.europa.eu/maes/glossary-of-terms.

Buchan, David. The Rough Guide to the Energy Crisis. London: Rough Guides, 2010.

California ISO. "What the Duck Curve Tells Us about Managing a Green Grid." 2013. Accessed June 15, 2015. https://www.caiso.com/Documents/FlexibleResourcesHelpRenewables_FastFacts.pdf.

Carpenter, Steve, Brian Walker, J. Marty Anderies, and Nick Abel. "From Metaphor to Measurement: Resilience of What to What?" Ecosystems 4 (2001): 765–781.

Carrera, Fabio. "City Knowledge: An Emergent Information Infrastructure for Sustainable Urban Maintenance, Management and Planning." Department of Urban Studies and Planning, Massachusetts Institute of Technology. September 2004. http://dspace.mit.edu/handle/1721.1/28790.

Ceruzzi, Paul. Computing: A Concise History. Cambridge, MA: MIT Press, 2012.

Chi, Kelly Rae, James MacGregor, and Richard King. Fair Miles: Recharting the Food Miles Map. London: Published by IIED in association with Oxfam GB, 2010.

City of Baltimore. "Planning / Baltimore Food Policy Initiative." Accessed November 20, 2015. http://archive.baltimorecity.gov/Government/AgenciesDepartments/Planning/BaltimoreFoodPolicyInitiative.aspx.

Convention on Biological Diversity. "What's the Problem?" Accessed December 17, 2014. http://www.cbd.int/agro/whatstheproblem.shtml.

Corbitt, Robert A. Standard Handbook of Environmental Engineering. New York: McGraw-Hill, 1999.

Costanza, Robert, Ralph d'Arge, Rudolf de Groot, Stephen Farber, Monica Grasso, Bruce Hannon, Karin Limburg, Shahid Naeem, Robert V. O'Neill, Jose Paruelo, Robert G. Raskin, Paul Sutton, and Marjan van den Belt. "The Value of the World's Ecosystem Services and Natural Capital." Nature 387 (15 May 1997): 253–260.

Data Revolution Group. "A World That Counts: Mobilising the Data Revolution for Sustainable Development." UN Report. Accessed April 2014. http://www.undatarevolution.org/report/.

Department for Transport (UK). "Factsheets, UK Transport Greenhouse Gas Emissions." Accessed November 22, 2015. https://combehavendefenders.files.wordpress.com/2014/01/climatechangefactsheets-1.pdf.

Dobbs, Richard, Corinne Sawers, Fraser Thompson, James Manyika, Jonathan Woetzel, Peter Child, Sorcha McKenna, and Angela Spatharou. "Overcoming Obesity: An Initial Economic Analysis." Discussion paper, McKinsey Global Institute. November 2014.

Dormehl, Luke. The Formula: How Algorithms Solve All Our Problems—and Create More. New York: Penguin Group, 2014.

Dunnett, Nigel, and Andy Clayden. Resources: The Raw Materials of Landscape. London: Spon Press, 2000.

Ecofys. Accessed January 13, 2014. www.ecofys.com.

Economist Intelligence Unit. "A Summary of the Livability Ranking and Overview." August 7, 2012. www.eiu.com.

Elkington, John. Cannibals with Forks: The Triple Bottom Line of 21st Century Business. Gabriola Island, BC: New Society Publishers, 1998.

Energy Storage Association. "Energy Storage Technologies." Accessed June 20, 2015. http://energystorage.org/energy-storage/energy-storage-technologies.

Energy Transition. "Glossary." Accessed January 16, 2015. http://energytransition.de/glossary/.

Engelhaupt, Erika. "Do Food Miles Matter?" Environmental Science and Technology 42, no. 10 (2008): 3482–3482.

Environmental Protection Agency (US). "Advancing Sustainable Materials Management: Facts and Figures." 2013. http://www.epa.gov/osw/nonhaz/municipal/pubs/MSWcharacterization_508_053113_fs.pdf.

Environmental Protection Agency (US). "A Handbook for Water and Wastewater Utilities." Accessed November 22, 2015. http://water.epa.gov/infrastructure/sustain/upload/EPA-s-Planning-for-Sustainability-Handbook.pdf.

Environmental Protection Agency (US). "EPA Strategic Plan Charts Direction for Next Four Years." October 2014. Accessed March 30, 2014. http://yosemite.epa.gov/opa/admpress.nsf/bd4379a92ceceeac8525735900400c27/75143c6974e6b05f85257cb60051d435!OpenDocument.

Environmental Protection Agency (US). "Municipal Solid Waste Generation, Recycling, and Disposal in the United States: Facts and Figures for 2010." http://www3.epa.gov/epawaste/nonhaz/municipal/pubs/msw_2010_rev_factsheet.pdf.

Environmental Protection Agency (US). "Municipal Solid Waste (MSW) in the United States: Facts and Figures." February 2014. http://www.epa.gov/solidwaste/nonhaz/municipal/msw99.htm.

Environmental Protection Agency (US). "Polluted Runoff (Nonpoint Source Pollution)." Accessed October 30, 2015. http://www.epa.gov/owow/nps/nps-conserve.html.

Environmental Protection Agency (US). "Water: Public Water Systems. Definition of a Public Water System." Accessed February 20, 2014. http://water.epa.gov/infrastructure/drinkingwater/pws/pwsdef2.cfm.

Ernst Strüngmann Forum. "Water." In Thomas E. Graedel and Ester van der Voet, eds., Linkages of Sustainability. Cambridge, MA: MIT Press, 2010.

European Commission. "Mapping and Assessment of Ecosystems and their Services." April 2013. http://ec.europa.eu/environment/nature/knowledge/ecosystem_assessment/pdf/MAESWorkingPaper2013.pdf.

European Commission. "National Renewable Energy Action Plans." Accessed January 10, 2014. http://ec.europa.eu/energy/renewables/action_plan_en.htm.

European Environment Agency (EEA). "Trends and Projections in Europe 2014." EEA Report No 6/2014, October 2014. Accessed February 10, 2015. http://www.eea.europa.eu/publications/trends-and-projections-in-europe-2014.

Faculty of Public Health (UK) in association with Natural England. "Great Outdoors: How Our Natural Health Service Uses Green Space to Improve Wellbeing." 2010, http://www.fph.org.uk/uploads/bs_great_outdoors.pdf.

Farmingfirst.org. 2015. "Green Economy | Farming First." Accessed October 29, 2015. http://www.farmingfirst.org/green-economy/.

Folke, Carl, Steve Carpenter, Brian Walker, Marten Scheffer, Thomas Elmqvist, Lance Gunderson, and C. S. Holling. "Regime Shifts, Resilience, and Biodiversity in Ecosystem Management." Annual Review of Ecology, Evolution, and Systematics 35 (2004): 557–581.

Food and Agriculture Organization of the United Nations (FAO). "Belo Horizonte: Growing Greener Cities in Latin America and the Caribbean." Accessed November 20, 2015. http://www.fao.org/ag/agp/greenercities/en/GGCLAC/belo_horizonte.html.

Food and Agriculture Organization of the United Nations (FAO). "Food Agriculture and Cities." Accessed October 29, 2015. http://www.fao.org/fileadmin/templates/FCIT/PDF/FoodAgriCities_Oct2011.pdf.

Food and Agriculture Organization of the United Nations (FAO). "Food and Nutrition in Numbers." 2014. Accessed October 29, 2015. http://www.fao.org/3/a-i4175e.pdf.

Food and Agriculture Organization of the United Nations (FAO). "Food Wastage Footprint: Impacts on Natural Resources." 2013. Accessed October 29, 2015. http://www.fao.org/docrep/018/i3347e/i3347e.pdf.

Food and Agriculture Organization of the United Nations (FAO). "How to Feed the World in 2050." Accessed October 29, 2015. http://www.fao.org/fileadmin/templates/wsfs/docs/expert_paper/How_to_Feed_the_World_in_2050.pdf.

Forman, Richard T. T. Landscape Mosaics: The Ecology of Landscapes and Regions. New York: Cambridge University Press, 1995.

Forman, Richard T. T. Urban ecology: science of cities. Cambridge, Mass.: Cambridge University Press, 2014.

Forman, Richard T. T., et al. Road Ecology: Science and Solutions. Washington, DC: Island Press, 2003.

Gleick, Peter H., ed. The World's Water: The Biennial Report on Freshwater Resources. 7th ed. Produced biennially by the Pacific Institute for Studies in Development, Environment, and Security. Washington, DC: Island Press, 2011.

Gleick, Peter H., ed. The World's Water: The Biennial Report on Freshwater Resources. 8th ed. Produced biennially by the Pacific Institute for Studies in Development, Environment, and Security. Washington, DC: Island Press, 2014.

Gold, M. "Sustainable Agriculture: Information Access Tools." United States Department of Agriculture, Alternative Farming Systems Information Center, May 2015. Accessed October 29, 2015. http://www.nal.usda.gov/afsic/pubs/agnic/susag.shtml#define.

Goldsmith, Stephen, and Susan P. Crawford. The Responsive City: Engaging Communities through Data-Smart Governance. San Francisco: Jossey-Bass, 2014.

Goldstein, Brett. Beyond Transparency: Open Data and the Future of Civic Innovation. San Francisco: Code for America Press, 2013.

Gordon, Alex, Christopher Akinrele, and Reed Smith. "Right to Water." Legal Guide, June 2011. Updated March 2012. Published by Advocates for International Development (A4ID). Accessed April 5, 2014. http://a4id.org/sites/default/files/user/%5BA4ID%5D%20Right%20to%20Water.pdf.

Green Infrastructure Evidence Base. "2 Green Infrastructure: Concepts and Definitions." Accessed August 10, 2014. http://gievidencebase.botanicgardens.sa.gov.au/contents/green-infrastructure-concepts-and-definitions.

Green Municipal Fund. "Getting to 50% and Beyond: Waste Diversion Success Stories from Canadian Municipalities." http://www.fcm.ca/Documents/tools/GMF/Getting_to_50_percent_en.pdf.

Grey, David, and C. W. Sadoff. "Sink or Swim? Water Security for Growth and Development." Water Policy 9 (2007): 545–571. http://cip.management.dal.ca/publications/Water%20security%20for%20growth%20and%20development.pdf.

GRID-Arendal. "Vital Climate Graphics." GRID-Arendal publications 2014. Accessed April 26, 2014. http://www.grida.no/publications/vg/climate/page/3090.aspx.

Halweil, Brian. Home Grown: The Case for Local Food. Washington, DC: Worldwatch Institute, 2002.

Harvard Leadership Conference 2014. "Policymakers in Cyber Security: War Gaming HTTPS and the DigiNotar Attack." http://hgc.harvard.edu/hlc/2014/agenda.

Hellinikon SA. "The Company." Accessed March 01, 2015. http://www.hellinikon.com/en/.

Hitachi, Ltd. 2012. "Hitachi's Vision of the Smart City." SCW-E001, Hitachi's Social Innovation Business. http://www.hitachi.com/products/smartcity/.

Hoornweg, Daniel, and Perinaz Bhada-Tata. "What a Waste: A Global Review of Solid Waste Management." World Bank, Urban Development & Local Government Unit, No. 15, March 2012. http://siteresources.worldbank.org/INTURBANDEVELOPMENT/Resources/336387-1334852610766/What_a_Waste2012_Final.pdf.

Hutchinson, B. G. Principles of Urban Transportation Systems Planning. New York: McGraw-Hill, 1974.

IBM Corporation. "IBM Intelligent Transportation." IBM Industry Solutions, Solution Brief. http://files.cnblogs.com/files/menghe/Intelligent_Transportation_GWS03009USEN.pdf.

IBM Corporation. "IBM Intelligent Water." IBM Industry Solutions, Solution Brief. http://www-935.ibm.com/services/multimedia/Intelligent_Water.pdf.

IBM Corporation. "Smarter, More Competitive Cities." IBM Smarter Cities, Point of View, 2014. https://www.ibm.com/smarterplanet/global/files/smarter_more_competitive_cities.pdf.

IBM Institute for Business Value. "Smarter Cities for Smarter Growth: How Cities Can Optimize Their Systems for the Talent-Based Economy." IBM Global Business Services, Government Executive Report. http://www.zurich.ibm.com/pdf/isl/infoportal/IBV_SC3_report_GBE03348USEN.pdf.

IEA Bioenergy. "Benefits of Bioenergy." 2005. http://www.seai.ie/Renewables/Bioenergy/Benefits_of_bioenergy.pdf.

Independent Evaluation Group. 2011. "Capturing Technology for Development, An Evaluation of World Bank Group Activities in Information and Communication Technologies." Independent Evaluation Group, World Bank Group, 2011. http://ieg.worldbankgroup.org/Data/reports/ict_evaluation.pdf.

INSEAD and World Economic Forum. "The Global Information Technology Report 2009–2010, ICT for Sustainability." http://www.ifap.ru/library/book466.pdf.

INSEAD and World Economic Forum. "The Global Information Technology Report 2010–2011 Transformations 2.0." 10th Anniversary Edition. http://www.ifap.ru/library/book494.pdf.

INSEAD and World Economic Forum. "The Global Information Technology Report 2013, Growth and Jobs in a Hyperconnected World." http://www3.weforum.org/docs/WEF_GITR_Report_2013.pdf.

INSEAD, World Economic Forum, and Johnson Graduate School of Management, Cornell University. "The Global Information Technology Report 2014, Rewards and Risks of Big Data." http://www3.weforum.org/docs/WEF_GlobalInformationTechnology_Report_2014.pdf.

Institute for Sustainable Infrastructure (ISI). "Envision® Sustainable Infrastructure Rating System." Accessed April 29, 2013. http://www.sustainableinfrastructure.org/rating/index.cfm.

International Energy Agency (IEA). "Climate and Electricity Annual 2011." 2011. http://www.iea.org/publications/freepublications/publication/Climate_Electricity_Annual2011.pdf.

International Energy Agency (IEA). "Energy Outlook 2013." 2013. http://www.worldenergyoutlook.org/weo2013/

International Financial Corporation. "IFC Performance Standards on Environmental and Social Sustainability." International Financial Corporation, World Bank Group, January 2012. http://www.ifc.org/wps/wcm/connect/115482804a0255db96fbffd1a5d13d27/PS_English_2012_Full-Document.pdf?MOD=AJPERES.

International Society for Ecological Economics. "Research Archives." Accessed April 26, 2014. http://www.isecoeco.org/category/resources/research-2/.

International Telecommunication Union (ITU). "Monitoring the WSIS Targets: A Mid-term Review." World Telecommunication/ICT Development Report 2010. http://www.uis.unesco.org/Communication/Documents/WTDR2010_e.pdf.

International Water Management Institute. "Regions." 2012. Accessed April 26, 2014. http://www.iwmi.cgiar.org/regions/.

Jollands, Nigel. "Cities and Energy." In Competitive Cities and Climate Change. Paris: OECD Publishing, 2009.

Kafantaris, Tasos. "Hellinikon: A Vision for a New City." Accessed February 23, 2015. http://www.tovima.gr/science/article/?aid=492710.

Khalil, Mohsen A., Philippe Dongier, Valerie D'Costa, Christine Zhen-Wei Qiang, Peter L. Smith, Randeep Sudan, Eric Swanson, and Bjorn Wellenius. "Information and Communications for Development: Extending Reach and Increasing Impact." The World Bank Group, 2009. http://documents.worldbank.org/curated/en/2009/01/10647109/2009-information-communications-development-extending-reach-increasing-impact

King, Michael. "Community Energy: Planning, Development and Delivery." International District Energy Association, 2012. http://www.districtenergy.org/assets/pdfs/Community-Energy-Dev-Guide-US-version/USCommunityEnergyGuidehi.pdf.

Koc, Mustafa, Rod MacRae, Jeniffer Welsh, and Luc J. A. Mougeot. For Hunger-Proof Cities. Ottawa: International Development Research Centre, 1999.

Konsbruck, Robert Lee. "Impacts of Information Technology on Society in the New Century." http://www.zurich.ibm.com/pdf/news/Konsbruck.pdf.

Kyrtsis, Alexandros-Andreas, ed. Constantinos A. Doxiadis: Essays, Plans, Settlements. Athens: Ikaros, 2006.

Lappe, Frances. 2014. "The City That Ended Hunger." YES! Magazine, spring 2009. Accessed November 20, 2015. http://www.yesmagazine.org/issues/food-for-everyone/the-city-that-ended-hunger.

Le Courtois, Alexandra. "Municipal Solid Waste: Turning a Problem into Resource." In Waste: The Challenges Facing Developing Countries, no. 15, October 2012. Private Sector and Development, PROPARCO.

Littlefair, Paul J. Environmental Site Layout Planning: Solar Access, Microclimate and Passive Cooling in Urban Areas. London: Building Research Establishment, 2000.

Lyle, John Tillman. Regenerative Design for Sustainable Development. New York: Wiley, 1994.

Makropoulos, Christos K., and David Butler. "Distributed Water Infrastructure for Sustainable Communities." Water Resources Management 24, no. 11 (2010): 2795–2816.

Mayer, Peter W., and William B. DeOreo. Residential End Uses of Water. Denver: AWWA Research Foundation and American Water Works Association, 1999.

McDougall, Forbes R., Peter R. White, Marina Franke, and Peter Hindle. Integrated Solid Waste Management: A Life Cycle Inventory. Oxford: Blackwell Publishing, 2001.

McHarg, Ian L. Design with Nature. 2nd ed. New York: John Wiley & Sons, 1992.

McKinsey & Company. "How to Make a City Great." McKinsey Cities Special Initiative 2013. http://www.mckinsey.com/insights/urbanization/how_to_make_a_city_great.

Meiklejohn, Katie, Rob Ament, and Gary Tabor. "Habitat Corridors & Landscape Connectivity: Clarifying the Terminology." Center for Large Landscape Conservation. http://largelandscapes.org/media/publications/Habitat-corridors-and-landscape-connectivity1.pdf.

Metz, B., O. R. Davidson, P. R. Bosch, R. Dave, and L. A. Meyer, eds. Climate Change 2007: Mitigation of Climate Change. Contribution of Working Group III to the Fourth Assessment Report of the Intergovernmental Panel on Climate Change. Cambridge: Cambridge University Press, 2007.

Millennium Ecosystem Assessment. Ecosystems and Human Well-Being: A Framework for Assessment. Washington, DC: Island Press, 2003.

Millennium Ecosystem Assessment. Ecosystems and Human Well-Being: Health Synthesis. Washington, DC: World Resources Institute, 2005.

Ministry of Environment and Energy (Greece). "Greek State Law 4062/2012 (FEK A'70/30.03.2012)." Accessed March 02, 2015. http://www.ypeka.gr/Default.aspx?tabid=506. English version site address, accessed January 09, 2015, http://www.hradf.com/sites/default/files/attachments/Law_4062_2012_HELLINIKON.pdf.

Mitchell, William J. Me++: The Cyborg Self and the Networked City. Cambridge, MA: MIT Press, 2003.

Mobbs, Paul. "A Practical Guide to Sustainable IT 2012." Commissioned by the Association for Progressive Communications (APC), with the support of the International Development Research Centre (IDRC). http://www.apc.org/en/system/files/APC_a_practical_guide_web.pdf

Naumann, S., M. Davis, et al. "Design, Implementation and Cost Elements of Green Infrastructure Projects." Final report to the European Commission. Brussels, 2011.

Nestle, Marion. What to Eat. New York: North Point Press, 2006.

Newman, Peter, and Timothy Beatley. Resilient Cities Responding to Peak Oil and Climate Change. Washington, DC: Island Press, 2009.

New York City Economic Development Corporation. "Food Retail Expansion to Support Health (FRESH)." Accessed November 20, 2015. http://www.nycedc.com/program/food-retail-expansion-support-health-fresh.

Ng, Edward. Designing High-Density Cities for Social and Environmental Sustainability. London: Earthscan, 2010.

NYC Special Initiative for Rebuilding and Resilience. "A Stronger, More Resilient New York." NYC Report. http://www.nyc.gov/html/sirr/html/report/report.shtml.

Office for United Nations High Commission for Human Rights (OHCHR). "OHCHR Study on Human Rights Obligations Related to Equitable Access to Safe Drinking Water and Sanitation: Contributions Received." Accessed April 20, 2014. http://www2.ohchr.org/english/issues/water/contributions.htm.

O'Neill, Katie, and P. J. Rudden. "Good Practice Report." http://ec.europa.eu/environment/europeangreencapital/wp-content/uploads/2011/04/MDR0763Rp00013_Good-Practice_Final2.pdf.

Paarlberg, Robert L. Food Politics. What Everyone Needs to Know. Oxford: Oxford University Press, 2010. 2nd ed. 2013.

Pacific Institute. "The World's Water: Water Conflict." 2013. Accessed April 26, 2014. http://worldwater.org/water-conflict/.

Passive House Institute US. Accessed January 13, 2014. http://www.passivehouse.us/passiveHouse/PHIUSHome.html.

PennWell Corporation. Accessed March 30, 2014. http://www.waterworld.com/.

Piotrowicz, Wojciech, and Richard Cuthbertson. "Sustainability—A New Dimension in Information Systems Evaluation." Said Business School, University of Oxford, 2009. http://eureka.sbs.ox.ac.uk/771/1/Piotrowicz_Cuthbertson,_sustainability_-_a_new_dimensionin_in_IS_evaluation.pdf.

Pollalis, Spiro N. "About Hellinikon." Accessed February 15, 2015. http://pollalis-hellinikon.com/wp/?page_id=11.

Pollalis, Spiro N. "Innovative Design Strategies." Presentation submitted to the Greek newspaper "TO BHMA." Accessed February 15, 2015. http://pollalis-hellinikon.com/wp/wp-content/uploads/2013/09/2012-12-22_%CE%95%CF%86%CE%B7%CE%BC%CE%B5%CF%81%CE%AF%CE%B4%CE%B1-%CE%92%CE%97%CE%9C%CE%91-%CE%A4%CE%BC%CE%AE%CE%BC%CE%B1-%CE%95%CF%80%CE%B9%CF%83%CF%84%CE%AE%CE%BC%CE%B7%CF%82.pdf.

Pollalis, Spiro N, et al. "'Hellinikon SA' Urban Development Model Technical Report, Part A." Accessed February 15, 2015. http://fds.lib.harvard.edu/fds/deliver/46593785/013728962.pdf.

Pollalis, Spiro N, et al. "The Urban Development of the Former Athens Airport." Paper presented at AESOP-ACSP Joint Congress, Dublin, July 15–19, 2013.

Pollalis, Spiro N, Andreas Georgoulias, Stephen Ramos, and Daniel Schodek. Infrastructure Sustainability and Design. New York: Routledge, 2012.

Qiang, Christine Zhen-Wei. "Information Communications Technology for Development." The World Bank. Accessed April 15, 2014. http://live.worldbank.org/information-communications-technology-development.

Rights to Water and Sanitation. "The Rights to Water and Sanitation in National Law." Accessed April 20, 2014. http://www.righttowater.info/progress-so-far/national-legislation-on-the-right-to-water/.

Rocky Mountain Institute. Accessed January 13, 2014. http://www.rmi.org/.

Rogers, Peter P., Kazi F. Jalal, and John A. Boyd. An Introduction to Sustainable Development. London: Earthscan, 2008.

Rogers, Richard George, and Anne Power. Cities for a Small Country. London: Faber, 2000.

Rottle, Nancy, and Ken Yocom. Basics Landscape Architecture 02: Ecological Design. United Kingdom: AVA Publishing, 2010.

Rowe, Peter G., and Hashim Sarkis, eds. Isopolis: Addressing Scales of Urban life in Modern Athens. Cambridge, MA: Harvard University Graduate School of Design, 1997.

Sarker, Ashutosh, Helen Ross, and Krishna K. Shrestha. "A Common-Pool Resource Approach for Water Quality Management: An Australian Case Study." Ecological Economics 68 (2008): 461–471.

Selman, P. "Planning for Landscape Multifunctionality." Sustainability: Science, Practice, and Policy 5, no. 2 (2009). http://www.google.gr/archives/vol5iss2/communityessay.pselman.html.

Serero Architectes. "Hellinikon Park." 2003. Accessed October 20, 2014. http://serero.com/projet/hellenikon-park/.

Sherriff, Graeme, and Catalina Turcu, under the guidance of Yvonne Rydin and Simon Guy. "Energy: Looking to the Future: A Tool for Strategic Planning." The CLUES project (Challenging Lock-in through Urban Energy Systems), University College London, October 2012. http://www.ucl.ac.uk/clues/files/CLUES_Tool_2013.

Sniffer Project UEUB01. "Urban Networks for People and Biodiversity—Form and Function." May 2008. http://www.sniffer.org.uk/search?q=Multifunctional+green+networks+.

Spivack, Nova. "Minding the Planet: The Meaning and Future of the Semantic Web." http://lifeboat.com/ex/minding.the.planet.

State of South Australia. "Sustainable Landscapes Project." Accessed June 05, 2014. http://www.environment.sa.gov.au/botanicgardens/Learn/Green_Infrastructure/Sustainable_Landscapes_Project.

Stockholm International Water Institute. "2008 World Water Week in Stockholm." Accessed November 25, 2015. http://www.worldwaterweek.org/

Tester, Jefferson W. Sustainable Energy: Choosing among Options. Cambridge, MA: MIT Press, 2005. 2nd ed., 2012.

Thorne, J. F. "Landscape Ecology: A Foundation for Greenway Design." In D. S. Smith and P. Cawood Hellmund, eds., Ecology of Greenways: Design and Function of Linear Conservation Areas. Minneapolis: University of Minnesota Press, 1993.

Townsend, Anthony M. Smart Cities: Big Data, Civic Hackers, and the Quest for a New Utopia. New York: Norton, 2013.

Unicef and World Health Organization. "Progress on Drinking Water and Sanitation 2012 Update." Accessed October 30, 2015. http://www.unicef.org/media/files/JMPreport2012.pdf.

United Nations. "General Assembly Adopts Resolution Recognizing Access to Clean Water, Sanitation as Human Right, by Recorded Vote of 122 in Favour, None against, 41 Abstentions." Accessed April 10, 2014. http://www.un.org/News/Press/docs/2010/ga10967.doc.htm.

United Nations Conference on Trade and Development. "Information Economy Report 2012: The Software Industry and Developing Countries." United Nations, New York and Geneva 2012. http://unctad.org/en/PublicationsLibrary/ier2012_en.pdf.

United Nations Environment Programme (UNEP). "Solid Waste Management, Volume I." December 2005. http://www.unep.org/ietc/Portals/136/SWM-Vol1-Part1-Chapters1to3.pdf.

United Nations Environment Programme (UNEP). "UNEP 2003 Annual Report." 2003. http://www.unep.org/annualreport/2003/.

United Nations Environment Programme (UNEP). "Waste: Investing in Energy and Resource Efficiency." 2011. http://www.fcm.ca/Documents/tools/GMF/Getting_to_50_percent_en.pdf.

United Nations Environment Programme (UNEP). "Water Withdrawal and Consumption: The Big Gap." Accessed April 21, 2014. http://www.unep.org/dewa/vitalwater/article42.html.

University of Washington. "Grocery Delivery Service Is Greener than Driving to the Store." Last modified April 29, 2013. http://www.washington.edu/news/2013/04/29/grocery-delivery-service-is-greener-than-driving-to-the-store/.

Urban Environment Laboratory of the National Technical University of Athens. "Hellinikon Metropolitan Park of Athens." Accessed May 15, 2015. http://courses.arch.ntua.gr/elliniko.htm.

Verebes, Tom, ed. Masterplanning the Adaptive City: Computational Urbanism in the Twenty-First Century. London: Routledge, 2014.

Water Footprint Network. Accessed November 25, 2015. http://www.waterfootprint.org/.

Waterlex. 2013. Accessed April 20, 2014. http://www.waterlex.org.

Whole Building Design Guide of the National Institute of Building Sciences. "Optimize Operational and Maintenance Practices." October 2014. Accessed December 03, 2014. http://www.wbdg.org/design/optimize_om.php.

Wolf, A. T., J. Natharius, J. Danielson, B. Ward, and J. Pender. "International River Basins of the World." International Journal of Water Resources Development 15, no. 4 (1999): 387–427.

Wolf, Kathleen L. "Ergonomics of the City: Green Infrastructure and Social Benefits." In Engineering Green: Proceedings of the 2003 National Urban Forest Conference. Washington, DC: American Forests, 2003.

World Business Council for Sustainable Development. "Water Facts and Trends." 2005. Accessed October 30, 2015. http://www.unwater.org/downloads/Water_facts_and_trends.pdf.

World Food Programme. "Climate Change." Accessed November 25, 2015. http://www.wfp.org/climate-change.

World Health Organization. "Health Topics: Diet." Accessed November 25, 2015. http://www.who.int/topics/diet/en/.

World Nuclear Association. "Comparison of Lifecycle Greenhouse Gas Emissions of Various Electricity Generation Sources." Accessed December 10, 2014. http://www.world-nuclear.org/WNA/Publications/WNA-Reports/Lifecycle-GHG-Emissions-of-Electricity-Generation/.

World Nuclear Association. "Renewables 2014 Global Status Report." Accessed December 10, 2014. http://www.ren21.net/Portals/0/documents/Resources/GSR/2014/GSR2014_full%20report_low%20res.pdf.

World Resource Institute. "Topics: Water. Mapping, Measuring, and Mitigating Global Water Challenges." Accessed April 20, 2014. http://www.wri.org/our-work/topics/water.

Wuppertal Institut für Klima, Umwelt, Energie. "Eco-innovation—Putting the EU on the Path to a Resource and Energy Efficient Economy." Accessed January 13, 2014. http://epub.wupperinst.org/frontdoor/index/index/docId/3433.

Yerra, Bhanu Murali Krishna. "The Emergence of Hierarchy in Transportation Networks." M.S. thesis, University of Minnesota, 2003.

Zero Waste Alliance. "ZW Definition." Accessed February 5, 2014. http://zwia.org/standards/zw-definition/.

ZOFNASS PROGRAM CORE RESEARCH TEAM FOR THE PLANNING GUIDELINES

Professor Spiro N. Pollalis is Professor of Design, Technology and Management at the Harvard Design School. Since 2008, he has been the Director of the Zofnass Program for Sustainable Infrastructure that has led to the Envision® Rating System. He is also the Principal Investigator of the Gulf Sustainable Urbanism project studying 10 cities in the Arab Gulf, sponsored by the Qatar Foundation. He has taught as a visiting professor at the ETH-Zurich, Switzerland; TU-Delft, Holland; Uni-Stuttgart, Germany; U-Patras, Greece; and has offered joint courses with the Harvard Business School on planning and development. He serves as the co-chair of the Advisory Committee on Future Cities for the Singapore-ETH Center. Professor Pollalis received his first degree from the University in Athens (EMP) and his Master's and PhD from MIT. His MBA in high technology is from Northeastern University. He has an honorary Master's degree in Architecture from Harvard.

Evgenia Hagistavrou is a Researcher at the Harvard Graduate School of Design for the Zofnass Program for Sustainable Infrastructure. She has 10 years of professional experience in landscape planning and design, landscape integration impact analysis, as well as urban planning. She holds a diploma of Architecture from the School of Architecture of the National Technical University of Athens and a Master's degree in Landscape Architecture from the Barcelona School of Architecture, Technical University of Catalonia.

Angela Kouveli is a Research Associate at the Harvard Graduate School of Design for the Zofnass Program for Sustainable Infrastructure and previously worked on the Gulf Sustainable Urbanism project. During 2011–2012, she was the coordinator and urban planner of the Hellinikon Urban Development Model project at Hellinikon SA, a Greek state company responsible for the management of the former Athens Airport area at Hellinikon. As a master plan designer she has collaborated with Professor Spiro N. Pollalis in large-scale projects in Pakistan and South Korea. She holds a Master's degree (MArch) in Territorial, Research & Planning from IAAC Barcelona and a Diploma in Architecture as well as a Master's degree (MPhil) in Architecture Design-Space-Culture from the School of Architecture of the National Technical University of Athens.

Dimosthenis Lappas is an architect and urban designer with expertise in sustainable urban planning. Dimosthenis is currently a researcher at the Zofnass Program for Sustainable Infrastructure at the Harvard Graduate School of Design. Before that he was a Research Associate for the Gulf Sustainable Urbanism project. He has worked in architectural offices in London and Athens and is accredited as an Envision® Sustainability Professional (ENV SP) by the Institute for Sustainable Infrastructure (ISI). Dimosthenis studied architecture at the Technical University of Athens and received a Master of Architecture from University College London (UCL).

Eleonora Marinou is an architect and currently a Researcher at the Harvard Graduate School of Design for the Zofnass Program for Sustainable Infrastructure. Since 2008 she has participated in multiple international urban planning and research projects. As a researcher she worked on the development of a new model settlement of 25,000 inhabitants for the victims of the 2008 wildfires in Peloponnese, Greece. She also contributed as research assistant to the holistic, cross-disciplinary and cross-border research project Gulf Sustainable Urbanism. As an urban planner she worked for the preparation and development of a detailed master plan and sustainable infrastructure solutions for the new DHA City Karachi now under construction. She holds a Diploma in Architecture from the University of Patras, Greece.

Yannis Orfanos is a Research Associate at Harvard's Graduate School of Design, where he is a Core Group member of the Zofnass Program for Sustainable Infrastructure and a Computational Design Lead at the Health and Places Initiative. His work focuses on applied research, computational urbanism, and using data to address urban sustainability challenges. Yannis has worked within a number of international design practices in London, Barcelona, and Athens, including KPF Associates and Prof. Pollalis Inc. where he was the Project Manager for the new DHA City Karachi for 600,000 people. He was a key collaborator as Consultant for the Harvard research project Gulf Sustainable Urbanism. Yannis is accredited as an Envision® Sustainability Professional & Trainer (ENV SP) by the Institute for Sustainable Infrastructure and is a Registered Architect in the UK (ARB). His education includes a Diploma in Architectural Engineering and MPhil in Architecture from National Technical University of Athens, and an MArch in Architecture and Urbanism from Architectural Association Design Research Lab.

Vicky Sagia is an architect and planner specializing in sustainable planning and design strategies. As a Research Associate for the Harvard Graduate School of Design she has worked on the development of the Zofnass Program Planning Guidelines and the Gulf Sustainable Urbanism project, investigating the correlations between urban development, building techniques, and the natural environment in the region. She has worked for large-scale urban planning projects, such as the new DHA City Karachi for 600,000 people, currently under construction, and the mixed-use development of the 1,500-acre former Athens International Airport site. Vicky is an Envision® Sustainability Professional (ENV SP) and holds a Diploma of Architecture from the University of Thessaly and an MSc in Sustainable Environmental Design from the Architectural Association Graduate School.

Richa S. Vuppuluri has been a Research Associate at the Harvard Graduate School of Design since 2010, where she has been lead researcher and program manager for the Gulf Sustainable Urbanism project and the Zofnass Program for Sustainable Infrastructure. Her primary interests are in design research, integrative methods for urban practice, environmental management, and international development. Richa was a recipient of a Community Service Fellowship Award from Harvard's GSD and the Penny While Travel Fellowship to study greenhouse agriculture in Almeria, Spain. She has worked in architectural offices in Mumbai and held a teaching position in KRV Institute for Architecture, Mumbai. She is accredited as an Envision® Sustainability Professional & Trainer (ENV SP) by the Institute for Sustainable Infrastructure, and is a member of the American Water Works Association and the Council of Architects in India. Richa holds the degrees of Bachelor of Architecture from Mumbai University and Master of Architecture in Urban Design from Harvard University.

Olga Tzioti is an architect specializing in sustainable environmental design. She has worked in several architectural offices in Athens and Geneva and has participated in a variety of projects as a sustainable design consultant. Since 2009 her work has focused on sustainable urban planning, master planning, and infrastructure. She has participated as a Research Associate in several sustainability projects including Harvard's Gulf Sustainable Urbanism project and the Zofnass Program for Sustainable Infrastructure. Olga holds a Diploma in Architecture from the National Technical University of Athens and has obtained a Master of Science in Sustainable Environmental Design from the Architectural Association School of Architecture.

CONTRIBUTORS

Gary Adamkiewicz is Assistant Professor of Environmental Health and Exposure Disparities at the Harvard T. H. Chan School of Public Health, where his work focuses primarily on environmental exposures within low-income communities. Dr. Adamkiewicz also serves as the Healthy Cities Program Leader at the Harvard Center for Health and the Global Environment where he directs studies examining the intersection between sustainability and environmental health. He co-created the Harvard Extension School course, "From Farm to Fork: Why What You Eat Matters," which focuses on the local and global implications of our dietary decisions. Dr. Adamkiewicz holds a Ph.D. in chemical engineering from the Massachusetts Institute of Technology and a Master of Public Health from the Harvard T. H. Chan School of Public Health.

Robert M. Beinstein (PE, ENV SP, EFCG) is Director of Sustainability for the Environmental Financial Consulting Group. He works closely with the Zofnass Program for Sustainable Infrastructure at Harvard and the Institute for Sustainable Infrastructure, roles that leverage his passion for developing and articulating the business case for realigning engineering and corporate practices around more sustainable outcomes. He has a B.S. in civil and environmental engineering and has held technical and managerial roles in the commercial nuclear industry, managed environmental programs in the Federal marketplace, and led a diverse array of client service-related work. Bob is an advisory board member at the Zofnass Program, a licensed professional engineer, a certified Envision® Sustainability Professional, and an approved Envision® trainer.

Terry D. Bennett (LS, LPF, MRICS, ENV SP, LEED AP) is the senior industry strategist for civil infrastructure at Autodesk. He is responsible for setting the company's future vision and strategy for technology serving the planning, surveying, civil engineering, and heavy construction industries, as well as cultivating and sustaining the firm's relationships with strategic industry leaders and associations. He currently serves as a Board Member of Harvard's Graduate School of Design's Sustainable Infrastructure Advisory Board, is a founding editorial board member of *Rebuilding America's Infrastructure magazine*, a charter member of the Institute for Sustainable Infrastructure and a member of its economics and technical committee, and a council member on the Urban Land Institute's Public Development and Infrastructure Council. You can find his publications and projects on his LinkedIn page at https://www.linkedin.com/in/terrydbennett.

William J. Bertera is President and CEO of the Institute for Sustainable Infrastructure. He is the former Executive Director of the Water Environment Federation, the Rebuild America Coalition, and the American Public Works Association. He has also held senior management positions with the International City Management Association, the National Association of Counties, and the National Solid Wastes Management Association.

Laura Bonich (PE, LEED AP) has worked for NV5 since 1994, where she directs the Sustainability Practice and provides strategic approaches to all aspects of infrastructure master planning, emphasizing an integrated process for water use and water resource planning. Laura has advanced goals for the sustainable optimization of infrastructure, recommending best practices that help reduce potable water use and wastewater generation as a complement to programs to reduce outdoor water use on large public and private projects throughout the United States. She further emphasizes the financial impacts of implementing sustainability to reduce traditional infrastructure and site development costs. Laura is a registered civil engineer, LEED AP, Chair of the Harvard Graduate School of Design Sustainable Infrastructure Advisory Board, Member of the Institute for Sustainable Infrastructure National Technical Committee for the Envision® rating system, full member of the Urban Land Institute, past branch president of ASCE, and holds a Master's in Business Administration.

Jonathan Buckley (BAHons, DipLA, CMLI, FFB) is a professional landscape architect with over 22 years of private-sector experience. He is a graduate in Landscape Architecture from Gloucestershire College of Arts and Design and then post-graduate Leeds Metropolitan University with Commendation. Jonathan has specialist experience in the development of urban green spaces, large-scale landscape planning, education, and historical projects. Jonathan also leads the Urban Development & Infrastructure practice at Golder, which operates out of 180 offices in 26 countries.

Michael Cant is a Principal and the Global Waste Practice Leader at Golder Associates in Whitby, Ontario, Canada. He has over 25 years' experience working in the waste management field. He has conducted presentations across North America on waste planning, diversion, and life cycle assessments. He was the Project Director for the Region of Waterloo's Residual Waste Management Planning Study and the City of Toronto business plan for a Mechanical Biological Treatment Facility. He assisted the City of Surrey on a business case for an 80,000-per-year-tonne Anaerobic Digestion Facility and was the Project Manager on the permitting application for Covanta's Energy from Waste facility in the regional municipality of Durham in Ontario, Canada. Michael was a contributing editor for *Solid Waste and Recycling* magazine and sits on the Board of Directors of the Ontario Waste Management Association. He is also an Adjunct Professor at the University of Toronto lecturing on waste management issues.

Melissa M. Carter (PE, PMP, Project Management Director, MWH Global) is a vice president with MWH, an international water and natural resources company. She also serves as the chair of the Committee on Sustainability for the Boston Society of Civil Engineers. Melissa has over 19 years of experience as a "wet infrastructure" engineer, including large-scale wet weather overflow control projects. She received both her Bachelor of Civil Engineering and Master of Science in environmental engineering degrees from the Georgia Institute of Technology, where she first developed an interest in sustainability principles in engineering. Melissa practiced engineering in the southeastern United States for over a decade before relocating to Massachusetts with her husband and two children, where she currently practices her profession and advocates for sustainability.

Richard Corolewski (PE, Electrical Engineer/Federal Division Director) is the director of the Power Engineers' Federal Division and has 30 years' experience in the A/E industry supporting projects in the U.S. and overseas. He is currently responsible for all business development, strategic planning, and operations of POWER's Federal Division, which specializes in energy engineering and design. Mr. Corolewski is a member of POWER's Sustainability Committee and is integral to shaping POWER's sustainable engineering solutions. Together with POWER's CEO, he serves on the Sustainable Infrastructure Advisory Board for the Zofnass Program for Sustainable Infrastructure at the Harvard University Graduate School of Design. With other industry leaders, he promotes topics of sustainability within today's global infrastructure. Mr. Corolewski offers a thorough understanding of all aspects of project management and execution, including programming, planning, design, bidding, and construction administration. He has managed federal, industrial, and commercial projects, ranging in size from $3 to $100 million, and he specializes in Department of Defense and Department of State programs.

Daniel Dietch (MBA, MPS, Client Service Manager CH2M) has more than 20 years of academic and professional consulting experience with integrated solid waste planning and management. He uses this experience to provide clients with strategic consulting services including integrated solid waste and recycling planning, procurement services, contract development and negotiations, ordinance development, rate structure/alternative financing analyses, and performance monitoring. Daniel has advanced degrees from Cornell University in Environmental Management and from the University of Miami in Business Administration. In addition, he is a Board Member of Recycle Florida Today and serves as a judge for SWANA's Integrated Solid Waste Management Excellence Award. In his spare time he is a three-term Mayor in his hometown of Surfside, FL.

Marty Janowitz (MES, ENV SP, Vice President & Practice Leader Sustainable Development, Stantec) has more than 30 years' experience consulting internationally in environmental and sustainability planning, policy, and engagement. He leads Stantec's efforts to become an exemplary model of sustainability in all its operations and to develop an integrated sustainability consulting practice. He's played key roles for the Institute for Sustainable Infrastructure as an Envision® Professional, Verifier, Trainer, Technical Sub-Committee Chair, and member of its Envision® Review Board. Marty serves on the Zofnass Program's Sustainable Infrastructure Advisory Board. He was advisor to the first two Envision® and Platinum Awarded projects in Canada, Stantec's Grand Bend Wastewater Treatment Plant and Port Metro Vancouver Low Level Road projects. Marty was selected a member of Canada's Clean 50—outstanding contributors to sustainable development and clean capitalism.

Roberto Mezzalama (MSc, Global ESIA Technical Community Leader, Golder Associates) has been Principal and Senior Project Director of Golder Associates since 2006 and is currently Leader of the Global ESIA (Environmental and Social Impact Assessment) community. From 2011 to 2014 he was Golder's Global Sustainability Advisor. From 2007 to 2011 he was President and Managing Director of Golder Associates Europe, coordinating the 14 European operating companies and managing a staff of about 800 people. Roberto holds a BSc in ecology and an MSc in environmental engineering. In his capacity as Senior Project Director and Leader of the Global ESIA community he worked on the development of power, mining, transportation, manufacturing, and oil and gas projects mostly in Europe, North America, and Africa, and he is now leading several ESIA international projects.

Erin Mosley (PE, ENV SP, Vice President, Director of Management Consulting at CH2M) currently leads a global management consulting practice that serves the water, transportation, facilities, urban environments, and energy sectors—helping clients reach their goals through strategic alignment, organizational development and capacity, management science and data-driven decisionmaking, meaningful public engagement, and commitment to sustainability. She is an Envision® Sustainability Professional and Trainer and a member of the Harvard Zofnass Program Sustainable Infrastructure Advisory Board. She also serves as a member of the Water Environment Federation Board of Trustees (2013–2015). Erin received a BS in Civil Engineering from Northeastern University and completed graduate coursework at Tufts University and the Harvard Extension School.

Douglas M. Owen (PE, BCEE, ENV SP) is an Executive Vice President with ARCADIS and is the Chief Technical Officer for the Water Division. A professional engineer in three states, he has led applied research projects on advanced technologies, consulted with utilities on treatment and facility planning for over 6 billion gallons per day of treatment capacity throughout the United States, and has provided technical and facilitation support to USEPA and other policymakers on a range of policy issues since 1990. He currently serves as the Chair of the WateReuse Research Foundation, as a member of the Zofnass Sustainable Industry Advisory Board at Harvard, and has served on USEPA's National Drinking Water Advisory Council. He has published widely on water treatment and planning topics in books, peer-reviewed journals, and at national and international conferences.

Jade Paul (PLA) is a Senior Landscape Architect at CH2M. Jade has more than 15 years of professional consulting experience, specializing in thoughtful planning and design that balances natural and built components within our urban landscapes. She has helped develop and implement award-winning green infrastructure strategies on city- and countywide efforts, and has served on municipal task force committees developing sustainable urban design guidelines for public spaces. Jade holds two degrees from North Dakota State University in Landscape Architecture and Environmental Design, is certified with the Council of Landscape Architectural Registration Boards, and is registered in multiple states.

Rick Phillips is an award-winning urban designer and planner who specializes in the visioning, planning, and design of complex engineering and public transportation projects. For HNTB Corporation, Rick serves as director of Urban Design for Northern California. He has been deeply involved in the planning and design of California's future high-speed rail system, as well as regional rail and urban public transit in the Bay Area and beyond. Rick played a central role in the creation of Downtown in Motion, Salt Lake City's innovative downtown transportation master plan, built on a foundation of multimodal transportation, walkability, and sustainable land use. In 2007, Rick authored the citation on "Transit Oriented Development" for the inaugural edition of the international reference work *Landscape Architectural Graphic Standards*. He presents regularly at conferences and academic meetings on transportation design and the intertwined roles of land use and transportation in supporting a sustainable urban future.

Lynda Sharkey has been in the GIS, planning, and sustainability space for the last 15 years. She has worked in South Africa, Northern Europe, and the US, in roles ranging from Technical Specialist to Technical Marketing Manager. Lynda's most recent role was as Sustainable Cities and Transportation Product Manager at Autodesk, responsible for developing, pro-typing, and bringing-to-market solutions that help make sustainability easier, more insightful, and more cost-effective for cities.
Right now, however, you can find her riding the roads of Southern Europe on a Triumph Explorer.

Cathy Smith (HBA, MA) has 19 years' experience in solid waste management. She has a Sustainability Planner's accreditation from the Institute for Sustainable Infrastructure at Harvard University and now actively utilizes this knowledge in planning and infrastructure projects. Cathy is Stantec's National Waste Sector Leader. As a former manager of the City of Guelph Solid Waste Division, she managed 100 staff through day-to-day operations of a city-operated landfill, MRF, organic waste processing facility, waste transfer station, and curbside collection services. Cathy now uses her extensive background in waste operations and management to assist in the development of long-term solid waste management plans, in assessing system improvements through the implementation of best practices and implementing recommended system changes and financing strategies.

Jeanette M. Southwood (MASc, PE, FEC, FCAE, QPESA, QPRA, Engineer and Global Sustainable Cities Initiative Leader / Canadian Urban Development & Infrastructure Leader / Principal, Golder Associates) is an award-winning engineer and a Principal at Golder, a global, employee-owned firm of more than 8,000, where she leads the Global Sustainable Cities team and the Canadian Urban Development & Infrastructure Sector. One area of focus for her teams is the strategic integration of cutting-edge international innovation and knowledge into value-creating solutions for urban resilience and revitalization. A dedicated volunteer on several boards over the years, Jeanette is also a Fellow of Engineers Canada and of the Canadian Academy of Engineering, a recipient of the Province of Ontario Leading Women Building Communities Award, one of Canada's 2014 Clean50, a recipient of the Ontario Professional Engineers Award for Engineering Excellence, and one of the WXN Top 100 for 2015.

Ana Maria Vidaurre is Principal Investment Officer with the Infrastructure Division of the Structured and Corporate Finance Department of the Inter-American Development Bank, based in Washington, DC. The Structured and Corporate Finance Department of the IDB fosters private-sector participation in sustainable economic growth in Latin America and the Caribbean. As a Principal Investment Officer, she leads the origination, analysis, structuring, and execution of project finance transactions in the infrastructure space. Ms. Vidaurre has worked on financing water and sanitation, gas transportation, port, and power utility projects for over 15 years, and is currently focused on financing renewable energy projects in Latin America. She is also responsible for the IDB Infrastructure 360 Sustainability Awards, which, in three editions, has received 148 project applications with investments amounting to US$120 billion in the Latin American region. Prior to joining the IDB, she worked in infrastructure project finance in Latin America's Andean Region at CAF development bank. Ms. Vidaurre has a BA degree in Business Administration from Universidad Metropolitana in Caracas, Venezuela, and an MBA from École Supérieure des Sciences Économiques et Commerciales (ESSEC), Paris, France.

John Wood (PE) is Global Practice Director for Integrated Waste Solutions at CH2M. John has more than 25 years' experience in the solid waste field. His background includes technology assessment, facility siting, landfill design, project development, and recycling alternatives. He has participated in more than 200 solid and industrial waste management projects for private and public clients throughout the United States and internationally. John leads a practice that provides comprehensive and sustainable solid waste management solutions to government and industry worldwide, with extensive experience in conversion technologies, landfill development and closure, landfill gas management, solar installations, solid waste planning, sustainability and climate change, and material transfer and recovery facilities.

Paul Zofnass (Zofnass Program Founder, President, Environmental Financial Consulting Group, Inc.) is President of the Environmental Financial Consulting Group (EFCG), a firm he founded in 1990 to provide financial, strategic, and investment banking services initially to environmental engineering and consulting firms, and now to the overall engineering/consulting (e/c) industry. EFCG is headquartered in New York City, and serves as a retained adviser to many of the major global e/c firms. EFCG has hosted for 25 years the annual Engineering/Consulting CEO Conference, which brings together the leaders of over 200 major engineering firms throughout the world. EFCG is also the leading merger and acquisition adviser to the e/c industry. Paul is a magna cum laude graduate of Harvard College, and an alumnus of both the Harvard Business School and the Harvard Law School. His wife is an attorney in New York, and both his daughters are alumnae of Harvard. He is a dedicated environmentalist, assisted in developing (in the early 1990s) the Harvard University Center on the Environment, and created (in 2008) the Zofnass Program for Sustainable Infrastructure at the Harvard GSD. He also funded the New England Forests Exhibit at the Harvard Museum of Natural History, and is responsible for creating and funding a 250-acre nature preserve near his weekend home in Pound Ridge, NY, just north of New York City, on which he has built a 10-mile hiking/nature trail known as the Westchester Wilderness Walk/Zofnass Family Preserve. He also serves on the board of Riverkeeper, of the Mount Auburn Cemetery, and as a trustee of Harvard's Organismic and Evolutionary Biology Department.

INDEX

E

F

G

N

O

P

Q

R

S

T

U

V

W

Y

Z

For Product Safety Concerns and Information please contact our EU representative GPSR@taylorandfrancis.com Taylor & Francis Verlag GmbH, Kaufingerstraße 24, 80331 München, Germany

Batch number: 10733588

Printed by Printforce, the Netherlands